OLIVER HEAVISIDE

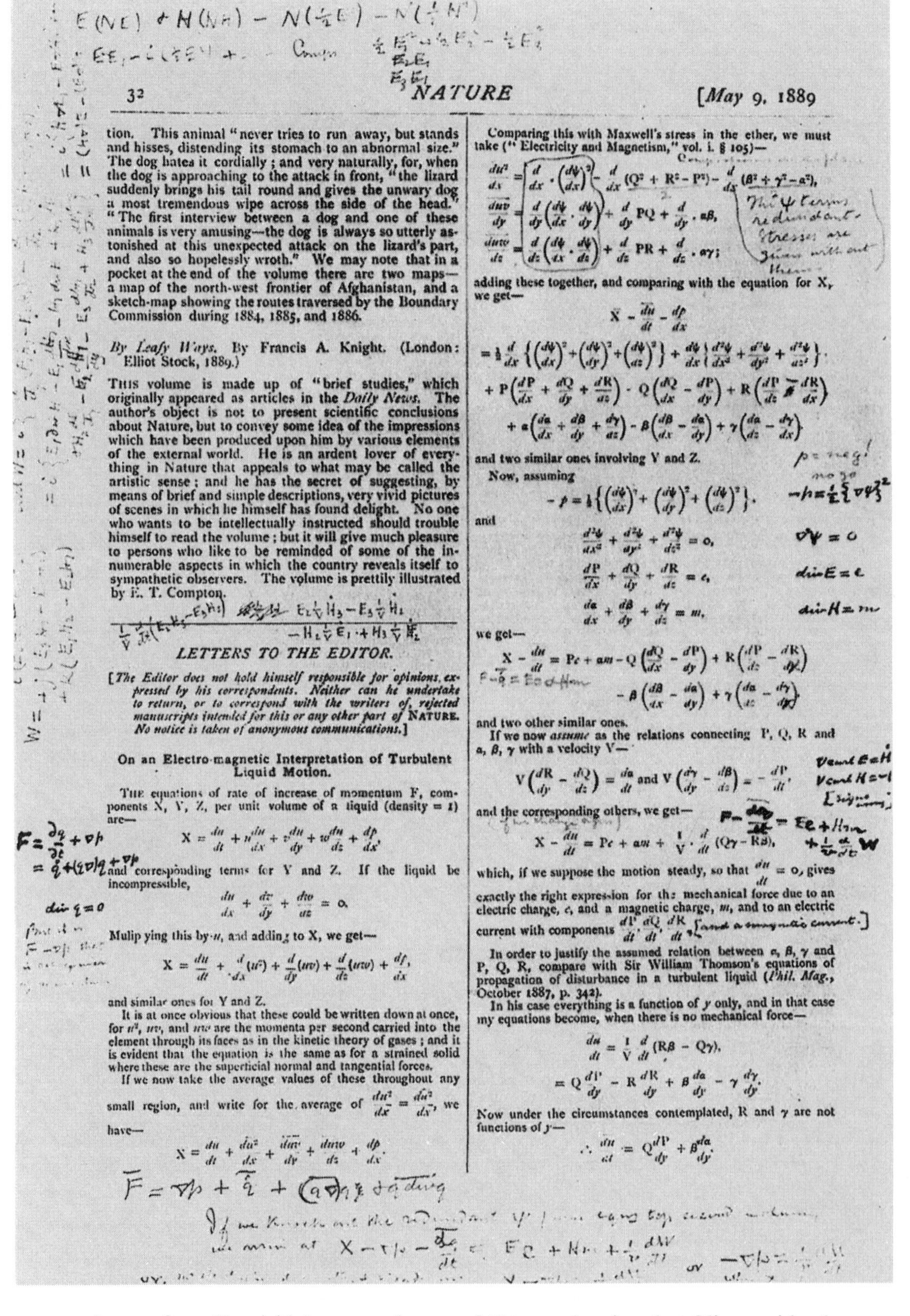

tion. This animal "never tries to run away, but stands and hisses, distending its stomach to an abnormal size." The dog hates it cordially; and very naturally, for, when the dog is approaching to the attack in front, "the lizard suddenly brings his tail round and gives the unwary dog a most tremendous wipe across the side of the head." "The first interview between a dog and one of these animals is very amusing—the dog is always so utterly astonished at this unexpected attack on the lizard's part, and also so hopelessly wroth." We may note that in a pocket at the end of the volume there are two maps—a map of the north-west frontier of Afghanistan, and a sketch-map showing the routes traversed by the Boundary Commission during 1884, 1885, and 1886.

By Leafy Ways. By Francis A. Knight. (London: Elliot Stock, 1889.)

THIS volume is made up of "brief studies," which originally appeared as articles in the *Daily News*. The author's object is not to present scientific conclusions about Nature, but to convey some idea of the impressions which have been produced upon him by various elements of the external world. He is an ardent lover of everything in Nature that appeals to what may be called the artistic sense; and he has the secret of suggesting, by means of brief and simple descriptions, very vivid pictures of scenes in which he himself has found delight. No one who wants to be intellectually instructed should trouble himself to read the volume; but it will give much pleasure to persons who like to be reminded of some of the innumerable aspects in which the country reveals itself to sympathetic observers. The volume is prettily illustrated by E. T. Compton.

LETTERS TO THE EDITOR.

[*The Editor does not hold himself responsible for opinions expressed by his correspondents. Neither can he undertake to return, or to correspond with the writers of, rejected manuscripts intended for this or any other part of* NATURE. *No notice is taken of anonymous communications.*]

On an Electro-magnetic Interpretation of Turbulent Liquid Motion.

THE equations of rate of increase of momentum F, components X, Y, Z, per unit volume of a liquid (density = 1) are—

$$X = \frac{du}{dt} + u\frac{du}{dx} + v\frac{du}{dy} + w\frac{du}{dz} + \frac{dp}{dx},$$

and corresponding terms for Y and Z. If the liquid be incompressible,

$$\frac{du}{dx} + \frac{dv}{dy} + \frac{dw}{dz} = 0.$$

Mulip ying this by u, and adding to X, we get—

$$X = \frac{du}{dt} + \frac{d}{dx}(u^2) + \frac{d}{dy}(uv) + \frac{d}{dz}(uw) + \frac{dp}{dx},$$

and similar ones for Y and Z.

It is at once obvious that these could be written down at once, for u^2, uv, and uw are the momenta per second carried into the element through its faces as in the kinetic theory of gases; and it is evident that the equation is the same as for a strained solid where these are the superficial normal and tangential forces.

If we now take the average values of these throughout any small region, and write for the average of $\frac{du^2}{dx} = \frac{d\overline{u^2}}{dx}$, we have—

$$X = \frac{du}{dt} + \frac{d\overline{u^2}}{dx} + \frac{d\overline{uv}}{dy} + \frac{d\overline{uw}}{dz} + \frac{dp}{dx}.$$

Comparing this with Maxwell's stress in the ether, we must take ("Electricity and Magnetism," vol. i. § 105)—

$$\frac{d\overline{u^2}}{dx} = \frac{d}{dx}\cdot\left(\frac{d\psi}{dx}\right)^2 - \frac{d}{dx}\frac{(Q^2 + R^2 - P^2)}{2} - \frac{d}{dx}\frac{(\beta^2 + \gamma^2 - \alpha^2)}{2},$$

$$\frac{d\overline{uv}}{dy} = \frac{d}{dy}\left(\frac{d\psi}{dx}\cdot\frac{d\psi}{dy}\right) + \frac{d}{dy}PQ + \frac{d}{dy}\cdot\alpha\beta,$$

$$\frac{d\overline{uw}}{dz} = \frac{d}{dz}\left(\frac{d\psi}{dx}\cdot\frac{d\psi}{dz}\right) + \frac{d}{dz}PR + \frac{d}{dz}\cdot\alpha\gamma;$$

adding these together, and comparing with the equation for X, we get—

$$X - \frac{d\bar{u}}{dt} - \frac{dp}{dx}$$

$$= \frac{1}{2}\frac{d}{dx}\left\{\left(\frac{d\psi}{dx}\right)^2 + \left(\frac{d\psi}{dy}\right)^2 + \left(\frac{d\psi}{dz}\right)^2\right\} + \frac{d\psi}{dx}\left\{\frac{d^2\psi}{dx^2} + \frac{d^2\psi}{dy^2} + \frac{d^2\psi}{dz^2}\right\}$$

$$+ P\left(\frac{dP}{dx} + \frac{dQ}{dy} + \frac{dR}{dz}\right) - Q\left(\frac{dQ}{dx} - \frac{dP}{dy}\right) + R\left(\frac{dP}{dz} - \frac{dR}{dx}\right)$$

$$+ \alpha\left(\frac{d\alpha}{dx} + \frac{d\beta}{dy} + \frac{d\gamma}{dz}\right) - \beta\left(\frac{d\beta}{dx} - \frac{d\alpha}{dy}\right) + \gamma\left(\frac{d\alpha}{dz} - \frac{d\gamma}{dx}\right),$$

and two similar ones involving Y and Z.

Now, assuming

$$-p = \frac{1}{2}\left\{\left(\frac{d\psi}{dx}\right)^2 + \left(\frac{d\psi}{dy}\right)^2 + \left(\frac{d\psi}{dz}\right)^2\right\},$$

and

$$\frac{d^2\psi}{dx^2} + \frac{d^2\psi}{dy^2} + \frac{d^2\psi}{dz^2} = 0,$$

$$\frac{dP}{dx} + \frac{dQ}{dy} + \frac{dR}{dz} = e,$$

$$\frac{d\alpha}{dx} + \frac{d\beta}{dy} + \frac{d\gamma}{dz} = m,$$

we get—

$$X - \frac{d\bar{u}}{dt} = Pe + \alpha m - Q\left(\frac{dQ}{dx} - \frac{dP}{dy}\right) + R\left(\frac{dP}{dz} - \frac{dR}{dx}\right)$$

$$- \beta\left(\frac{d\beta}{dx} - \frac{d\alpha}{dy}\right) + \gamma\left(\frac{d\alpha}{dz} - \frac{d\gamma}{dx}\right),$$

and two other similar ones.

If we now *assume* as the relations connecting P, Q, R and α, β, γ with a velocity V—

$$V\left(\frac{dR}{dy} - \frac{dQ}{dz}\right) = \frac{d\alpha}{dt} \text{ and } V\left(\frac{d\gamma}{dy} - \frac{d\beta}{dz}\right) = -\frac{dP}{dt},$$

and the corresponding others, we get—

$$X - \frac{d\bar{u}}{dt} = Pe + \alpha m + \frac{1}{V}\cdot\frac{d}{dt}(Q\gamma - R\beta),$$

which, if we suppose the motion steady, so that $\frac{d\bar{u}}{dt} = 0$, gives exactly the right expression for the mechanical force due to an electric charge, e, and a magnetic charge, m, and to an electric current with components $\frac{dP}{dt}, \frac{dQ}{dt}, \frac{dR}{dt}$.

In order to justify the assumed relation between α, β, γ and P, Q, R, compare with Sir William Thomson's equations of propagation of disturbance in a turbulent liquid (*Phil. Mag.*, October 1887, p. 342).

In his case everything is a function of y only, and in that case my equations become, when there is no mechanical force—

$$\frac{d\bar{u}}{dt} = \frac{1}{V}\frac{d}{dt}(R\beta - Q\gamma),$$

$$= Q\frac{dP}{dy} - R\frac{dR}{dy} + \beta\frac{d\alpha}{dy} - \gamma\frac{d\gamma}{dy}.$$

Now under the circumstances contemplated, R and γ are not functions of y—

$$\therefore \frac{d\bar{u}}{dt} = Q\frac{dP}{dy} + \beta\frac{d\alpha}{dy}.$$

A page from Heaviside's personal copy of *Nature,* showing that Oliver evidently liked to read with a pencil at hand.

OLIVER HEAVISIDE

The Life, Work, and Times of an Electrical Genius of the Victorian Age

Paul J. Nahin

Department of Electrical and Computer Engineering
University of New Hampshire

The Johns Hopkins University Press
Baltimore and London

Printed in the United States of America on acid-free paper

Originally printed in a hardcover edition by The Institute of Electrical and Electronics Engineers, Inc., 1988
Johns Hopkins Paperbacks edition, 2002
2 4 6 8 9 7 5 3 1

The Johns Hopkins University Press
2715 North Charles Street
Baltimore, Maryland 21218-4363
www.press.jhu.edu

Library of Congress Cataloging-in-Publication Data

Nahin, Paul J.
Oliver Heaviside : the life, work, and times of an electrical genius of the Victorian age / Paul J. Nahin.
p. cm.
Originally published: New York : IEEE Press, 1987.
Includes bibliographical references and index.
ISBN 0-8018-6909-9 (pbk. : alk. paper)
1. Heaviside, Oliver, 1850–1925. 2. Electric engineers—England—Biography. I. Title.

TK140.H43 N34 2002
621.3′092—dc21
[B] 2001038916

A catalog record for this book is available from the British Library.

Contents

For My Family.

Preface to the Johns Hopkins Edition

England takes great pride and pleasure in her eccentrics; as the Spanish-born philosopher and poet George Santayana wrote in his 1922 *The British Character,* "England is the paradise of individuality, eccentricity, heresy, anomalies." She has never lacked for men (and women) with those qualities, but the nineteenth century was particularly well stocked with "mad dogs and Englishmen" who became famous for their unusual lives. They came in all possible shapes and sizes, with every occupation represented by at least one slightly odd (or perhaps very odd) individual. Names from that time, like Oscar Wilde, General Gordon of Khartoum, Florence Nightingale, and Lewis Carroll are instantly recognized, even by those whose interest in Victorian times is mainly limited to Jack the Ripper.

Even in anonymity the nineteenth-century Englishman can leave us astonished. We will, for example, sadly never know anything about the man (or possibly the woman) whose long abandoned box at London's Chancery Lane Safe Deposit Company was finally forced open in the 1940s. Found inside, according to Peter Bushell's *London's Secret History,* was only a pair of Victorian lady's knickers, along with an attached note declaring them to be "My life's undoing." In another such box was a packet of six live bullets; written on the packet, in a faded hand, was the single sentence: "One for each of the directors." I think I am on safe ground to believe that those notes certainly must excite your imagination; but not nearly as much, I think, as will the amazing story of Oliver Heaviside.

Heaviside was one of the most important yet least appreciated contributors to physics and electrical engineering during the second half of the nineteenth century. His very name conjures up the image of a Charles Dickens' character. Indeed, he was born at mid-century only a short distance from the shoe-blacking factory where the author of *Oliver Twist* and *Hard Times* learned firsthand the horrors of being down and out in Victorian London. After a mostly self-taught technical education in advanced electrical and mathematical matters, Heaviside engaged in a stupendously nasty and lengthy battle with William Preece, the engineering head of the Post Office, which was England's official agency for the development of electrical communications technology. Preece was a powerful government official, enormously ambitious, and, in some remarkable ways, an utter blockhead. He was everything Heaviside, a poor man with no weapons at his disposal but his brains, was not. Their conflict was of such an intense personal nature that it had all the makings of a Victorian melodrama. In the end, although it did him precious little good, Heaviside's brain proved to be the superior to Preece's politics.

Some time after the original 1988 edition of this book appeared I came across a review in *Time* magazine of Donald Spoto's biography *The Dark Side of Genius: The Life of Alfred Hitchcock.* The reviewer wrote "The word genius is good for starting arguments but bad for book titles, unless the book is about Mozart." Having committed that very "sin," using the word "genius" in this book's subtitle, was I contrite? Absolutely not—Heaviside *was* a genius, and I see no reason to hide that fact from readers. Indeed, that fact needs to be made explicit since, even among mathematicians, physicists, and electrical engineers (those who would be expected to know him best), he is still a mysterious figure. I feel fortunate that my long-time editor, Trevor Lipscombe, an Englishman himself and Oxford physicist to boot, loved the story of Heaviside, too. Trevor asked if I would consider the Johns Hopkins University Press for reprinting, and I enthusiastically replied, "Why, of course!"

Over the years I have received a large amount of correspondence from readers of the original edition, providing me with bits and pieces of Heavisidean lore that I had not known about. I have carefully saved all that material, so when Trevor asked for an essay to update the book, I knew my treasure-trove of "new" Heaviside stuff would at last come into its own.

A natural issue to take up immediately is that of other biographical writings on Heaviside that appeared after the 1988 publication of this book. Actually, as I mention in a footnote in the original Preface, a Heaviside biography did appear *earlier,* in 1985, in Russian, but at the time I wrote I knew nothing more of that work. It wasn't until some time after 1988 that I learned the same author, Boris M. Bolotovsky, had published an English language paper on Heaviside in the journal *International Studies in the Philosophy of Science* in April 1987 called "The Birth and Development of New Knowledge: Oliver Heaviside." In his paper Bolotovsky gives a nice summary of Heaviside's life and work, but it contains nothing that isn't also in this book.

In 1991, unaware of Bolotovsky's book and apparently not totally happy with this one, Arnold C. Lynch (a long-time employee of the British Post Office Engineering Department from the mid-1930s to the mid-1970s) published an interesting essay in the journal *History of Technology* titled "The Sources for a Biography of Oliver Heaviside." Lynch wrote that my book "is the best of the biographical work we have, and unlikely to be superseded," but also that "It is open to some criticism." Lynch feels I "play down the importance of Heaviside's mathematical methods," and makes it clear that he thinks my opinion is that those methods are obsolete (see Chapter 10). *But,* that doesn't mean I think Heaviside's methods historically unimportant; if I gave that impression then I am truly sorry. The German historian Susann Puchta has recently written ("Why and How American Electrical Engineers Developed 'Heaviside's Operational Calculus,'" *Archives Internationales d' Histoire des Sciences,* June 1997) in agreement with Lynch. She feels that I have too readily brought down the curtain in my discussion of the decline of Heaviside's operational calculus. I think she is correct. Despite the development of the Laplace transform the use of Heaviside's method did *not* cease overnight, and if I were to write Chapter 10 fresh, today, I would soften my position on that issue.

However, I do not agree with Puchta's speculation, at the end of her paper, that the occasional appearance of a historical paper on the operational calculus is evidence for a renewed interest in the subject by electrical engineers. I write this even though at least one electrical engineer had written in agreement with Puchta *before* Puchta's paper appeared; Per A. Kullstam presents his views in two papers displaying great mathematical virtuosity ("Heaviside's Operational Calculus: Oliver's Revenge" and "Heaviside's Operational Calculus Applied to Electrical Circuit Problems," *IEEE Transactions on Education,* May 1991 and November 1992, respectively). Still, while I admire Kullstam's skillful mathematics, I think his central conclusion that undergraduate electrical engineering students should be taught the operational calculus is wrong.

Every undergraduate engineering textbook I am familiar with, in *all* of the engineering disciplines, uses the Laplace transform to solve linear, ordinary differential equations; there is virtually no chance of the standard undergraduate electrical engineering curriculum replacing the Laplace transform with the Heaviside operational calculus. Indeed, that decision was made decades ago. When the first edition of Cornell University mathematics professor Ralph Palmer Agnew's classic *Differential Equations* was published in 1942, he included both the classical techniques and Heaviside's differential operator approach. The Laplace transform was nowhere to be seen. Agnew, then, had no bias against Heaviside's mathematics. But when the second edition appeared in 1960, after calling both the Laplace and Heaviside methods "expeditious," Agnew went on to say that "The Laplace transformation method seems to have completely won the battle for attention because it is more interesting, it is easier to apply to routine problems, and . . . it is more versatile." Nobody in engineering education has seen fit to argue with Agnew's assessment since he wrote those words more than forty years ago.

While I don't believe Kullstam's papers had any impact on electrical engineering educators, they did prompt the German historian Susann Hensel to write the very interesting essay "Ernst Julius Berg—Educator and Proselytizer of Heaviside's Calculus," *IEEE Potentials,* August/September 1994. You'll read more about Berg and Heaviside in Chapter 12 of this book, but Hensel's essay does mention one amusing anecdote that I had not known; in 1936 Berg sent Albert Einstein a copy of the new edition of his book *Heaviside's Operational Calculus,* and received a thank you note from the great man saying that he appreciated the gift because now he would be able to learn Heaviside's "peculiar mathematical witchcraft"!

Returning to Lynch's "Sources," in addition to this book he also mentions two other specific biographical works on Heaviside; a short 1987 book based on material prepared by Heaviside's friend G. F. C. Searle (*Oliver Heaviside, The Man,* St. Albans, edited by one Ivor Catt about whom I'll say more, soon), and an unpublished typescript by the late Henry J. Josephs (also a British Post Office Engineering Department employee, from even earlier times than those of Lynch). As Lynch observes, and as I stated in my "A Note on References," the Searle/Catt book appeared just as this book was going to the printer, so I did not have it available as a source. I have, however, read it since. It is an interesting amalgamation of letters written by Heaviside to Searle from 1892 on, with occasional commentary by Searle on the origins of particular letters. The Searle/Catt book, though, is limited to letters written sporadically to one individual over the last three decades of Heaviside's life, and it presents little discussion of Heaviside's life before 1892 and none at all about his work. It is mostly an expansion of a contribution Searle made to the *Heaviside Centenary* volume published in 1950 by the Institution of Electrical Engineers (IEE), which I used when writing this book.

In his "Editor's Preface," Catt writes "The behavior of the Scientific Establishment when faced by the brilliance of Heaviside was intensely political and destructive . . . A century later, I myself met with exactly the same kind of behaviour . . . when in my turn I produced major advances in electromagnetic theory. Like Heaviside's, my work was suppressed." This assertion is strange. Heaviside's work certainly was *not* suppressed by the scientific world (it would be better to say that for a long time it was *ignored* by the *engineering* establishment)—*The Electrician* and *Philosophical Magazine* regularly published him, and the papers appearing in those journals were widely read, admired, and publicly praised by such scientific stars as Oliver Lodge, William Thomson (Lord Kelvin), and John Strutt (Lord Rayleigh). It was through their support that he was elected a Fellow of the Royal Society, which is hardly a mark of scientific rejection. Heaviside is even mentioned in Maxwell's masterpiece *A Treatise on Electricity and Magnetism* (see note 28 of Chapter 2 of this book).

It is true that in 1888 Preece succeeded *for a while* in pressuring *The Electrician* into not publishing Heaviside, and that in 1894 the Royal Society refused to continue publishing his paper on operators, but neither act rises to the level of suppression. Heaviside by then had other avenues for disseminating his work—and he used them. Heaviside's "suppression" has not continued into modern times; his work has been almost continually cited. For example, Sir Edmund Whittaker's classic two volume *History of the Theories of Aether and Electricity* (the first volume of which was originally published in 1910 then reprinted in 1950; the second

Certificate of a Candidate for Election.

☞ Attention is especially called to the directions given on the other side.

(Name) Oliver Heaviside

(Title or Designation)

(Profession or Trade*) None

(Usual Place of Residence) Paignton Devon

Qualifications

Learned in the science of Electromagnetism, having applied high mathematics with [illegible] power & success to the development of Maxwell's theory of electromagnetic propagation, & having [illegible] extended our knowledge of facts & principles in several directions & into great detail. He is the author of the following papers among many others.

On Electromagnetic induction & its propagation 48 parts 1885, 6, 7 [illegible]

The induction of currents in cores 15 parts 1884, 5

Some electrostatic & magnetic problems 6 [illegible] 1883

Current energy 19 parts 1883

On the Electromagnetic effects due to the motion of electrification through a dielectric Phil Mag 89

The general solution of Maxwell's equations Phil Mag

On Electromagnetic waves 6 parts Phil Mag 88

On Resistance & conductance operators Dec 87

On the self induction of wires 7 parts Phil Mag 1886, 7

On the Electromagnetic wave surface [illegible] June 85

On the electromagnetic effects of a moving charge

The [illegible] of an electromagnetic [illegible] by [illegible] of the medium

On the working of cables with condensers [illegible]

On the extra current [illegible] 76

On [illegible] through telephone [illegible] Phil Mag 77

On the effect of faults on the speed of working cables

On Electromagnetic [illegible] S.T.E

On induction between parallel wires, S.T.E

being desirous of admission into the ROYAL SOCIETY OF LONDON, we, the undersigned, propose and recommend him as deserving that honour, and as likely to become a useful and valuable Member. Dated this 19th day of February 1890

From General Knowledge. From Personal Knowledge.

Oliver J. Lodge

William Thomson

Geo. Fras. Fitzgerald

J. H. Poynting

R. T. Glazebrook

J. T. Bottomley

A. W. Rücker

A. W. Reinold

John Perry

J. Hopkinson

J. A. Ewing

G. Carey Foster

C. V. Boys

Shelford Bidwell

Elected June 4th, 1891

Read to the Society on the 1st day of May 1890

Some of England's greatest scientists certify that Oliver Heaviside is their intellectual (if not social) equal.

volume was first published in 1953), has numerous references to Heaviside's work. The 1938 polemic against Einstein and relativity theory by Alfred O'Rahilly, the still occasionally cited book *Electromagnetics,* contains numerous critical references to Heaviside, both of him personally and of his work. As the old saying among actors and writers goes, however, "say what you want about me, just spell my name right." At the University of New Hampshire's Physics Library I found Julius Adams Stratton's classic 1941 book *Electromagnetic Theory,* which contains multiple references to Heaviside's work. And I also unearthed Sir James Jeans' 1951 *The Mathematical Theory of Electricity and Magnetism,* which refers the reader to Heaviside's seminal work on electric signaling over inductive cables. Heaviside certainly did not "disappear" from the view of those who read technical books.

Lynch writes in his essay that I "may not have seen the Josephs draft." Well, in fact I did, during a March 1985 visit to the Heaviside Collection at the IEE in London. Before starting to write this essay I consulted the research notebook I kept during that visit, and found a fair number of unhappy comments that I wrote to myself as I read the Josephs draft. Most material in the draft was available elsewhere in published documents with a proper citation of sources. Josephs claims to have had access to letters and papers that nobody else has, but to date these have not been produced for independent study. As such, there is no support for some of his assertions. I do not share Lynch's opinion that the "Josephs draft has value for the historian." You'll find more on Josephs in Chapter 13.

Not to be totally at odds with Lynch's essay, I did learn one quite interesting fact from it. As I write in Chapter 2 of this book, Heaviside's uncle was the famous scientist Charles Wheatstone, having married the sister of Oliver's mother. I had not known, however, that the marriage was an unusual one for the rigid, class-conscious Victorian age; Wheatstone married his cook.

Also in 1991, the year of Lynch's paper, in a book that is about not only Heaviside's work but also that of several other Victorian researchers, Bruce J. Hunt (a professor of history at the University of Texas, Austin) published *The Maxwellians,* Cornell University Press. Hunt's book is admirable; it is based on his 1984 doctoral dissertation written at the Johns Hopkins University, a copy of which I purchased and used extensively while writing this book, and it is cited in this book's notes. A few years later Ido Yavetz (a professor at the Cohn Institute for the History of Science at Tel Aviv University) wrote the 1995 book *From Obscurity to Enigma: The Work of Oliver Heaviside, 1872–1889,* Birkhäuser. Like Hunt's, Yavetz's book is a smoothly written, technically accurate and scholarly work. While it contains some biographical information, its view on Heaviside's life as opposed to his work is much narrower (as you might expect from the title) than is that of this book.

Finally, let me return for one last visit with Catt. In an audio tape (dated February 2, 1997) that he placed in the IEE archives in London (a transcript is available on the Web at *www .electromagnetism.demon.co.uk*), Catt discusses, with Lynch, what he calls "The International Incident/The Josephs-Gossick-IEE Clash." Gossick was Ben Roger Gossick who, at the time of his death in 1977, was professor of physics and electrical engineering at the University of Kentucky, Lexington. (Catt incorrectly states that Gossick was, of all things, a professor of music.) During the last years of his life Gossick worked on a biography of Heaviside (originally to be a dual biography of Heaviside and Wheatstone, which was partly preempted by Brian Bosworth's definitive 1975 biography of Wheatstone); much of that work made use of discussions Gossick had with Josephs during several research visits he made to England. The book was never published.

Josephs certainly must have been an interesting person; despite my severe reservations about his understanding of the need to properly document an historical work, it seems clear he possessed a good deal of mathematical ability. In 1946, for example, he published a highly readable monograph on Heaviside's physics and mathematics (*Heaviside's Electric Circuit Theory*), based on "after-hours" lectures Josephs had given to the engineering staff of the British Post Office. Later, Josephs contributed a lengthy commentary on Heaviside's mathematics to the 1950 IEE *Heaviside Centenary* volume. Both are erudite works, although the line between what is Heaviside and what is Josephs is often a blurry one.

Catt claims that Josephs' mathematical skills were such that all of the *technical* sections of Gossick's aborted book were due to Josephs, for Gossick was, after all, just a "professor of music." Gossick, however, was a Phi Beta Kappa at Pomona College (one of the Claremont Colleges in California) and had a Ph.D. in physics (Purdue University, 1954); his doctoral dissertation was titled "On the Transient Behavior of Semiconductor Rectifiers." During his career he worked as an electrical engineer at the University of Minnesota at Minneapolis, for RCA, and at the Oak Ridge National Laboratory, Tennessee, as well as holding academic positions *as a physicist* at Purdue, Arizona State University, and at Harpur College of SUNY, Binghamton. He conducted research on radiation effects in semiconductors, the dynamic behavior of semiconductor rectifiers, and p-n junction particle detectors. He published numerous scientific papers in refereed journals and wrote two books in advanced physics, both published by Academic Press (*Potential Barriers in Semiconductors,* 1964, and *Hamilton's Principle and Physical Systems,* 1967). He certainly did not need "help" in understanding either Heaviside's circuit theory or his mathematics. Indeed, in his 1967 book Gossick makes use of Heaviside's mathematics (as well as of the Laplace transform).

After his death Professor Gossick's research papers were donated by Jean Gossick to the Niels Bohr Library of the Center for History of Physics at the American Institute of Physics (then in New York City and now at College Park, Maryland). One need only study the Gossick papers a short time to appreciate how central a role his stillborn book on Heaviside must have played in Gossick's life. I never knew Ben Gossick, but I greatly admire his published work and hope that this book on Heaviside, while not his, would nonetheless have pleased him.

In the following section I have included discussion to serve as a supplement to and/or expansion on the original book. Much of the new material is due to the kindness of strangers, to readers of the original book who sent me letters of personal knowledge on various issues, often including Heaviside-related documents, as well. I have arranged these "updates" by order of original chapter.

Chapter 4: Heaviside's Early Telegraphy Work

In April 1994 I received a letter from David W. Kraeuter, a reference librarian at Washington and Jefferson College in Pittsburgh. An active member of the Pittsburgh Antique Radio Society, Kraeuter had an obvious interest in Heaviside and had written to tell me that I had missed an important detail of Heaviside's work. While preparing his own book (*British Radio and Television Pioneers: A Patent Bibliography,* Scarecrow Press, 1993), he had come across a patent in Heaviside's name, something that I believe has not been mentioned before in any historical treatment of Heaviside. The discovery of this patent resolves a long-standing puzzle mentioned in Ido Yavetz's book, *From Obscurity to Enigma.*

Kraeuter kindly sent me a copy of the entire patent document, and it made extremely interesting reading. Dated April 6, 1880, (when Heaviside was living at 3 St. Augustine Road, in London) it is British Patent 1407 with the title "Improvements in Electrical Conductors, and in the Arrangement and Manner of Using Conductors for Telephonic and Telegraphic Purposes." This patent is interesting because it describes two ways for "zeroing-out" all inductive coupling between separate but adjacent circuits. It is the first method that is the meat of the patent. Heaviside opens his claim by stating the problem as follows:

> When a number of wires run parallel to one another, either suspended or otherwise, any change in the current flowing in one wire causes currents in all the rest by induction, and the effect may be so great as to seriously interfere with the working of telephonic circuits, and to a less degree of ordinary telegraphic circuits also.
>
> It is a common practice to complete the circuit by means of a second wire instead of the earth and it is well known that the inductive interference is thereby reduced in magnitude, the induced electro-motive forces in one wire canceling those in the other to a certain extent. The nearer the wires are brought together the less does the inductive interference become, but it cannot be altogether eliminated in this way, because the axes of the two wires cannot be made to coincide so that they shall be both at the same distance from any disturbing wire.

Next comes Heaviside's solution:

> My improvements have for object to obtain perfect protection, and to render a circuit completely independent under all circumstances of external influences. For this purpose I use two insulated conductors for the circuit, and place one of them inside the other; thus one conductor may be a wire, and the other a tube or sheath, thus forming a compound conductor consisting of a central wire surrounded by an insulating covering, which is in its turn surrounded by a conducting tube or sheath, which must also be insulated. When the tube and inner wire are electrically connected at both ends of the line, as through apparatus in the usual manner, the circuit as thus described is completely independent of other circuits, and any number of such circuits, each containing an insulated tube and inner wire, may be laid side by side and worked without any mutual inductive interference, and without interference from other wires worked in ordinary manners.

That is, Heaviside's patent is for nothing less than the modern, ubiquitous coaxial cable, of which a light-year or two (metaphorically speaking) must be used each year around the world. This was the geometry of the Atlantic Cable, discussed in Chapter 3 of this book, but apparently nobody patented it. That cable was used at signaling speeds so slow that inductive effects were not evident (or, at least, William Thomson's theoretical analyses ignored such effects), and it was Heaviside who first saw the coaxial cable as the solution to avoiding *magnetic coupling interference* between the adjacent, high-speed telephone circuits of the late nineteenth century that far outran the slower mid-century submarine cabling speeds. The second method described in the Heaviside patent is separate and distinct from the first; it shows how to position two pairs of wires, with each pair forming a conventional circuit, so as to cancel the inductive coupling between the two circuits.

As I mentioned earlier, this patent now clears up a historical puzzle, namely the nature of a letter Heaviside's brother Arthur wrote on June 29, 1881, (just a year after the patent date). In it Arthur advises Oliver to "sign the blooming [royalty] agreement and take . . . the cheque for [100 pounds]." In his book Yavetz states (p. 13) that it has been speculated that Arthur was referring to an invention by one or both of the brothers, and that the invention concerned means of "neutralizing disturbances in cables"; it would now seem that we do indeed know that this was the case, as well as the details of exactly what was proposed, and that Oliver was the sole inventor.

Chapter 5: The Scienticulist

A scholarly paper dealing with Preece appeared *before* 1988, but I missed it in my original bibliographic search. It is extremely interesting reading, addressing in part (as the author writes) Preece's "almost arrogant self-confidence which led him into serious error on occasions. This comes out most strongly in his work on telephone transmission, where he pushed forward his own erroneous views in opposition to the work of Oliver Heaviside, which he could not understand."

Read by Professor D. G. Tucker at the Science Museum in London, March 1982, you can find it in print as "William Preece (1834–1913)," *Transactions of the Newcomen Society for the Study of the History of Engineering Technology* 53, 1981/82, pp. 119–138.

Chapter 6: Maxwell's Electricity

I say in note 18 of this chapter that there is no extensive biographical treatment of George Green. That was true in 1988, but in 1993 a really very nice little book by D. M. Cannell appeared: *George Green, Mathematician and Physicist, 1793–1841: The Background to His Life and Work*, Athlone Press. As I write, the Society for Industrial and Applied Mathematics (SIAM) has just published the second edition. The new edition contains a fascinating essay by physicists Freeman Dyson of the Institute for Advanced Study and the 1965 Nobel laureate Julian Schwinger, describing the historical and contemporary applications of Green's work.

Chapter 10: Strange Mathematics

I have already discussed Kullstam's modern-day championing of Heaviside's mathematics. In this note I want to direct your attention to a fascinating essay by Bruce J. Hunt on the *politics* of that mathematics; in particular, the refusal of the Royal Society in 1894 to continue publishing Heaviside's multi-part paper "On Operators in Physical Mathematics." You can find Professor Hunt's analysis ("Rigorous Discipline: Oliver Heaviside Versus the Mathematicians") in the book *The Literary Structure of Scientific Argument: Historical Studies*, ed. Peter Dear, University of Pennsylvania Press, 1991.

Hunt's argument is essentially that Heaviside simply appeared on the scene at the wrong time, when English mathematicians were attempting to develop an identity for their discipline that would be separate and distinct from the *applied* mathematics of physics and engineering. The very nature of the Mathematical Tripos examination, with its increasing emphasis on mathematical physics, was increasingly disturbing to Cambridge mathematicians in the 1880s and 1890s. As I discuss in Chapter 10, Heaviside was definitely an applied mathematician of the wildest sort (as he wrote to Hertz in 1889, "Physics is above mathematics, and the slave must be trained to work to suit the master's convenience"), and when his unrigorous use of fractional operators and divergent series ran wild in the "On Operators . . ." paper, the new breed of "mathematical rigorists" reacted in a hostile way; not to Heaviside the man, but to Heaviside as an icon of "sloppy" mathematics that would simply no longer be tolerated.

This attitude can be found expressed in the classic book by G. N. Watson, *A Treatise on the Theory of Bessel Functions*, Cambridge, 1922; in the midst of a discussion in his chapter on asymptotic expansions, Watson writes "A number of extremely interesting symbolic investigations of the formulae are to be found in Heaviside's papers [Watson cites the first two parts of "On Operators . . ." that the Royal Society did publish], but it is difficult to decide how valuable such researches are to be considered when modern standards of rigor are adopted." Heaviside was still alive when that appeared; I don't know if he saw Watson's book but I can imagine what he said if he did.

Hunt reports that it was Andrew Forsyth (1858–1942) who raised the initial objections that eventually resulted in the Heaviside manuscript being sent to its doom at the hands of William Burnside, as discussed in Chapter 10. (Hunt doesn't mention it, but I think it an amusing irony that Forsyth, both the son of an engineer and the Senior Wrangler of the 1881 Mathematical Tripos, was the one who started the chain of events that stopped the "unpure, too practical" Heaviside.) Burnside had little patience for the "intuitive genius" of Heaviside; as Hunt reports, Burnside was later equally harsh with the self-trained Indian genius S. Ramanujan, despite his being the protégé of G. H. Hardy, the purest of the pure in English mathematics.

The Laplace transform has come up several times before in this essay, but there remains one additional historical aspect I learned of only after the original publication of this book. The English mathematician Harry Bateman (1882–1946) used the Laplace transform in a 1910 paper (the year he moved permanently to the United States, where he taught at Bryn Mawr, Johns Hopkins, and, from 1917 until his death, Caltech). Titled "The Solution of a System of Differential Equations Occurring in the Theory of Radio-active Transformations," it appeared in the *Proceedings of the Cambridge Philosophical Society*—and in it he used contour integration in the complex plane to invert transforms, the same technique Bromwich would later use to justify Heaviside's operational calculus.

The author of Bateman's obituary (as a Fellow of the Royal Society) wrote that Bateman felt his pioneering work on the Laplace transform had been ignored (Doetsch's 1937 book fails to mention Bateman). In a 1945 letter Bateman explained that, in fact, his own interest in the transform had been sparked by lectures on it that he had heard at Cambridge in either 1903 or 1904, and that he had written John R. Carson (the AT&T engineer who proselytized Heaviside's operational calculus) to point out to him that the Laplace transform was yet another way to attack circuit problems. By 1950 others had learned that lesson. See, for example, the paper by Louis A. Pipes, "The Summation of Fourier Series by Operational Methods," *Journal of Applied Physics,* April 1950, in which Heaviside and his *Electromagnetic Theory* are referenced but the actual mathematics is the Laplace transform.

Chapter 12: The Final Years of the Hermit

After this book appeared in 1988, the editors at *Scientific American* invited me to write an article about Heaviside for the magazine; that essay appeared in the June 1990 issue and, with one exception, contained no new information. But that exception is an interesting one. In December 1988 I received a letter from James L. Hunt, professor of physics at the University of Guelph in Ontario, Canada, bringing to my attention that about fifteen years earlier the British novelist Beverley Nichols (1899–1983) had published a nonfiction book titled *Father Figure* (Simon & Schuster, 1972). It is a very dark work describing the horrors Nichols' abusive, alcoholic father had inflicted upon his family throughout the entire marriage. This is relevant because the Nichols family had been Heaviside's next-door neighbors at Torquay, and the young Beverley had been one of the boys that Heaviside had frequently recorded as having "tormented" him. Nichols had a symmetrical relationship with Heaviside on that score, calling him a "singular character" and just the first of "an ever weirder horde of eccentrics" that had lived all about the Nichols' home.

Nichols recollection of Heaviside, in particular, is shocking. He wrote

> Professor Oliver Heaviside. He was our immediate neighbor on the west. . . . At the time of my boyhood Professor Heaviside was engaged in certain intricate researches into the nature of etheric waves. I cannot describe his discoveries, but they must have been of some importance; when he died *The Times* published a long obituary about him, in which his name was coupled with Edison and Marconi.

> But I could have told *The Times* a great many things about Professor Heaviside which, though they might not have been of interest to the scientists, would have provided the student of eccentric psychology with abundant food for thought. *The Times,* for instance, did not mention that he seldom dressed, and was usually attired in a kimono of pale pink silk. Nor did *The Times* see fit to mention that Heaviside, in a moment of pique, had caused most of the furniture of his house to be removed, and had replaced it by large granite rocks, which stood about in the bare rooms like the furnishings of some Neolithic giant. Through those fantastic rooms he wandered, growing dirtier and dirtier, and more and more unkempt—with one exception. His nails were always exquisitely manicured, and painted a glistening cherry pink.

This passage certainly gives some credibility to the idea that Heaviside was, in some sense, mentally unbalanced during his last years. Ido Yavetz mentions in *From Obscurity to Enigma* (note 194 on his page 281) that B. R. Gossick claimed in a 1977 paper Heaviside (see note 17 of the Epilogue of this book) to have been in fact mad. Yavetz thinks that claim to have "marred" Gossick's paper, but Nichols' words seem to support Gossick. Nichols himself was actually not prepared to go quite as far as had Gossick, but his own additional commentary on this issue *is* nevertheless damning:

> Was Professor Heaviside mad? Presumably not—his scientific record shows him to have had a brain of exceptional range and delicacy; he was probably one of the few men of this century who could have argued intelligently with Einstein. But to me he was just another figure of whom to be afraid; he was surrounded with a childish aura of terror. Sometimes I used to creep under the bushes and peer through a gap in the wall, and watch him prowling about his ragged, thorny garden. Through the unwashed windows I could see the rocks, standing against the walls of his drawing-room. Now and then he would pause, and glare in my direction; he had a habit of suddenly pulling his pink dressing-gown very tight, and flicking his glistening fingers above his head.

Apparently Nichols and his brothers did rather more than just watch Heaviside because, as he goes on to write,

> occasionally, during the summer months, my father would receive a letter from [Heaviside] on the subject of tennis balls which I, or my brothers, had inadvertently knocked into his gardens. These letters always began as follows:
>
> > *His Wormship, Professor Oliver Heaviside, (W.O.R.M.), presents his compliments to Mr. John Nichols, and wormfully requests permission to bring certain matters to his attention.*
>
> Then came the substance of the complaint. The letters were always signed Oliver Heaviside, W.O.R.M. It is an ironic thought that the author of these letters was largely responsible for bringing radio technique to its present pitch of perfection.

Nichols wrote that his father thought Heaviside to be simply a "grand joke," but to the young boy he was far more threatening. Nichols ends his remembrance of his scary neighbor with these words:

> to me, as a child, he was a bogy, a witch-doctor, another figure to fear in a world in which fear was the guiding force. Never would I venture into his garden, to fetch a ball that had been lost; I felt that once inside those walls I should never come out again . . . or if I did, that I should be changed, and have a light of madness in my eye, and should wander down the hill, and out into the world, a lost boy, whom nobody knew.

I learned of Nichols too late for the original edition of this book, but I was able to include an abbreviated summary of his recollections in the *Scientific American* essay.

CHAPTER 13: EPILOGUE

In October 1988 I received a letter from Professor Dirk Struik (1894–2000), professor emeritus of mathematics at MIT and a well-known historian of mathematics. He had been given a copy of this book by "my grandson the electronics engineer" for his 94th birthday, and wrote to tell me about his friendship with his MIT colleague Norbert Wiener. He also included some comments about Wiener's novel *The Tempter* (starring Heaviside as the thinly disguised hero). Struik told me that he and Wiener had had many conversations over the relative merits of using differential operators and the Laplace transform to solve differential equations; Struik believed that Wiener was "one of the first to see the connection [between the Laplace transform and the Heaviside operational calculus]." Indeed, when Wiener's electrical engineering colleague at MIT, Vannevar Bush, published his 1929 book *Operational Circuit Analysis,* Wiener contributed an appendix. It was on Fourier series and integrals, and of asymptotic series expansions of differential operators. Perhaps it was the writing of that appendix that originally brought Heaviside to Wiener's attention.

To quote from Struik's letter, concerning *The Tempter:* "he was quite excited when it appeared, went every day to the [campus bookstore] to see how many copies were sold, and was utterly depressed when a review appeared that was somewhat critical. 'Even Einstein cannot be a Dostoyevsky at the same time,' I told him."

There is one more interesting story concerning *The Tempter* that I learned about after this book appeared in 1988.

In 1993, nearly thirty years after Wiener's death, the MIT Press published a book by him (a preliminary draft had been written in 1954, for Doubleday, and then abandoned), titled *Invention: The Care and Feeding of Ideas.* Wiener had lost interest in *Invention* and had decided to put all of his energies towards his novel. As Steven J. Heims writes in his introduction to the MIT Press publication of the *Invention* manuscript (found among Wiener's papers in the MIT Library Archives),

> He was . . . starting to contemplate a more literary work [than his earlier two-volume autobiography, or his famous 1948 book *Cybernetics*]. He had for some time been thinking about the theatrical or novelistic possibilities of a dramatic tale in the history of invention in which Oliver Heaviside, a pioneer in the theory of telephonic communication, would be the hero while the American Telephone and Telegraph Company and the engineer Michael Pupin of Columbia University would be the villains. In 1941 Wiener had sent an outline of the plot and the major characters to Orson Welles, inviting him to use it as a basis for a film, but nothing ever came of this offer.

That last sentence *would* have been a tantalizing enigma to me except for a letter I had received in 1990, from John W. Verity of Brooklyn, New York. He had just read my *Scientific American* article on Heaviside and thought I might find a copy of Wiener's letter to Welles interesting, and indeed I did (the original is in the MIT Institute Archives).

Writing from South Tamworth, New Hampshire on June 28, 1941, Wiener began his letter (addressed only as to "Orson Welles, Esq., c/o Hollywood, California) with the heading

> Oliver Heaviside and the ATandT
> The spectacle of a single poor, deaf
> man negotiating with equal strength
> with a company greater than many nations
> and successfully defying it
> A movie proposal by Norbert Wiener

Wiener began his letter by telling Welles that he had just seen *Citizen Kane,* and thought it so well done that perhaps Welles could do the same with Wiener's proposal. In the proposal Wiener makes it clear that he despised Pupin (he called Pupin's 1923 Pulitzer prize autobiography *From Immigrant to Inventor* a "particularly nauseating panegyric"), and that he thought little better of Preece ("a fool"). Wiener respected AT&T's staff engineer George Campbell, however; although he did not specifically name him, he wrote of "Mr. C—(Still alive and well). The *real* inventor of the wave filter and originator of the details of modern loading-coil technique, as all his colleagues in the profession recognize. Not known outside the profession."

Wiener went on in his letter to give Welles a rather well-written crash course on the inductive loading of telephone lines by discretely positioned coils; he also explained why AT&T found that particular invention absolutely essential to support its growing web of *long* lines in those days before the invention of the vacuum tube amplifier. Wiener argued that AT&T needed to own and control the patent rights for loading, and no doubt would have paid Heaviside for such rights *if Heaviside had filed for a patent*—which he had not. So, as Wiener wrote, "since Heaviside had no legal papers to show his fatherhood of his intellectual child, it was necessary to find another daddy for the baby."

It was Wiener's thesis that AT&T took two parallel paths to "find a new daddy." First, it assigned its employee George Campbell to the problem of determining the proper spacing of the loading coils, a problem Campbell did indeed solve (and which Heaviside had not). Second, it somehow involved an outsider (Pupin) in the matter—Wiener claimed not to know how—from whom AT&T could then buy *his* patent rights. As Wiener wrote to Welles,

> C---'s work is the real patent . . . On the other hand, Pupin put in a claim . . . at the patent office [and] the matter came into interference proceedings in the U.S. Patent Office:—proceedings to which Heaviside was naturally not a party. Pupin won these, and AT&T paid for his rights . . . The company had a valid patent, established as valid by proceedings which had gone triumphantly against them . . . Pupin had a fortune, and the job of convincing himself that he had really deserved it.

Heaviside, of course, had nothing.

The invention of loaded cables is actually a bit more complex than Wiener's version; the first patent for a loaded cable was granted to the French inventor Lazare Weiller (1858–1928) in 1888. Since the original publication of this book a very nice paper on the history of that invention has appeared. Heaviside does indeed play an important role, to be sure, but there were others, too—see Helge Kragh, "The Krarup Cable: Invention and Early Development," *Technology and Culture,* January 1994. (I mention Krarup only very briefly in this book, in note 38 of Chapter 12, in connection with continuously loaded cables.) Wiener's understanding of the complex technical issues of an electrical invention may be the result of a pure mathematician trying to apply the laws of clean logic to decidedly messy human politics—in his *Invention,* for example, Wiener completely garbles, *twice,* the nature of the Edison effect (the existence of a current, under certain conditions, between two electrodes in a vacuum). One should, I think, take Wiener's explanation of hardware inventions not of his doing with a grain (or two) of salt.

Whatever the correctness of his views, it was his outraged sense of justice denied that prompted Wiener to write to Welles. He concluded his movie proposal in the strongest possible words against Pupin, writing (correctly) that "Heaviside despised him as a fraud," and that (perhaps incorrectly) "[AT &T] despised him as a stooge." Wiener declared that Pupin's name might be enshrined at Columbia University on the physical laboratory that bears his name, "but Pupin's real monument is in the hearts of his colleagues, and it is built of contempt." These are very harsh words, and it would be interesting to explore the possible reasons behind such extreme feelings—did Wiener and Pupin have a bad personal interaction at some time?

At the end of Chapter 6 of his *Invention,* Wiener continued to vent his anger at Pupin with almost unbelievable passion, comparing Pupin to Marlowe's Doctor Faustus who sold his soul to the devil:

> Heaviside may have been a very snuffy lower-middle-class Prometheus, but at least he had snatched a piece of fire for mankind. If the vultures of poverty and the sense of persecution were gnawing at his liver, he shared with Prometheus the sense of having performed a godlike feat. Pupin, on the other hand, had wrapped his soul as part and parcel of a commercial bargain. When a soul is bought by anyone, the devil is the ultimate consumer. Even a public penance was denied Pupin. Although he was unable to contain himself in silence, the lies and bluster which he was forced to resort must have echoed hollowly in the empty space where his soul had been.

One last comment on Wiener's letter to Welles; he wrote in a postscript "If you wish to verify my own *bona fides* or the authenticity of the incidents here related, I suggest you turn to Professor Eric Temple Bell, of the California Institute of Technology in Pasadena." E. T. Bell (1883–1960) was a remarkable and somewhat mysterious character; he was an accomplished mathematician (he made important discoveries in number theory), a best-selling author (his 1937 *Men of Mathematics* has never gone out of print), and a pioneering writer of science fiction (his 1931 novel *The Time Stream* is a recognized classic in the time travel genre). But why should he have been able to confirm the story of Heaviside to Welles? Did Bell perhaps learn of Heaviside from his Caltech math colleague, Harry Bateman, whose interest in the Laplace transform would have surely led him to Oliver's operators? Equally interesting is the question of why Wiener would refer Welles to Bell as a personal reference? In Wiener's two-volume autobiography (*Ex-Prodigy* and *I Am a Mathematician*) Bell's name does not appear even in passing. And nowhere in the 1993 biography of Bell by Constance Reid (*The Search for E. T. Bell*) is there even one mention of Wiener. One might wonder if the two men actually ever met.

It would seem from Wiener's postscript that they must have been acquainted (and in 1925–26 Bell did indeed spend a sabbatical semester at Harvard, just "down the street" from MIT); but why, of all people, did Wiener refer Welles to *Bell* concerning a project so obviously important to Wiener? Was it perhaps Welles' famous radio prank of 1939 (of a Martian invasion of Earth) and Bell's writing of science fiction that caused Wiener to think Welles would look more favorably upon Bell's words than words from others in the world of academic science? Well, I honestly don't know the answers to any of these questions; I'll simply end with the hope that some historian reading this essay will look into the Bell/Wiener connection (if any) and write a paper on it.

Newmarket, New Hampshire
June 22, 2001

Preface to the Original Edition

Why am I the author of this book?

What an odd question, you may think, for an author to ask his readers. Surely if anyone knows it should be the author, and if *he* doesn't know then God help us all! But my question isn't meant to be cryptic. Rather, why have I, an electrical engineer, a technical man who should feel more at home with a tough equation (but not *too* tough) or a page of FORTRAN computer code than with a magnifying glass trying to read century-old letters, why has such a person as myself written this book? Why not, instead, a professional historian?

It certainly isn't because historians of science are unappreciative of Oliver Heaviside. In the biographies of other scientists, and in the journal literature discussing the history of various technological innovations, Heaviside's name appears over and over again. And yet this is the first major treatment[1] of his life and work. Perhaps it is, as William Berkson wrote in his admirable *Fields of Force* (New York, NY: Wiley, 1974): "Because the bulk of his work is extremely technical and difficult, it is unlikely that a full scientific and personal biography of him will be written."

To unravel that "technical difficulty" was, indeed, no easy task. In our modern times, in which nearly everything has to be packaged for the "person on the go," including *Reader's Digest* essays (and even some so-called books) that can be consumed during the brief time one allows for washing down a hamburger with a diet soft drink, who is going to take the time to plow through Heaviside? There are mere *paragraphs* in his papers that can literally take hours to figure out! And the decipherment isn't just the reading of cramped handwriting. Heaviside's work is full of monstrously long and involved mathematics, enough to give even the technically trained a headache. It is no wonder that for an historian, trained to pursue *literary* and *social* threads through the fabric of his subject's life, the sight of all that intricate math and physics is more than sufficient to create a tidal wave of despair. To hell with Heaviside, they might think, and then turn their scholarly attention to Lincoln, Hitler, Faraday, or even Einstein. Lytton Strachey wrote in *Eminent Victorians* (New York, NY: Putnam, 1918), his biographical masterpiece that debunked the values of the well-to-do classes of Heaviside's 19th century England, that "it is perhaps as difficult to write a good life as to live one." He was right.

To write this book required that I be willing to spend hours and days and weeks and months and *years* reading Heaviside (often right off the filthy pages of century-old copies of *The Electrician*), and to be honest and frank with you, it *was hard to do*. It almost did me in, and I now understand why so many of his readers tore their hair out trying to read him. I do not sneer at historians who may have once thought of doing a book like this but then thought the better of it once they realized what they would have to go through. Perhaps, in fact, they showed good

[1] There was a book published in 1985, in Russian, that mainly treated Heaviside's controversial mathematics. I know of this book only through the very brief review of it that appeared in *Mathematical Reviews,* issue 87h, p. 4001, August 1987 (review 87h: 01063).

sense, and my decision to go ahead and do the book was really the result of what my mother and father always told me was my chief fault—a stubborn and occasionally unreasonable tendency to do the first most damn foolish thing that zipped through my mind as soon as I woke up in the morning. My wife assures me that I have not outgrown this fault, so perhaps that *is* the reason for this book.

But as Heaviside said after the Royal Society refused to publish his third paper on operational mathematics, "it is not sympathy that is particularly wanted"—I just want to make the point that a non-shallow book on Heaviside *must* treat his most technical and complex work, and that this almost automatically dictates that the author have an advanced technical background. There are, of course, historians of science with such backgrounds, but not many. So perhaps I just was the first person foolhardy enough to "do Heaviside."

In his brilliantly done history (*Lives and Letters,* New York, NY: Knopf, 1965) of literary biography in England and America, Professor Richard Altick wrote of the not-uncommon shallow literary biography, "...one could make a good story out of the life of an author without saying much about his works: it was what he was, not what he wrote, that count. By regarding authors solely as colorful human beings who led eventful lives...popularized literary biographies appealed to countless readers who had little interest in literature as such and were not especially eager to acquire any knowledge of it. For every reader of a vividly written biography of Byron who went on to sample *Don Juan,* there were thousands who were content to have enjoyed a guided tour through his love life."

To a much lesser extent this same misrepresentation has been committed in scientific biographies (Tesla and Einstein are examples), but not in this book. Heaviside certainly was a colorful character, even a mysterious one, and I have tried hard to get the flavor of those aspects of his life into the pages that follow, but it is his *work* that is of primary concern here. And that's the problem, of course—his work is so damn much work to figure out!

But no matter, it has been a labor of love. It has been the most enjoyable and satisfying intellectual accomplishment of my life. Indeed, as the job draws to a close I am bedeviled by the question, what do I do *now*? I have spent years reading old books, crawling through dusty, gloomy library stacks, writing letters to interesting people whom I no doubt would otherwise never have had the pleasure of knowing, and peering through my five-inch magnifying glass at Oliver's letters[2] from 1890, trying (but not always succeeding) to turn an ink-blob back into the word it once was. And I have loved every single minute of it! It has made me feel as close as I'll ever come to my hero, that intrepid professor of action and romance, Indiana Jones. (Is there a college professor *anywhere* whose heart doesn't pound with pride every time that scholarly teacher of archaeology bullwhips, shoots, and outsmarts the Nazis to the secret of the Ark of the Covenant?) Certainly *my* actual experiences have been more like those described in Richard Altick's *The Scholar Adventurers* (New York, NY: Macmillan, 1950)—the discovery of "mouse-chewed papers of an old family in a dormant English hamlet"—but the *possibility* of greater discoveries was always there, too.

And now that it is done, I am sad. Saying good-bye to the task that has occupied nearly all my free time for over six years is not easy.

Of course, how I'll feel about things a year or two from now will depend to some extent on how the book is received. I am well aware that historical books by non-historians, and

[2] I have learned, somewhat to my dismay, that the Victorians did *not* routinely draft letters to each other in beautiful handwriting. This myth, which I recall being drummed into my head sometime between the fourth and seventh grades, vanished in the reality of the uncrossed *t*, the undotted *i*, and the misshapen everything else in the correspondence found in the Heaviside archives.

particularly *technical* history books by technicians, are often at extra jeopardy.[3] How often I have read the review by an historian, of a technical history, that begins with the fateful sentence: "This is a book by an engineer." And we all know what *that* means, don't we? Nothing good, I assure you.

Well, I have done the very best I know how, and while my primary concern has been to present the story of Heaviside in a nontrivial yet interesting way to engineers, mathematicians, and scientists, I do hope that historians will find much in the book to interest them, too. I certainly owe a great deal to several historians who have willingly and graciously corresponded with me. To paraphrase Will Rogers, I have never interacted with an historian who didn't treat me with professional courtesy and scholarly collegiality. I hope I can at least partially justify their interest in my book by not now making them wish they'd saved their stamps.

One question I am often asked by those who hear I have been working on this book[4] is: "*Heaviside?* What did *he* do?" My own children are even more direct: why not write a book, they ask, that will make a lot of money, like one of the trashy horror novels that seem to come and go at bookstores and airport paperback stands like spots of breakfast on a white shirt? Or, perhaps, "The College Professor's Two-Week Wonder Diet Plan" book. In response I usually just give a boyish grin, shuffle my feet, and mumble something inane. Actually, of course, I have nothing at all against making money, and to be paid for a hot-selling trashy horror novel wouldn't bother me one bit. What *would* bother me is the sheer and utter transient shallowness of the doing of such a book. There could be no joy, for me, in producing what would be, in the end, today's trashy novel and tomorrow's recycled paper.

What a naive, romantic position, many will think, and that's just my point. Heaviside was one of the most romantic, naive characters of 19th century science, and I have found that to be an irresistible attraction. Who, for example, could *not* be fascinated by a man that, in the same letter to a technical journal, managed to discuss electrical science and to toss out a sneer or two at Archbishops (see Chapter 7)?

One of the romantic aspects of Heaviside and his colleagues is, to me, how they are at one and the same time so close and yet so far away from us. Kellow Chesney made this same point in his wonderful book on the Victorian underworld (*The Anti-Society,* Boston, MA: Gambit, 1970). In his opening words, "Much of the fascination of the Victorian age derives from its strange familiarity." There are people almost surely still alive today who saw Oliver in the flesh. If he had lived only fifteen years more (to age ninety, a not outlandish figure) he could have bounced *me* on his knee. His closest friend, who had once met the great Maxwell, lived until the end of 1954, when I was in high school. I could have shaken the hand of a man who shook hands with Clerk Maxwell! And yet Heaviside is *so far* away from us, too. When he was a teenager it would still be forty years *before* the Wright brothers would first stagger through

[3] For those who may not quite believe this I recommend reading the dual book reviews, written by Professor L. Pearce Williams, which appeared under the incredible title "Should philosophers be allowed to write history?", *British Journal for the Philosophy of Science,* vol. 26, pp. 241–253, September 1975. Williams is a Cornell University historian whose reputation rests primarily on his 1965 biography of Faraday. He claimed to be outraged at what he called "serious, indeed crippling flaws" in the two books under review (one dealt with Faraday in its entirety, and the other in significant part). Williams wrote savage reviews of both. The assaulted authors replied with equal ferocity in the same journal (vol. 29, pp. 243–252, September 1978), with Williams' final comments appearing immediately afterward (p. 252). There is a bit of irony in all this—a few years before, in the same journal, these roles had been reversed when Williams' Faraday biography was mauled by a famous *historian.* Reading the book review pages of a history journal can be more exciting than watching a Clint Eastwood movie!

[4] One particularly funny incident happened when I mentioned to a nontechnical acquaintance that I was writing a book on Oliver Heaviside. His eyes brightened in recognition and he said "Oh, you mean the old-time movie comedian!" It took me a few seconds before I realized how he had arrived at this astonishing conclusion. He was thinking of Oliver Hardy, the larger half of Laurel & Hardy, who was of course quite a bit on the *heavy side.*

the air a few hundred feet. Now all this may not send shivers up and down *your* spine, but it does mine. Ah, well, you must keep in mind I'm a college professor and that's a profession that *requires* a certain odd twist to the brain. Let me give you another example.

One day while in London to study at the Heaviside Collection, lovingly maintained by the Institution of Electrical Engineers, I took a bit of time off to stroll down Fleet Street. The editorial offices of *The Electrician* had been on Fleet Street, and while Heaviside apparently never actually visited there during business hours, he might well have walked down the same street some evening a century or more ago, just to see where his manuscripts were going. It was, at any rate, fun to *imagine* I might be retracing his steps. And as I walked along Fleet Street on to Ludgate Hill and then up to the magnificent St. Paul's Cathedral, I realized that the old General Post Office on St. Martin's-le-Grand, the lair of William Henry Preece (Oliver's great nemesis), had been just a few blocks over on my left. Preece, a hundred years ago, had surely walked right where I then walked. I was thrilled!

I recall how the thought of Preece turned my mind to imagining how the plump figure of that self-confident bureaucrat must have appeared as it sailed majestically among the throng of London-others, most far less sure than he of their place in the sun. Preece does not come off well in this book, but that is entirely his own fault. This is not a revisionist history; my views are essentially those of Sir Edmund Whittaker who more than a half a century ago wrote, "If, as seems likely, Preece is known to posterity only by what is said about him in the works of Heaviside, time will have brought an ample revenge."

Some may feel it mean-spirited to kick a man now expired (this silly dictum would allow every scoundrel with the luck to be dead to escape the pen of the historian, making life inestimably duller both for the historians and their readers). Others may leap to the defense of the departed with the cry of "Don't judge the man by our times, but by his own." By the measure of either time, I claim Preece has no defense, and since Preece is no longer here to dispute the issue, which he would surely do with enough bluster and bombast to refloat the *Titanic*, I have gone to great lengths in this book to support my general charge of a *malicious spirit*. To those who disagree with my position, I would reply *read Preece*, himself. The man was quite open in revealing his true mental state, and I have used *his* pen, not just Heaviside's, to draw the noose around his neck.

I *am* a romantic person (please understand, I do *not* mean I read nurse stories[5]), even if the last paragraph sounds a bit bloodthirsty, and I have tried very hard to get a flavor of that into this book. I would trade almost everything I have (with the exceptions, *perhaps*, of my health and the inexplicable affection a few dear friends seem to have for me) to experience just one day in the Victorian London of 1887. How I would love to hear, from Oliver's own lips, just what he thought of William Preece. I can guess what he would say, in essence, but just imagine *how* he might put it! The thought that it will never happen makes me almost faint with frustration. But I have not let this blind me to my historical obligations.

I started this book an uncritical admirer of Heaviside. He was, quite literally, my hero. And why not? I accepted the standard story of the impoverished, genius nobody who came out of nowhere and took on all the pompous, know-it-all government officials and snobbish university

[5] The sort of books I *do* consider to be romantic are the love stories told in the time-travel novels of Richard Matheson (*Bid Time Return,* New York, NY: Viking, 1975) and Jack Finney (*Time and Again*, New York, NY: Simon & Schuster, 1970). Matheson's masterpiece moved me so much I made a pilgrimage (from New Hampshire) to the old Hotel Coronado off the coast of San Diego proper so that I too could stand in the Hall of History. See Finney's wonderful preface to his more recent book, *Forgotten News* (New York, NY: Doubleday, 1983), for the best description I've ever read of what it means to be a *romantic*, in the historical sense in which I have used the word.

mathematicians. Good old Oliver! He showed them you can't keep a good man down! Sock it to 'em, Ollie!

Alas, as I learned more, old Ollie started to show his spots. I still admire Heaviside. He did the very best he could do, and that was very good, indeed. But I am not so sure that I would have *liked* Oliver. Maybe, but maybe not. This, however, is not, as I said earlier, a revisionist biography, and Oliver is not revealed to have been secretly Jack the Ripper, or to have been a collector of Victorian pornography. It is, instead, as true and historically accurate as I could make it. In a letter written near the end of his life to the President of the Institution of Electrical Engineers, Heaviside declared, "I don't want compliments, but just Facts and Justice." I have tried to honor that request.

In the Introduction to his scholarly and entertaining anthology, *The Historian as Detective* (New York, NY: Harper & Row, 1969), Robin Winks wrote, "In history perhaps more than in any other discipline the book is the man." That is certainly true in the case of this book, and my authorship is reflected all through it by a marked tendency to insert chatty, first-person commentary. This may offend the truly academic, but I really see no intellectual merit in striving for the dry and ponderous tone for its own sake. There is already plenty of that sort of stuff in the journal literature.

There is, of course, the concern for a biographer being too visible in a work that rightfully belongs to the subject. Even less welcome than children at an adult party, modern biographers are traditionally expected to be unseen as well as unheard. I admit I am well aware of early Victorian historian Lord Thomas Babington Macaulay's assertion that Boswell appeared in his own *Life of Johnson* as a conceited fool, as "a man of the meanest and feeblest intellect...servile and impertinent, shallow and pedantic, a bigot and a sot... ." Well, you get the idea. And I also recall Professor Altick's acid observation more than a century later (*Lives and Letters*) that even if it is true Boswell was conceited he "might at least have had the sense to remove himself from [the pages of his book] so that the fact would not have been so obvious." Ah, well, I do take some comfort in the fact that Macaulay also called Boswell "the first of biographers." For those who don't like my style, I am sorry. But what you read here *is,* for better or worse, the way I do things and I am far too old, much like the proverbial dog, to change my ways now.

Finally, I would like to address a question that historians will certainly find odd (as I do), but one which I hear all too often from a certain kind of technologist. "Who cares," asks this breed of electrical engineer as he glances up from his integrated circuit breadboard, "about what happened a hundred years ago? It's all obsolete, dead, and gone. But if you want to talk state-of-the-art, well then let's boogie!" The hardcore "state-of-the-art junkie" has no doubt not even read *this* far, of course, but for the broader technical type who does think it is important to know about the origins of his or her profession (but perhaps still feels a bit guilty taking the time to read books like this), let me offer a little solace. You are not alone. George Francis FitzGerald, for example, wrote to Oliver in 1898 and asked for a photograph. To his surprise he actually got one, and wrote back to say of Heaviside's well-known reluctance to seek publicity, "I can see no good reason for depriving your fellow creatures of an opportunity for satisfying a *natural and innocent curiosity* [my emphasis]." FitzGerald did, I must in all honesty add, conclude his letter with, "This craving for information as to the private affairs of individuals is a sort of intellectual itch from which the nineteenth century suffers severely." FitzGerald's itch is still here in the 20th century, though, and I say that the *right* thing to do with an itch is to scratch it!

There is a long and honorable tradition for finding enlightenment (as well as entertainment) in other people's lives. As James Boswell recorded in his biography of Johnson, Dr. Johnson enthusiastically declared in 1763 "the biographical part of literature is what I love most." Or as Boswell wrote twelve years later in his own journal, in a curious twist on the words of Shakespeare's Othello just before he murders his wife Desdemona, "Be there a thousand lives, My great curiosity has stomach for 'em all."

With all this now said, let me conclude with the hope that you enjoy reading this book even just half as much as I loved writing it. As a great man (whose name I've forgotten) once said, "Sometimes all we need to make this life a happy one is for someone to admire our mudpies."

A Note on Mathematics

To write a history of the life and work of a man like Oliver Heaviside *without* the appearance of mathematics would be to defraud readers. Exactly *how* to present the mathematics, of course, depends greatly on the intended audience. This book is frankly not written for those seeking a way to "kill time," but rather for those who understand mathematics at the level of the first two years of college (integral calculus and differential equations) and who don't faint at the sight of a square root. A good number of bright high school seniors are at that level these days!

This is *not* a "gee-whiz" book, the sort of biography that Nikola Tesla has, for example, unfortunately received. Tesla has attracted mathematically shy (in my opinion) biographers precisely because he was strictly an *intuitive* genius whose greatest insight, the rotating magnetic field, came to him in finished form (as far as *he* took it) literally "in his head." He neither used nor needed analytic reasoning. If Tesla ever performed a mathematical analysis, or displayed even the slightest comprehension of Maxwell's theory, I am unaware of it. This is not meant as a criticism of Tesla (he was, indeed, a genius, and the fact that he lived his life promoting a self-aggrandizing[1] fantasy world just makes him all the more interesting), but it goes far in explaining some of his biographers, who wouldn't know a rotating magnetic field from the stationary football kind. How much easier it must have been for those writers to ramble on about Tesla's crazy talk of "rings around the Earth," "death rays," and the "telegeodynamic oscillator to destroy the Empire State Building," than to get into the real scoop on multiphase power! When such writers take a look at what *Heaviside* did, why they must slam his books shut in horror and try to recover from "math anxiety" by reading an issue of *The National Enquirer*.

My position on mathematics is not as extreme as Robert Heinlein's, who had his immortal (in every sense of the word) character, Lazarus Long, utter these brutal words: "Anyone who cannot cope with mathematics is not fully human. At best he is a tolerable subhuman who has learned to wear shoes, bathe, and not make messes in the house." Strong words, yes, and maybe just a little too rough, but I do sympathize with the intended message. As Lord Kelvin's famous aphorism[2] makes the point, "when you can measure what you are speaking about, and express it in numbers, you know something about it; but when you cannot measure it, when you cannot express it in numbers, your knowledge is of a meager and unsatisfactory kind: it

[1] Readers who think I am being too hard on Tesla should read his strange article "The problem of increasing human energy" in the June 1900 issue of *Century Magazine*. In this very long and very peculiar article he claimed (among many claims) to have discovered errors in Hertz's research and so he "long ago ceased to look upon his [Hertz's] results as being an experimental verification of the poetical conceptions of Maxwell." Hertz's big mistake? He neglected the effect of air on the oscillation frequency of his circuits! Tesla also claimed in this same piece that crystals are "living beings."

[2] For a very interesting *social* history of this, see: R. K. Merton, D. L. Sills, and S. M. Stigler, "The Kelvin dictum and social science: An excursion into the history of an idea," *Journal of the History of the Behavioral Sciences*, vol. 20, pp. 319–331, October 1984.

may be the beginning of knowledge, but you have scarcely, in your thoughts, advanced to the stage of *science*.''

But, of course, *too much* math can be just as deadly and so I've attempted to seek the middle ground with the device of end-of-chapter *Tech Notes*. The chapters themselves are nearly all prose, with mathematical *results* stated when appropriate. For readers who wish to see more detail (along with even more prose, speculation, and historical information), the *Tech Notes* are designed to help satisfy that curiosity. Realizing that math if not used on a regular basis gets rusty (just like any other tool), the *Tech Notes* have been written in a tutorial style that is in no way intended to be patronizing. The *Tech Notes* can be skipped, but even those who would rather tread lightly on the math will, I think, find in them much of interest.

A Note on References

Citations from Oliver Heaviside's five books (the two volumes of *Electrical Papers* and the three volumes of *Electromagnetic Theory*) are so numerous in this book that I have adopted the abbreviated notations, EP and EMT. For example, EP 1 and EMT 3 denote, respectively, volume 1 of *Electrical Papers* and volume 3 of *Electromagnetic Theory.* There have been several reprints of these books over the years, but my sources have been the very excellent editions by Chelsea Publishing Company (New York, NY: 1970, 1971), and I thank Chelsea for permission to quote from its works.

The Institution of Electrical Engineers (IEE) in London has an extensive collection of Heaviside's research notebooks, his copies of *Nature* with many interesting comments written in the margins, and personal letters. When referencing an annotated *Nature* I have used the notation AN; e.g., AN, May 9, 1889, p. 32 refers to Heaviside's personal copy of *Nature* owned by the IEE. There are also 21 surviving notebooks in the IEE archives, numbered 1, 1A, 2, 2A, 3, 3A, 4, 5, ..., 18. For example, NB 13:313 refers to Notebook 13, page 313. Unless otherwise noted, all letters from which quotations have been taken are in the IEE Heaviside Collection.

Much (but not all) of the IEE material is also available on microfilm at the Niels Bohr Library of the Center for History of Physics (American Institute of Physics) in New York City. In addition, Microfilms International Marketing Corporation has announced they have microfilm copies for sale of the entire IEE Heaviside Collection.

And finally, coming to my attention literally as the last words of this book were being set into type, was a book of recollections by Heaviside's close friend, G. F. C. Searle of Cambridge University. I had access to all of the surviving Heaviside/Searle correspondence, but this short (79 pages) book (*Oliver Heaviside, The Man,* I. Catt, Ed., St. Albans, Herts: CAM, 1987) may include additional insights. I have not seen the book, but an interesting review of it by A. C. Lynch can be found in *Electronics & Power,* vol. 33, p. 469, July 1987.

A Note on Money

To understand the ''worth'' of money is a confusing business under the best of circumstances, even when it is your own currency of the present day. But what if it is money from a hundred years ago, and perhaps (for non-British readers) *foreign* money to boot? Just *what,* for example, does it *mean* to say a telegram could be sent across England for sixpence in 1885, or that beginning in 1896 Oliver received an annual Government stipend of 120 pounds? Questions like these come up in the book from time to time, and I have tried to put the issue of the ''worth'' of Victorian money in a proper context at most such occurrences. For the curious reader, however, I can recommend the delightful book *Busy Times* by E. Royston Pike (New York, NY: Praeger, 1969), which devotes an entire chapter of thirty pages to this single topic.

A major point to keep in mind, in any case, is that the British now have a modern decimal currency, with 100 pence (or pennies) to the pound sterling. In Heaviside's day, however, there were 240 pence to the pound (i.e., 20 shillings, each the equal of 12 pence). And as a curious testament to the nature of the financial minds behind the Victorian mint, I remind readers that the now obsolete *guinea* was a gold coin worth precisely one shilling more than a pound.

Acknowledgments

My name alone appears as author, but this book is really the result of the efforts of so many other, generous people who have shared their considerable talents with me. The several library staffs that assisted me have opened my eyes to how academic professionalism and scholarly curiosity exist far beyond the province of the teaching faculty that all too often think these virtues belong only to themselves. The staff members of the Dudley Knox Library of the U.S. Naval Postgraduate School in Monterey, California, where I spent 1981–1982 as a visiting faculty member, were most kind and helpful. The Library's complete run of the *Journal of the IEE* was invaluable. Similar aid was efficiently rendered by Michael Kohl (Head of Special Collections at Clemson University's Robert Muldrow Cooper Library), Karl Kabelac (Manuscripts Librarian in the Department of Rare Books and Special Collections at the Rush Rhees Library of the University of Rochester), Ellen Fladger and Ann Seemann (Archivist and Librarian, respectively, at the Schaffer Library of Union College), and John Aubry (Niels Bohr Library, Center for History of Physics, at the American Institute of Physics in New York City).

The Inter-Library Loan (ILL) system in this country was of inestimable value, providing me with ready access to out-of-print books, long-forgotten articles, and the complete run of the now decaying *Electrician*. The Center for Research Libraries in Chicago, and the Baker Memorial and Kresge Libraries of Dartmouth College, in particular, provided special assistance. This book would have been unthinkable and impossible without ILL. At my own campus library, Dimond Library of the University of New Hampshire (UNH), very special thanks must go to Library Specialist Jane Russell and Head Reference Librarian Professor Deborah Watson, both of whom provided professional counsel of the highest caliber as I progressed with my writing. In addition, Professor Watson did the German translations, made sure I dotted my umlauts, kept my English spelling of Russian names honest, read the entire book in galley proofs (a grubby but necessary task, as was observed by the new father while changing diapers: "The *fun* is in the creation!"), and helped prepare the index. Professor Robert Morin helped me overcome my faltering French. Library Assistants Karen Fagerberg, Susan Metcalf, and Diane Webb would deny they gave me any special treatment, but somehow I always seemed to get my ILL microfilm, book, and photocopy requests processed in record time. Their super service and friendly good cheer did not go either unnoticed or unappreciated. And I must not forget Associate University Librarian Dr. Diane Tebbetts, who tracked down several incredibly obscure 19th century literary references after I had given up the chase in despair. I still don't know how she did it. "All in a day's work," she would tell me, but I *know* there was magic involved!

For his friendship and shared love for the "scholarly hunt" which made me feel so at home in Dimond Library (as well as for a bottle of sherry, a magnificent dinner, and marvelous conversation at the *Crystal Quail* in the depths of a spooky New Hampshire forest on a frosty Halloween night), I owe much more than I can ever hope to acknowledge to UNH University Librarian Dr. Donald Vincent. He, along with his colleagues Reina Hart, J. C. Kapoor, Constance Reik, Frank Adamovich, Nathalie Wall, Arthur Lichtenstein, Robin Lent, Barbara Lerch, Robert Reed, Joan Griffith, Kevin Coakley-Welch, and all the other Dimond Library staffers mentioned before, "took in" this somewhat gone astray electrical engineer who wanted to write a history book, a stranger to their literary world, and made me feel a welcome part of it.

In London, Eleanor Symons, the Archivist of the Institution of Electrical Engineers (IEE) has been *for years* a willing correspondent, as well as a gracious host during a research visit in 1985 to the IEE Heaviside Collection. It was Eleanor who taught me that while the English *really do* take their afternoon tea seriously, there could be no munchies allowed while reading historical documents. Thanks to her there are no strange jam spots on Oliver's letters (at least not *mine*) to mislead future historians. Her able staff, Allyson Whyte and Jean Robertson, also did much to make my stay at the IEE as productive as possible. And I *did,* so very much, enjoy Allyson's excitement as she helped me discover in the IEE Archives an ancient, dusty photo album containing a long-lost, never before published photograph of Charles Henry Walker Biggs. My pleasure was magnified by hers.

I am especially indebted to Jean K. Gossick of Lexington, Kentucky, who graciously allowed me to study her late husband's photocopy collection of correspondence written by Heaviside to George Francis FitzGerald. B. R. Gossick, Professor of Physics at the University of Kentucky, died in 1977 before he could significantly use these letters in his own research. I greatly regret not having had the opportunity to know Ben Gossick. The holder of the literary rights to the letters is the Royal Dublin Society (RDS), Ireland, and I thank Alan Eager, Librarian at the RDS for permission to quote from the Gossick copies. The microfilm version of the IEE Heaviside Papers at the American Institute of Physics (AIP) in New York was originally owned by Professor Gossick, and its present availability to all Heaviside scholars is, again, due to the generosity of Jean Gossick.

For whatever reasons, I found it impossible to obtain from the traditional historical research funding sources the necessary travel money to visit archival holdings. Fortunately for me, however, the University of New Hampshire has an enlightened group of administrators who saw nothing odd about an electrical engineering professor writing a history book, and so provided me with the needed funds. For two travel grants (from the Faculty Development Program and the Central University Research Fund), as well as a 1986 Summer Faculty Research Fellowship which paid the rent, I am most grateful. My former Department Chairman, Professor Ronald Clark, also did the best he could for me out of a thin department travel budget.

Ever since I began this book, my present Chairman, Professor John Pokoski, encouraged me to continue by his willingness to let me barge into his office at any time to tell him breathlessly of my latest astonishing discovery about Ollie. His eyes would glaze over now and then, but he never once threw me out (although I am sure he *thought* about it). I must also thank all the rest of my fellow EE professors who, with very little complaining, put up with the debris left on the department photocopy machine after each of my sessions with a disintegrating volume of *The Electrician*. I cleaned up as best I could, but a thin layer of 19th century dirt I had overlooked seemed always to be clinging to something.

The illustrations in this book play an important role, and I devoted not a small amount of energy to finding what I feel are just the right ones. Some of them have never been published before. I was efficiently and pleasantly aided in this quest by Joyce Bedi, Curator of the IEEE Center for the History of Electrical Engineering, Louise Hillard at the AIP Center for History of Physics, and Dorothy Nelhybel at the Burndy Library, Norwalk, Connecticut. UNH photographers Ronald Bergeron and Yvette Croteau made perfect copy negatives and prints of every illustration I brought to them, including those from volumes of *Punch* so ancient they had to hold their breath to keep from inhaling the ink from right off the pages.

Academic historians who helped me in some way or another, often in what might well have seemed like minor ways to them but not to me, include Richard Altick (Ohio State), Bruce Hunt (University of Texas/Austin), James Brittain (Georgia Tech), Richard Kremer (Dartmouth), Edward Layton (University of Minnesota/Minneapolis), and Stewart Gillmor (Wesleyan).

For all my students who have had to listen to countless "Oliver stories" for the last decade, what *can* I say? *Thanks for listening,* and *I am sorry* are perhaps both appropriate.

W. Reed Crone, Managing Editor of the IEEE PRESS, and Randi Scholnick, the book's editor, have together seen this book through from its initial proposal, to paper at last running swiftly through inky presses. Dr. Ronald Kline (former Director of the IEEE Center for the History of Electrical Engineering and now on the faculty at Cornell) has given me the benefit of his own extensive research into the relationship between Charles Steinmetz, Heaviside, and Professor Ernst Berg of Union College (who met and corresponded with Heaviside near the end of Oliver's life), as well as shared his insights into the evolution of Heaviside's operational calculus. He also alerted me to the availability of Heaviside's letters to and from Berg at Union College, and once went beyond mere professional courtesy by taking time from his own duties to look up and send me a nearly forgotten, ancient journal article from the IEEE archives that I desperately needed on short notice.

The final revisions of the book were written in several places, but in particular I wish to thank the staff of the Chief Petty Officers' Mess of the aircraft carrier USS FORRESTAL. The CPO Mess of CV-59 provided me with a congenial work place for an uninterrupted week, as well as allowed me to verify that U.S. Navy coffee is, *indeed,* the thinking man's brew!

And, of course, I must thank Alice Greenleaf, who typed the entire manuscript *at least* ten times. Nan Collins, head of my college's Word Processing Center, wisely assigned the book exclusively to Alice to maximize revision efficiency, and the result was that Alice is now among the world's authorities on Oliver Heaviside. I hope this expression of gratitude to, and affection for, one of the nicest ladies I know, will be some compensation for all the red-ink manuscript drafts I dumped on her desk for so many years.

Acknowledgments for the Johns Hopkins Edition

All of the photographic illustrations in this book have been reset for this new edition. Producing new prints, under a tight deadline, from dozens of old negatives that had spent the last fourteen years rather casually tossed into an office drawer was the task of Lisa Nugent, Douglas Prince, and Beverly Conway, staff photographers at the University of New Hampshire (UNH). They were more than equal to the job, and their efforts are much appreciated. My friend Donald Vincent, late University Librarian at UNH, liked the first edition of this book very much. His good opinion meant a great deal to me, and he encouraged me to continue my historical writing. Don died in 1994, but, even now in the Other Place with Oliver, I hope he would be pleased with this printing, too.

OLIVER HEAVISIDE

1
The Origins of Heaviside

To the majority of people the unusual and arresting name of Oliver Heaviside conveys nothing.

— Richard Whiddington, from his essay
"Oliver Heaviside" in *The Post-Victorians*

Heaviside's life was in some respects a great tragedy, but he emerges from it a man of tremendous intellectual stature....

— Willis Jackson, *Nature*

Heaviside's whole life can be read as a kind of allegory on the ultimate triumph of virtue.

— D. W. Jordan, *Annals of Science*

The Man

Who was Oliver Heaviside?

That question has many answers, so let me give you a few of them straightaway. He was a man born in mid-Victorian times at a low social and economic level, who, with no formal education after the age of sixteen, eventually came to be accepted as the intellectual equal of the finest scientific minds of the day. He was a man who lived as a recluse among his relatives, having resigned from his one and only job at age twenty-four. He then devoted the next thirty-five years of his life to first-rate scholarly research, the publication of technical papers of astonishing achievement, and the carrying-on of some of the most interesting correspondence this side of the secret love letters (if only they had ever been written) between Oscar Wilde and Lola Montez.

He was a man who often was incapable of conducting himself properly in the most elementary social interactions (meeting a local townsperson while on a walk around the grounds of his home inevitably led to a note in his diary about what a nasty business he had just endured). As far as is known, his only continuing contacts with women were limited to his mother, nieces, and housekeepers. He was a man who knew the power of money and desired it, but refused to work for it, preferring to live off the sweat of his family and long-suffering friends whom he often insulted even as they paid his bills.

In short, Oliver Heaviside was a highly complex man, with some rather rough edges. He often was an uncivil, insensitive, boorish individual. But he was much more, too. He was a mathematical physicist of the first rank, an electrical engineer with amazing powers of analytic and physical insight, and an accomplished artist at the game of literary invective never before or since equaled in the scientific literature. He could also be an extremely funny fellow in print and didn't hesitate, even in the midst of a swirl of equations, to insert an hilarious jab at some individual who had displeased him. In one of his copies of *Nature*, in the margin next to the printed observation "The virtues of science are exactness, impartiality, candor," Heaviside wrote[1] "Candor should come first." Clearly, then, this is a man worth learning about!

The Nature of His Work

There is a strange paradox involved in trying to understand just what it was Oliver Heaviside did. To understand Einstein's special theory of relativity, for example, one needs mathematics no more advanced than algebra. To truly delve into Heaviside's major works, however, requires a bit more, indeed some fairly heavy mathematical artillery. The paradox in this is that Einstein's theory applies almost everywhere in the universe (one notable exception is the inside of a black hole where nobody knows *what* happens), while Heaviside's efforts are mostly applicable to much more restrictive situations, e.g., the insides of an electric cable. It seems superficially that more mathematics is required to understand the details of a telephone circuit than to comprehend the most intimate secrets of Mother Nature.

Another part of the paradox is that no matter how much theoretical physicists may grumble about it, Oliver Heaviside's work on how to make a decent telephone cable plays a vastly greater role in our everyday lives than does Einstein's work (knowledge of the equation behind The Bomb, $E = mc^2$, is *not* essential for its making,[2] i.e., Hiroshima and Nagasaki would have been doomed even had Einstein never lived). We all make several telephone calls nearly every day, and we expect those on the other end of the line to be able to understand the sounds that emerge from their receivers. How often, on the other hand, does the conduct of your life depend on the nuances of relativity?

To many the study of telephone cables may seem a bit pedestrian, smacking just a little too much of shirt-sleeve engineering (with its common public image of brawny arms all black with dirt and wet with sweat), and rather unworthy of being compared to Einstein's cerebral work. Even if that were true (and it isn't, as we'll see with Heaviside's theory of the distortionless circuit), we could alternatively ponder one of Heaviside's more abstract studies, one that touches on the very foundations of quantum mechanics, a subject even Einstein never quite came to grips with, psychologically. Heaviside, to be sure, is not to be placed in the same class of intellect as Maxwell and Einstein (who, with Newton, by those who like to make rankings, are considered the supreme physicists of all time). But with this example we see Heaviside operating at a very high level, indeed.

First, a basic fact. An accelerated electric charge radiates energy. This statement is verifiable by mathematical manipulation of Maxwell's electromagnetic field equations (about which I'll have a great deal more to say, since it was Heaviside, *not* Maxwell, who first[3] expressed the theory in modern mathematical form using vector notation), and is the cornerstone of radio, television, radar, and microwave ovens, too, for that matter. Indeed, the theoretical prediction by the equations of the possibility of electromagnetic radiation predated *by over twenty years* the experimental discovery by Heinrich Hertz in 1888.[4] What happens, in greatly simplified terms, is that the negatively charged electrons in the metal of a television station's antenna are made to oscillate back and forth to the tempo of the evening news broadcast, and since the electrons are then constantly changing their velocity they are, by definition, *accelerating* (or, in what amounts to the same thing, decelerating). This results in the radiation of energy, and the news, into space which can be intercepted by the antenna of your home television receiver, possibly many miles distant.

Now, just one more basic fact. Great progress was made in the late 19th century in understanding the nature of matter and, in particular, the now familiar image of the atom as a heavy nucleus (with positive electric charge) orbited by electrons was developed. There was a paradox associated with this image, however—an orbiting electron is an *accelerated* charge since even if it moves at constant speed it is continuously changing its direction (and thus its velocity). Therefore an orbiting electron should radiate energy and in so doing spiral in toward the central nucleus; i.e., according to 19th century theory all atoms should *collapse*, an event difficult to overlook! The radiation rate is predicted to be so huge, in fact, that no atom should be able to survive much longer than one ten-thousandth of a millionth of a second. Physicists call this (erroneous) prediction the "violet death of the universe" as the entire cosmos should collapse in one blinding flash of high-frequency radiation.[5]

Heaviside was among the first[6] to calculate this enormous (and luckily for us, unobserved) rate of energy loss in 1904, as a special case of a more general analysis he published in 1902, thus drawing attention to the fact that something was not quite right with classical electromagnetic theory, at least when applied to atoms. As he bluntly put it, "The radiation of energy is very rapid—so great as seeming to shut out the possibility of anything more than momentary persistence of revolution [of the electron]." Heaviside was tremendously frustrated by this failure of his beloved electromagnetic theory, and page after page in his surviving notebooks show he was never able to resolve the issue, try as he did. It would be the 1920s, in fact, before this embarrassing situation would be fully explained by the invention of a new physics (quantum mechanics) based on revolutionary new concepts introduced by such intellectual giants as Max Planck, Niels Bohr, Erwin Schrodinger, Paul Dirac, and Werner Heisenberg.

Quantum mechanics tells us the accelerated electrons *bound* to an atom do not radiate except during an orbital transfer. The energy radiation behavior of *free* electrons, on the other hand, is nicely explained by Heaviside's equations. One exotic application of them, indeed, is of major concern today (and is so bizarre that Heaviside could not possibly have imagined it). High-altitude nuclear explosions emit powerful gamma rays that transfer their energy to electrons in air molecules via a physical process called Compton scattering. These newly energetic, relativistically fast electrons are then forced into curved (accelerated) trajectories by the Earth's magnetic field, and thus they radiate energy. The net effect is to transfer a significant fraction of the bomb's explosive energy into a powerful electromagnetic pulse (EMP) of radio energy that can damage Earth-based electronic and orbiting satellite systems many miles distant. Indeed, the dangers of EMP were first discovered during an H-bomb test in July 1962, some hundreds of miles above Johnston atoll in the Pacific. Electrical systems, much to everybody's shocked surprise, were damaged in Hawaii, 800 miles away, by the blast's EMP.[7]

The calculation of the energy loss by an accelerated charge at any speed is now a routine exercise in college physics courses, but I know of not one textbook writer that gives Heaviside credit for it—it is just one of those things most students believe "has always been known." As a final comment on this bit of forgotten lore it is amusing to read Heaviside's last word on the topic. After presenting a simplified derivation of his earlier 1902 radiation results, he concluded with a sharp dig at the needlessly turgid writing so often found in the technical literature, an act very few modern scientists would dare copy (which is regrettable): "I hope this will be satisfactory. If not, there are lots of other more complicated ways of doing the work." A study of his research notebooks suggests, however, that this remark may also have been made in relief as well as fun. Written[8] among the rough calculations (showing all the dead ends he first explored) of the work that eventually appeared in *Nature*, are the words "It is wonderful how little work there is when you know how to do it."

The Grim World of Heaviside's Youth

Oliver Heaviside lived through the first quarter of the 20th century, but he was born at the beginning of the mid-Victorian age and his character was set in concrete (his hardheaded behavior makes this almost the literal truth) as that period drew to a close in 1875. He was definitely a man of his time and social class, and to understand how Heaviside became Heaviside it is essential to say a few words about the society that molded (one is tempted to say *warped*) him.

The year he was born Queen Victoria had been on the throne thirteen years. It was a time when "the wealth of Britain was regarded as fabulous; and the prosperity of the wealthier of its citizens was all the more conspicuous for the sharpness of their juxtaposition with the many who appeared to remain in something like immemorial poverty."[9] A survey taken in 1872–73 showed, for example, that less than 0.025% of the population owned 50% of the land.[10] The grossly nonuniform distribution of wealth went hand in hand with a tremendous class consciousness, far beyond merely the difference between having and not having money. It was Heaviside's bad luck to start life as a new statistic among the poor, Godless masses.

The lower classes were actually made up of three layers—the "dregs of humanity," the merely poor, and skilled craftsmen who might one day (with luck) actually move up to the lower middle class. Heaviside's parents were one of the merely poor at the time of his birth, but some years later they had progressed a step or two. Heaviside's brothers were solid middle class later in their lives (and he shared in their prosperity as he sponged off them). There was a "cult of work" in those times, and Heaviside's lack of employment might seem inconsistent with that spirit until it is realized that the "cult" was mostly associated with the fascination by the middle and upper classes for visible "respectability." "Respectability" had a low priority in Heaviside's value system. Heaviside's brothers eventually rose above their low social class at birth, but Heaviside himself remained true, all his adult life, to the disinterest in steady employment for regular wages held by the lowest classes in Victorian England. This was to cause much friction and ill feelings in Heaviside's family relationships.

It is probably impossible to paint too dismal a picture of the appalling way of life for the poor in Heaviside's youth, as "the conditions of the poor masses both at work and at home were miserable beyond the imagination of most of their mid-20th century descendents."[11] The conditions of life then were probably much like they may be again if the world suffers the apocalypse of general nuclear war.

When Heaviside was born, for example, it was a time when even the upper classes rarely had a water closet in their homes and the privy reigned supreme, followed (not too closely, one can hope) by the open hole over a cesspit, and the wet-midden (an uncovered heap of human excrement, holder of the dubious honor of being the "greatest of all forms of sanitary barbarism"). Mid-19th century rural life in England is often romanticized these days, but consider the following words of Benjamin Disraeli (twice British prime minister) describing a small countryside town as he saw it in 1845: "Before the doors of the crumbling cottages of the workers ran open drains full of animal and vegetable refuse, decomposing into disease, or sometimes in their imperfect course filling foul pits or spreading into stagnant pools, while a concentrated solution of every species of dissolving filth was allowed to soak through, and thoroughly impregnate, the walls and ground adjoining."[12]

In 1859 Hippolyte Taine, a French writer and traveler, had much the same to say: "The

"The Black Country"—the industrial landscape of 1866 at Wolverhampton, Staffordshire.

whole [British] countryside seems to be a fodder factory. The mere anteroom to a ... slaughter house... ."[13]

Life in the cities was no better, and they could be extremely dangerous places. Typhus, small pox, diphtheria, scarlet fever (which would partially deafen Heaviside), and typhoid were common—East London was swept by an Asiatic cholera[14] epidemic as late as 1866–67. The source of all this misery was a lack of sanitation that we would, today, consider criminal. The rivers and canals were used as public sewage pipes, and into them were continually dumped raw excrement, dead animals, and the discarded bones and blood from slaughter houses. It was not a rare sight to see birds walking on the surface of streams, so thick was their pollution. In the summer of 1858, for example, the always terrible smell of the Thames reached such depths as to drive the M.P.s of the House of Commons from their deliberations; Charles Dickens bluntly called the river an open sewer, and Michael Faraday echoed him by calling it "a fermenting sewer."[15] One can speculate that if London was literally Hell on Earth (as that "merry old town" was often described in those awful times), then the Thames would surely have been an excellent choice for the Styx.

The degraded and sinister Thames was the worst, but not the only, offender of the public health. Consider, for example, the following iterative quote[16] from one of the more interesting books ever published on Victorian sanitation:

> The River Cam, like the Thames, was for many years not the best of places for a pleasure outing. Gwen Raverat, in her book about her Cambridge childhood, *Period Piece*, wrote: "I can remember the smell very well, for all the sewage went into the river, till the town was at last properly drained, when I was about ten years old. There is a tale of Queen Victoria being shown over Trinity by the Master, Dr. Whewell, and saying, as she looked down over the

PUNCH, OR THE LONDON CHARIVARI.—July 21, 1855.
FARADAY GIVING HIS CARD TO FATHER THAMES;
And we hope the Dirty Fellow will consult the learned Professor.

This illustration from *Punch* was inspired by a letter Faraday wrote to the London *Times*. In addition to the cartoon, the magazine wrote, ''Because we are losing brave men by war (the Crimean War), it is rather the more desirable than otherwise that we should not also lose useful citizens by pestilence, as we certainly shall if the Thames continues much longer to be an open sewer. We hope that Professor Faraday's publication will effect a saving of human life still greater than that which has resulted from his predecessor's (Sir Humphry Davy) safety-lamp.''

bridge: 'What are all those pieces of paper floating down the river?' To which, with great presence of mind, he replied: 'Those, ma'am are notices that bathing is forbidden.' ''

The year Heaviside was born was typical of the times with respect to housing—there wasn't nearly enough. London and the countryside were home to over 80,000 people crammed into what were called ''lodging houses,'' places often not much better than houses of ill-repute ''without ventilation, cleanliness, or decency, and with forty people's breaths perhaps mingling together in one foul choking steam of stench.''[17] When Heaviside was but a few months old the overcrowding was so bad as to be called ''a pestilential heaping of human beings.''[18]

The noise of city life (particularly in the London of Heaviside's youth) was an awful din. The steam engine was King as the prime mover for the factories and if a steam engine is anything, it *isn't* quiet. These factories burned coal to run the engines, and England's cities were forests of chimneys all belching forth a steady rain of coal dust.[19] Outside the factories was the constant racket of horses clattering along cobblestoned and wood-blocked streets. The uproar was enormous, but at least as vexatious must have been the pollution all those horses left daily in the streets, literally by the ton. To add its contribution to the thundering pandemonium was the ever expanding railroad network which by 1850 was already carrying nearly 70 million passengers annually (over 490 million by 1875).[20]

To be a poor man with a family, as was Heaviside's father, was a depressing business in those days. The social services we have all come to think our due were then still the fantasies of

''Over London—By Rail,'' an 1872 wood engraving by Gustave Dore.

"MAMMON'S RENTS"!!

"NOW, THEN, MY MAN; WEEK'S HUP! CAN'T 'AVE A 'OME WITHOUT PAYIN' FOR IT, YER KNOW!"

This 1883 cartoon from *Punch* illustrates how the poor despaired over the price of even just a hovel. It was inspired by the anonymous publication of a twenty-page pamphlet entitled *The Bitter Cry of Outcast London: An Inquiry into the Condition of the Abject Poor.* Its language was direct and blunt, with one passage reading:

> Whilst we have been building our churches and solacing ourselves with our religion and dreaming that the millenium was coming, the poor have been growing poorer, the wretched more miserable, and the immoral more corrupt... . Few who will read these pages have any conception of what these pestilential rookeries are... . To get into them you have to penetrate courts reeking with poisonous and malodorous gases arising from accumulations of sewage and refuse scattered in all directions... . You have to grope your way along dark and filthy passages swarming with vermin. Then, if you are not driven back by the intolerable stench, you may gain admittance to the dens in which these thousands of beings, who belong, as much as you, to the race for whom Christ died, herd together.

utopian dreamers. The reality then was a twelve-hour (or even longer) workday, and even a skilled craftsman measured his *annual* income at just a few hundred dollars. To lose your job was a terrifying thing, with no immediate future to look forward to except either humiliating (and insufficient) aid from private charity, or public relief *if* you were declared "worthy." If you were without a job too long you ran the risk of penury and the horrors of the workhouse (which were so awful that some people preferred death).

Even *with* a job, working conditions were often unpleasant. Employers did little for their workers and it was an accepted practice, for example, for men to bring their own coal to work each morning to burn in the company stove if they didn't want to freeze in winter. Those were times when the attitude of the factory owners was "the average wage-earner is not a man of flesh and blood. He is a most important figure, but not human."[21] It shouldn't be surprising then that such conditions of squalor and misery, both at home and work, drove many Victorian men to seek even the temporary relief of a drunken stupor at the local gin and beer shops. Violent crime and other forms of antisocial behavior were common in the cities, and to travel about the London slums without police escort (or knowing what you were doing) was a risky business. Mid-Victorian gentry were rightly terrified of the tenement districts, such as the Camden Town borough of St. Pancras (where Heaviside was born, and which has been described as a "ghastly slum"[22]).

If one was unfortunate enough to suffer a medical catastrophe, life and death could depend on social status. One typical tale is that of the last days of Charles Mansfield, the founder of coal tar chemistry. At one o'clock in the afternoon of a frigid February day in 1855, Mansfield and his eighteen-year-old assistant, George Coppin, blew themselves up in a benzene/air explosion of such force it collapsed the building in which they were working. The accident occurred on Regents Canal near St. Pancras Station (it is possible the Heaviside family could have heard the blast), and both men ran in flames toward the canal. Because the water was frozen over, however, there was a delay in putting the fires out and they were eventually transported to different hospitals "like shrivelled mummies rather than human beings."[23] Coppin, an Irish youth of "low status," was taken to the nearby Royal Free Hospital, while Mansfield had to endure a longer, agonizing trip to the distant Middlesex Hospital which was more befitting his loftier social level. Both men died of burns far beyond the day's medical capabilities,[24] but the point is clear—even in a situation of extreme emergency, social class was a major factor.

It was into this smelly, noisy, unhealthy,[25] and class-conscious world that Oliver Heaviside was born the son of a poor man on May 18, 1850. How he managed to persevere, and to finally triumph[26] in his odd way from such an impoverished beginning, is the story of the strength of the human spirit and of Heaviside's own remarkable personality.[27]

Notes and References

1. AN, October 8, 1908, p. 585.
2. It was believed well *before* Einstein that energy is related to mc^2. See, for example, *The Feynman Lectures on Physics,* vol. 2, Reading, MA: Addison-Wesley, 1964, pp. 28-1 to 28-4.
3. Much more will be said on this point later. It suffices here to say that nowhere in Maxwell's writings do the equations for the electromagnetic field appear as we write them today. Maxwell used an amalgamation of Cartesian component and quaternion notation, and it was Heaviside who first wrote the electromagnetic field equations in modern vector form.
4. Two interesting papers on the reason for this long delay in experimental verification of Maxwell's theory are by T. K. Simpson, "Maxwell and the direct test of his electromagnetic theory," *Isis,* vol. 57, pp. 411–439, 1966; and A. F. Chalmers, "The limitations of Maxwell's electromagnetic theory," *Isis,* vol. 64, pp. 469–483, 1973. To see how many others came close to predating Hertz but failed to understand what they had found, see the paper by C. Susskind, "Observations of electromagnetic-wave radiation before Hertz," *Isis,* vol. 55, pp. 32–42, 1964.
5. Readers should be careful not to confuse "violet death" with the similar sounding "ultraviolet catastrophe" for blackbody radiation, a separate and distinct failure of classical 19th century physics to explain observed behavior. This latter difficulty was also, curiously, a stimulus (indeed, the major one) for the development of quantum mechanics, and its resolution was due to Planck's conceptual discovery of energy "quanta" in 1901.
6. Heaviside was the first to find the radiation loss rate for *relativistic* (i.e., fast) electrons. See F. Rohrlich's "The electron: Development of the first elementary particle theory," in *The Physicist's Conception of Nature,* J. Mehra, Ed., Dordrecht: D. Reidel, 1973, p. 345. Sir Joseph Larmor had earlier (1897) calculated the loss rate for slow electrons, and his work reduces to a special case of Heaviside's. In addition, while Larmor realized the radiation loss should lead to the collapse of atoms, he tried to argue that the arrangement of the *several* electron orbits in molecules would result in a "null vector sum," i.e., as he put it, "The condition for no radiation is a natural law of the system." Heaviside, on the other hand, realized there was a problem and said so. Heaviside's work on radiation from an accelerated charge seems not to be widely known, even among physicists. In 1950 Richard Feynman (co-recipient of the 1965 Nobel prize in physics) independently rediscovered Heaviside's 1902

work. See J. J. Monaghan's "The Heaviside-Feynman expression for the fields of an accelerated dipole," *Journal of Physics,* vol. 1, pp. 112–117, 1968.

7. An outstanding, elementary (but not trivial) analytical treatment of EMP can be found in D. Hafemeister's article, "Science and society test VIII: The arms race revisited," *American Journal of Physics,* vol. 51, pp. 215–225, March 1983.
8. NB 13:313.
9. G. Best, *Mid-Victorian Britain,* New York, NY: Schocken Books, 1972, p. 3.
10. E. C. Black (Ed.), *Victorian Culture and Society,* New York, NY: Walker and Co., 1974, p. 3.
11. J. Roebuck, *The Making of Modern English Society from 1850,* New York, NY: Charles Scribner's Sons, 1973, pp. 27–28.
12. Quoted from Roebuck (Note 11), p. 2.
13. Taine was a shrewd and keen observer, who had a real flair for vivid imagery. He wrote, for example, after seeing the weather-stained statue in Trafalgar Square of Admiral Horatio Nelson, "That hideous Nelson, planted upon his column, like a rat empaled on the end of a stick!"
14. Cholera was a ubiquitous Victorian disease that claimed not a few lives of science. Sadi Carnot, the French engineer who laid the foundations of thermodynamics with his studies of steam engines, was killed by it in 1832, and Professor James Thomson (the father of William Thomson, later Lord Kelvin, a man with some influence in Heaviside's life) died of it in 1849.
15. F. S. Schwarzbach, *Dickens and the City,* London: The Athlone Press, 1979, p. 166; and L. P. Williams, *Michael Faraday,* New York, NY: Simon & Schuster, 1971, p. 496. Readers will be relieved to learn that some years later up to 50,000 tons of sodium manganate and sulfuric acid were being used, *daily,* to disinfect the London sewers (see *Nature,* vol. 32, pp. 415–416, September 3, 1885).
16. W. Reyburn, *Flushed with Pride: The Story of Thomas Crapper,* Englewood Cliffs, NJ: Prentice-Hall, 1971, pp. 24–25. This book (published with an impressive-looking toilet on an eye-catching pink cover) purports to tell the story of the man who, as a little thought will show, gave his name to the language *three* times—as a verb, a noun, and an adjective. It is a masterpiece of controlled wit, and of course is a complete fraud. Still, it does convey an accurate "flavor" of the sanitation of Victorian times, and is well worth the effort it may take to locate it today. Gwen Raverat, however, was no fraud (she was a granddaughter of Charles Darwin), and the quote is not fiction.
17. Quoted from Best (Note 9), p. 28.
18. Quoted from Best (Note 9), p. 60. H. G. Wells, whose own childhood was every bit as awful as Heaviside's, wrote, "... history fails to realize what sustained disaster, how much massacre, degeneration and disablement of lives was due to the housing of London in the nineteenth century." (From *Experiment in Autobiography,* London: Macmillan, 1934, p. 225.)
19. When mixed with one of London's infamous clammy fogs, coal dust could truly be a killer. See H. P. Dunn's essay, "What London people die of," *The Nineteenth Century,* vol. 34, pp. 875–898, December 1893.
20. Best (Note 9), p. 72. *The Times* (of London) saw an ancillary benefit to the coming of the railroads, picturing them as a means of slum clearance. In March 1861 it declared, "We accept railways with their consequences, and we don't think the worst of them for ventilating the City of London... . You can never make those wretched alleys really habitable, do what you will; but bring a railway to them, and the whole problem is solved." (From A. Welsh, *The City of Dickens,* London: Clarendon Press, 1971, p. 36.)
21. Quoted from Best (Note 9), p. 5. In the same vein, "A sick, crippled, or merely worn-out employee was an unregarded casualty of the system, cast out to suffer and die." (From R. D. Altick, *Victorian People and Ideas,* New York, NY: W. W. Norton, 1973, p. 43.)
22. Quoted from Best (Note 9), p. 31.
23. E. Ward, "The death of Charles Blachford Mansfield (1819-1855)," *Ambix,* vol. 31, pp. 68–69, July 1984.
24. For a fascinating (often stomach-wrenching, too) account of what medical care was like during Heaviside's youth, see M. F. Brightfield's "The medical profession in early Victorian England, as depicted in the novels of the period (1840-1870)," *Bulletin of the History of Medicine,* vol. 35, pp. 238–256, May–June 1961.
25. Many of Heaviside's London thought of it as literally a City of the Dead, a necropolis with Pluto as its mayor. Charles Dickens was highly aroused over this ghoulish aspect of London and often wrote about it in his novels (e.g., *Bleak House* and *Our Mutual Friend*), and in nonfiction essays in his weekly journal *Household Words.* In 1850, the year of Heaviside's birth, it was observed that "Under a surface of ground not amounting to 250 acres there had been interred within thirty years in the metropolis far more than 1,500,000 human beings. What must be the condition of the atmosphere affected by the exhalations from that surface?"
26. In a poll of its members, the Institute of Electrical and Electronics Engineers placed Heaviside among the top ten technical contributors to electrical engineering prior to 1900, and also for the period 1900–1939. See "Centennial Hall of Fame," *IEEE Spectrum,* vol. 21, pp. 64–66, April 1984.
27. As one writer put it in an essay comparing Heaviside to Lord Kelvin, "Being an outstanding student at Cambridge brought Thomson [William Thomson, Lord Kelvin's "real" name] in contact with scientists of the first rank, and he learned without effort how to get along with his peers. That pleasing quality was essentially absent in Heaviside, whose personality was as prickly as a porcupine, almost without grace." See B. R. Gossick's "Heaviside and Kelvin: A study in contrasts," *Annals of Science,* vol. 33, pp. 275–287, 1976. In another essay,

Heaviside is declared to have been "a wildly eccentric person and his work was notoriously difficult to understand" and "First-rank scientists who knew of his work respected him and learned much from his work, but in the main he was just not read. He had all the marks of a crank... ." See W. Berkson's highly recommended *Fields of Force: The Development of a World View from Faraday to Einstein,* New York, NY: John Wiley & Sons, 1974, p. 198. Berkson reported (p. 351) that London University's copy of Heaviside's two-volume classic *Electrical Papers* still had uncut pages nearly half a century after Heaviside's death! Michael J. Crowe put it in the following amusing way: "Heaviside's style can be placed somewhere between brilliant and obnoxious depending on one's point of view. At least he was never dull... ." Quoted from Crowe's *A History of Vector Analysis: The Evolution of the Idea of a Vectorial System,* Notre Dame, IN: University of Notre Dame Press, 1967, p. 169.

2
The Early Years

The following story is true. There was a little boy, and his father said, "Do try to be like other people. Don't frown." And he tried and tried, but could not. So his father beat him with a strap ... .

— Oliver Heaviside's opening words to EMT 3

The Beginning

A surviving photograph of the London house in Camden Town where Heaviside was born, on King Street (subsequently renamed Plender Street, between Camden High and Royal College Streets), near Regent's Park, shows a large, plain, three-story structure. It displays a portico that seems oddly out of place, hinting at a grandeur that is nowhere to be seen. It is a dreary-looking place (Heaviside called it[1] "horrid, with low neighbors and bad drains"). It is easy to look at it and shudder at what it must have been like to live in during the 1850s. The place has several windows in the front wall, but they are all clustered about the portico itself, leaving vast expanses of the building's side walls sealed solid and tight against the sunlight. It appears that each room must have had just one window, or even none at all.[2]

In a remarkable study[3] of London by Charles Booth, King Street is shown near a section color-coded for "Lowest class. Vicious, semi-criminal." The general area of Camden Town is described by Booth as having "many shops, most having lodgers above," "good deal of poverty among the labourers," and "A considerable amount of poverty exists in the centre, mixed with a bad element. Many families only occupy one room." Charles Dickens, who lived for a while as a boy in Camden Town (as did Tiny Tim in *A Christmas Carol*), occasionally used his memories of it in his stories. In *The Posthumous Papers of the Pickwick Club* (in the story of George Heyling's revenge), for example, he described Camden Town as "... whatever it may be now, was in those days a desolate place enough, surrounded by little else than fields and ditches." Heaviside was familiar with Dickens' works, and references to the novelist and his characters occasionally appeared in his papers.

Heaviside lived in the gloomy, dangerous dungeon on King Street until he was thirteen, and a miserable thirteen years they must have been to cause him to later declare[1] that they had "permanently deformed his future life." In a letter[4] written in 1897, he recalled his childhood in these bitter words:

> I was born and lived 13 years in a very mean street in London, with the beer shop and baker and grocer and coffee shop right opposite, and the ragged school just around the corner. Though born and raised up in it, I never took to it, and was very miserable there, all the more so because I was so exceedingly deaf that I couldn't go and make friends with the boys and play about and enjoy myself. And I got to hate the way of tradespeople, having to fetch the things, and seeing all their tricks. The sight of the boozing in the pub made me a teetotaller

Heaviside's birthplace at 55 King Street, Camden Town. With the exception of the portico, it bears an astonishing resemblance to Charles Dickens' house on Bayham Street, which was also the residence of Wilkins Micawber, in which David Copperfield lodged. See Plate VI in A. L. Hayward's *The Dickens Encyclopaedia*, Hamden, CT: Archon, 1968 (first published in 1924). Heaviside occasionally referred to Dickens' characters in his writings and, in an interesting coincidence, Dickens once inadvertently (but all too appropriately) returned the favor. A man with an odd sense of humor, Dickens had an ersatz book bound for his home library bearing the title *Heaviside's Conversations with Nobody*.

> for life. And it was equally bad indoors. A naturally passionate man [Heaviside's father], soured by disappointment, always whacking us, so it seemed. Mother similarly soured by the worry of keeping a school. Well, at 13, some help came, and we moved to a private house in a private street. It was like heaven in comparison and I began to live at once.
>
> C. Dickens, when he was at the blacking manufactory, lived in a lodging just around the corner and I know exactly how he got his most intimate knowledge of the lower middle class, as shown in his early papers. It was before my time, of course, though.

Heaviside's deafness (due to a bout with scarlet fever) was to bother him all through life. Even in his published work he referred to it; for example, in one article[5] when discussing how the human ear can decipher a voice in the midst of telephone circuit noise: "... some remarkable examples (of interpreting indistinct indication) may be found amongst partially deaf persons who seem to hear very well even when all they have to go by (which practice makes sufficient) is as like articulate speech as a man's shadow is like the man." In his private papers he was not so poetic, but bitter; for example,[6] "Got rid of deafness partly. ... Everything in this life that you want comes too late."

Perhaps being the youngest of four brothers[7] made his life a little less awful than it might have been, if only because his father, Thomas, had four targets over which to spread his anger. The opening words to Oliver's third volume of *Electromagnetic Theory* no doubt tell us a lot about his childhood with Thomas: "The following story is true. There was a little boy, and his father said, 'Do try to be like other people. Don't frown.' And he tried and tried, but could not. So his father beat him with a strap; and then he was eaten up by lions." Oddly enough, however, Oliver did not become estranged from Thomas, even though his father ruled the family as a despot, was subject to violent outbursts (one of which permanently drove Oliver's oldest brother, Herbert, from the home), and believed, as the previous quote indicates, in severe corporal punishment as an appropriate form of discipline.

It isn't hard to find the reason behind Thomas' anger. His was the frustration of a skilled artisan, a wood engraver with a fair amount of talent, who nonetheless was only marginally able to support his family. His was an occupation in transition to oblivion, with the death knoll for wood engraving having already been struck, even before Oliver's birth, by the invention of photography in 1839. One older history[8] of photography refers to it as "the greatest boon ever conferred on the common man," but it is doubtful Heaviside's father would have agreed. Wood engraving wouldn't be truly dead until the late 1890s, sometime after the development of the halftone process for the simultaneous printing of text and photographs, but even the 1850s were already very difficult times for Thomas Heaviside.

Thomas was frequently ill and often required the aid of his brothers (who were also engravers[9]) to complete his work. So often, in fact, did the family's finances drift near the edge of disaster that Oliver's mother, Rachel, decided to open a private school for girls. Before her marriage she had worked as a governess (one of her former students may well have been William Spottiswoode, in whose family service she had been employed, who later became a President of the Royal Society), and she believed her experience could be turned into badly needed money. It was at his mother's school, starting in 1855, that Oliver received his first formal instruction. He initially rebelled at being the lone boy in the midst of giggling girls, but after his father showed him the alternative by dragging him across the street to the regular neighborhood school (populated by a large number of young Camden Town roughnecks) he dropped his objections. In 1858 his mother placed him in the High Street School, St. Pancras, which provided education for boys of the "intermediate classes."

When Rachel's school eventually failed in 1862 she tried a new tactic—that of renting out part of their home to lodgers. This worked much better, to the extent that the family was finally

Oliver inherited some artistic ability from his father, as shown in these two pictures. The top one is inscribed in pencil in his own hand: "The Cart Horse. By Oliver Heaviside. Aged 11." The bottom one carries the legend "2nd Work by Oliver Heaviside. (No others preserved.)"

able to leave King Street for the new home Oliver found to be ''like heaven in comparison.'' He was now attending the Camden House School,[10] where his principal teacher was Mr. F. R. Cheshire, a man whom Oliver found worthy of his respect. Oliver was a good student, but never took anything as true merely because it came from an ''authority.'' In some ways he must have been a pain in the neck for Mr. Cheshire, particularly so when the class took up its lessons in geometry.

The traditional, axiomatic approach to geometry was originated by Euclid well over 2000 years ago and is commonly taught to this day in a singsong ''theorem–proof'' style. It is an approach that, while often helping to avoid sloppy reasoning, can be incredibly boring even to those who otherwise enjoy mathematics. The rigid nature of Euclid was repellent to Oliver's already developing pragmatic approach to scientific matters.[11] Mr. Cheshire, however, taught the subject according to Euclid, and it is not difficult to imagine his young student's loud and continual complaints. Mr. Cheshire, to his almost certain regret, had no awful alternative ''to drag Oliver across the street to'' as had the boy's father, and so Euclid continued to be ignored.

Many years later Oliver would write[12]:

> Euclid is the worst. It is shocking that young people should be addling their brains over mere logical subtleties, trying to understand the proof of one obvious fact in terms of something equally ... obvious, and conceiving a profound dislike for mathematics, when they might be learning geometry, a most important fundamental subject ---. I hold the view that it is essentially an experimental science, like any other, and should be taught observationally, descriptively and experimentally ... .

And then further on in the same essay, he continued:

> I feel quite certain that I am right in this question of the teaching of geometry, having gone through it at school, where I made the closest observations on the effect of Euclid on the rest of them. It was a sad farce, though conducted by a conscientious, hard-working teacher. Two or three [of Heaviside's classmates] followed, and were made temporarily into conceited logic-choppers, contradicting their parents; ... .

This attitude may explain why, when he took the 1865 College of Preceptors[13] Examination (with questions covering English, Latin, French, physics, chemistry, and mathematics), Oliver scored well above the minimum passing score of 600 with 1140 out of 2600, took fifth place overall out of a total of 538 candidates (of which he was the youngest), and scored first in the natural sciences, but could do no better than a dismal 15% in Euclidean geometry. Although he apparently performed satisfactorily in chemistry, this subject, too, was later recalled with sarcastic words[14]:

> Chemistry is, so far, eminently unmathematical (and therefore, a suitable study for men of large capacity who may be nearly destitute of mathematical talent) ... .

It is pretty certain that young Oliver's academic disdain wasn't limited to just Euclid and chemistry. Here's what he had to say[15] about a subject that probably has confused more young minds than has even poorly taught geometry:

> I always hated grammar. The teaching of grammar to children is a barbarous practice, and should be abolished. They should be taught to speak correctly by example, not by unutterably dull and stupid and inefficient rules. The science of grammar should come last, as a study for learned men who are inclined to verbal finnicking. Our savage forefathers knew no grammar. But they made far better words than the learned grammarians. Nothing is more admirable than the simplicity of the old style of short words, as in A sad lad, A bad

This Book has been awarded as a
Prize
to O. Heaviside,
pupil of F. R. Cheshire Esq A. C. P.,
of Camden House School,
Camden Town,
as being the First in Natural Science
of the Candidates examined
at Christmas 1865.

The bookplate in Oliver's 1865 prize award.

dog, of the spelling book. If you transform these to A lugubrious juvenile, A vicious canine, where is the improvement?

A Lucky Marriage

Oliver's struggling parents perhaps did the very best they could for their sons, given their limited resources. It was by an accident of fate, however, that they gave their boys, and in particular Oliver, a way to break free of the London slums. Rachel's sister, Emma, had in 1847 made a very good marriage to the then already well known electrical scientist and inventor Charles Wheatstone (after 1868, Sir Charles). Wheatstone today is actually better known among electrical engineers and physicists than is his nephew.[16] This is doubly ironic since, first, it is clear that Oliver's discoveries are of a far more fundamental and lasting nature than those of Wheatstone and, second, Wheatstone's continuing recognition rests primarily[17] on the ubiquitous "Wheatstone Bridge," a measuring instrument found in every high school and college physics lecture room, and electrical instrumentation laboratory in the world. The Bridge, however, is a device actually invented by someone else (Samuel Hunter Christie, a

Sir Charles Wheatstone (1802–1875)

Secretary of the Royal Society who knew Wheatstone quite well, first described the circuit in 1833), as Wheatstone himself always made abundantly clear!

Wheatstone was a highly successful scientific entrepreneur (at his death in 1875 his estate totaled over 70,000 pounds, a considerable amount of money even today, and a fortune then) who conducted his business affairs much like a lawyer or doctor in private practice. He was a highly visible role model of enormous attraction for the Heaviside boys (during Oliver's teenage years the Wheatstones lived in a house on Park Crescent, at the southern edge of Regent's Park, not far from the Heavisides, and the two sisters were on close terms), far more so than their debt-ridden father. Wheatstone's influence undoubtedly had a major impact on their lives.

Wheatstone was a man of extremely varied interests and talents; indeed, he had first arrived in London to enter the musical instrument-making business (his family was one of instrument makers, music publishers, and music teachers). In 1822 (at age twenty) he opened Wheatstone's Musical Museum[18] where, among other exhibits, he displayed the Enchanted Lyre, or Acoucryptophone, a cluster of supposedly automated horns driven by a fantastic clockwork mechanism to provide a half hour's program of robotic entertainment—in fact, it was a gentle fraud, with the music actually provided by hidden human performers!

With such a wonderful uncle as this it is no surprise that Oliver soon developed a lifelong interest in music, and learned how to play the piano[19] and the aeolian.[20] His brother, Charles, had his musical talents encouraged by Wheatstone, too (he became very proficient on the "English concertina," an instrument invented by Wheatstone[21]), and he eventually became a partner, and then sole proprietor, in a music store business.

Wheatstone's influence was most profound on Oliver, and at his suggestion the boy added German and Danish to his language studies. This last subject may[22] have been particularly helpful when later, with the aid of his uncle's connections, Oliver became a telegraph operator and spent some time on assignment in Denmark. His other brothers, Herbert and Arthur, also became involved with telegraphy.

First (and Last) Job

Very little more of Heaviside's earliest years is known,[23] except that at age sixteen he left school for good, his mathematical training extending certainly no further than algebra and trigonometry. There then followed a two-year period in which he remained at home pursuing his own self-designed course[24] of studies (which included teaching himself the telegraphic Morse code). Finally, in 1868, he ventured forth to take the first (and only) job he would hold during his entire life. This solitary period of employment would last just six years.

Oliver began his temporary stay in the working world as a telegraph operator with the Dansk-Norsk-Engelske Telegraf Selskab (Danish-Norwegian-English Telegraph Company), in Fredericia, Denmark, at the grand wage of 150 pounds per year. This was rather nice pay for a callow youth (and further supports the belief that Wheatstone must have provided very strong support for his nephew). By comparison, Herbert Heaviside earned an annual wage of just 90 pounds in 1870, working as a clerk in the telegraph offices of the Post Office in Newcastle-on-Tyne. Arthur was also in the telegraph business, having joined the Universal Private Telegraph Company at Newcastle-on-Tyne in 1861 (he stayed in its employ as a District Superintendent when the operation of all domestic telegraph services in England was transferred to the monopoly control of the British Post Office in 1870).

During his stay in Denmark Oliver observed that the rates at which intelligible telegraph signals could be sent across the 347 nautical mile Anglo-Danish cable (laid in September 1868) were highly dependent on the direction of transmission (England to Denmark in 1868 was 40% faster than the other way). And further, these rates varied with time, gradually decreasing (and becoming more nearly equal) during the two years Oliver observed them. The reason for this time-varying asymmetry was a mystery at the time (although it was suspected the occurrence of cable faults might be the explanation), and so it would remain until Oliver explained it later in his career (just how, we will see in Chapter 4). Until Heaviside presented his analyses, the state of telegraph cable theory remained as it had been left in 1855 by Professor William Thomson (later Lord Kelvin). Thomson's theory was a diffusion theory, modeling the passage of electricity through a cable with the same mathematics that describes how heat flows (or how an ink drop spreads in water), while Heaviside's final theory was a wave propagation theory. Thomson's analysis was adequate for the case of a very long, low-signal-rate cable where the effects of capacitance dominate those of inductance as was, for example, the situation for the 2000-mile-long transatlantic cable of 1866, joining Ireland and Newfoundland, but totally inadequate for the high-speed, relatively short, induction-dominated, overland telegraph lines that were rapidly coming into use. The high frequencies present in human speech (an order of magnitude higher than those in telegraphy) meant that Thomson's theory was also an inadequate one for long-distance telephony.

With the implementation of the Telegraph Acts of 1868 and 1869, which made the operation of telegraphs a government monopoly, came the disappearance of small, private companies,

Oliver's brother and occasional collaborator, Arthur (1844–1923).

and a shaking out and consolidation of the large international operations. The Dansk-Norsk-Engelske Telegraf Selskab was taken over by the Great Northern Telegraph Company in 1870, and Oliver was transferred to its Newcastle-on-Tyne office. There he was appointed Chief Operator in 1871 and given a raise to 175 pounds per year. One of his co-workers remembered him this way[25]:

> Oliver Heaviside was the principal operator at Newcastle—appointed no doubt by the influence of his uncle, Sir Charles Wheatstone. He was usually on day duty. He was a very gentlemanly-looking young man, always well dressed, of slim build, fair hair, and ruddy complexion.

From another account[6] we learn of Heaviside that "... he was a short (5′4 1/2″), red-headed Englishman of autocratic disposition, and of superb powers of mental penetration and intuition,... ."

During this phase of his employment he was not just a landlocked office telegraph operator. On September 11, 1871, for example, he wrote "Spent days waiting for 'Caroline'. It was always leaving tomorrow. A week spent in grappling, cutting and splicing cable." Other brief notes made at about the same time discuss various electrical experiments he performed in his free time, and also reveal broader interests. He devoted considerable thought and observation to the nature of human vision, earthquakes, and the weather (on April 9, 1871, for example, he noted that three earthquakes the previous month had apparently thrown so much dirt into the atmosphere that the sky's appearance made him think the "Day of Doom had come").

We can get an idea of his level of technical sophistication at this time from a letter he wrote near the end of his life, dated October 8, 1922. This letter was written to the parents of William Gordon Brown, a remarkably promising young theoretician who had been killed, at age twenty, in 1916 in France.[26] They had sent Heaviside a copy of their son's paper "On the Faraday-tube theory of electromagnetism," published in 1922 in the *Proceedings of the Royal Society of Edinburgh*, to ask his opinion of it. In his reply to the Browns, Heaviside wrote, "For a youth of 20 he was surprisingly advanced. Why, at his age I didn't know anything at all of Analysis, nor about Electricity, though I had made several inventions (telegraphic) at that time, and was trying to see my way."

In July 1872, Oliver began to give public evidence of his "superb powers" with the appearance of his first technical paper, "Comparing electromotive forces," in *English Mechanic*. This paper uses mathematics no more advanced than algebra, but his second publication in the February 1873 issue of *Philosophical Magazine* makes use of the differential calculus. This paper discusses an optimal arrangement of the Wheatstone Bridge and there is a note in one of Heaviside's notebooks[27] that Sir William Thomson discussed it with him during a visit to Newcastle. This same note includes the sentence "Sent Maxwell a copy of it, and Maxwell noted it in his 2nd Ed.″," which, indeed, he did.[28] It is clear that Heaviside was doing more with his evenings than idling the time away smoking a pipe by the fire. Indeed, on his own, he had by now mastered the work of Isaac Todhunter (calculus), the "damnable book"—as Heaviside called it—by George Boole (differential equations), and the work of Duncan Gregory and William Walton (solid geometry). Later, in 1874, he added George Airy's book on partial differential equations to his growing technical library. But as useful as these mathematical books were, none could approach the impact of the 1873 publication of James Clerk Maxwell's *A Treatise on Electricity and Magnetism*. This masterpiece by the genius Maxwell without doubt gave direction and inspiration to the genius of Heaviside, and Oliver was literally swept up by its power.

On May 31, 1874 he left his job as Chief Operator at Newcastle. He did this only a few months after joining the Society of Telegraph Engineers,[29] the next logical step in his developing professional career.[30] It has been speculated that increasing difficulties with his deafness had something to do with the decision, but much more likely reasons were bruised feelings and boredom. When he submitted his letter of resignation he gave no clue himself (it merely stated "I have obtained a situation elsewhere."), but two of his superiors wrote[31] revealing comments:

> Heaviside was a clever worker, but somewhat unruly and especially so since informed that he was not given an increase in salary. His application to leave the Company is therefore no great disaster, especially as it will mean a saving.

Oliver Heaviside Phil Mag. Feb 73
Great Northern telegraph company Newcastle on Tyne
on Best Arrangement of Wheatstone's Bridge

d resistance to be measured (given)
f battery resistance (given)
e galvanometer res (given)
then for best arrangement

$$a = \sqrt{ef}$$

$$b = \sqrt{de\,\frac{d+f}{d+e}}$$

$$c = \sqrt{df\,\frac{d+e}{d+f}}$$

tested J.C.M

$d = \gamma$ resistance to be measured

$f = a = B \qquad c = \sqrt{\alpha\gamma} \quad \sqrt{a\alpha} \quad \sqrt{BG}$

$e = \alpha \quad G \qquad b = \sqrt{a\gamma\,\frac{\alpha+\gamma}{a+\gamma}} \quad \sqrt{B\gamma\,\frac{G+\gamma}{B+\gamma}}$

$a = c$

$b = \beta \qquad \beta = \sqrt{\alpha\gamma\,\frac{a+\gamma}{\alpha+\gamma}}$

$c = b$

A note in Maxwell's hand on Oliver's Wheatstone Bridge paper. This note was found among Maxwell's papers after his death and was first reported on in the article by A. M. Bork, "Physics just before Einstein," *Science,* vol. 152, pp. 597–603, April 29, 1966.

and

> Very clever and reliable … . His conduct is exemplary, but in view of his age, high salary, and good qualifications, more interest in the daily work should be expected.

A Lifetime Decision

It is, perhaps, worth looking at why Heaviside elected to skip university work. Why, in fact, didn't he submit to a few years of extra academic work, which he surely could have done with little difficulty, and earn the degree that could have added "official" authority to his writings? A lack of money certainly isn't the answer, since Heaviside didn't have a job anyway—if his parents and brothers were willing to support him even as he scribbled mathematics all day with no obvious reward, then surely they wouldn't have objected to his attending classes! Indeed, they might have welcomed *that* as a respectable alternative, compared to what they actually had to put up with—an apparently idle relative whom the neighbors must have thought bordered on being a ne'er-do-well.

With the passage of more than a century, and because of Heaviside's lack of autobiographical interest, it is all speculation today as to what really motivated him. I believe that the answer may, however, be quite unmysterious and direct. Heaviside may have simply concluded in his pragmatic and stubborn way that there just wasn't anything for him to learn at a university. Heaviside's background and interests in the early 1870s were those of a practical engineer, and as one writer put it at the turn of the century,[32]

> It was not so very long ago that engineers became willing to recognize that technical training had an academic side at all.

Electrical engineering was a shirt-sleeve, workingman's job in those days, one sharply distinct from the atmosphere of scientific training provided by a university. As one ditty of the day put it:

> Some talk of millimetres, and some of kilogrammes,
> and some of decilitres to measure beer and drams;
> but I'm a British workman, too old to go to school;
> so by pounds I'll eat, and by quarts I'll drink and
> I'll work by my three-foot rule.

But let's suppose Heaviside *had* decided to attend a university, say Cambridge. There would have been the issue of having to take an oath on graduation, swearing to be a practicing member of the Church of England, and this I believe Oliver (not being much of a God-fearing man) would have refused to do. But suppose we ignore that. I don't believe he would have lasted very long! Certainly he would have rebelled at the educational philosophy of that institution's academic star, Isaac Todhunter, who wrote,[33] concerning his attitude toward a student who would dare put his faith in actual experiments and not just have blind faith in his teacher:

> If he does not believe the statements of his tutor—probably a clergyman of mature knowledge, recognized ability, and blameless character—his suspicion is irrational, and manifests a want of the powers of appreciating evidence, a want fatal to his success in that branch of science which he is supposed to be cultivating.

Heaviside, a man who energetically "took on" eminent representatives of the British Empire's electrical establishment, would no doubt have done something unforgivable if told this to his face in a professor's classroom. It probably was best, for Heaviside, for his potential professors, and most certainly for the independent spirit in Heaviside's papers, that he *didn't* go to Cambridge.[34]

It is clear that after he left his job there was never any question in his mind about what he would do next. What he did was to attain, again through unguided self-study, a mastery of the *Treatise* second to none, and then to devote almost all the remainder of his life to seeking out the implications hidden away in Maxwell's theory of the electromagnetic field.

In a letter[35] written near the end of his life (dated February 24, 1918) he recalled how he first learned of Maxwell's work one day while in a library:

> I remember my first look at the great treatise of Maxwell's when I was a young man. Up to that time there was not a single comprehensive theory, just a few scraps; I was struggling to understand electricity in the midst of a great obscurity. When I saw on the table in the library the work that had just been published (1873), I browsed through it and I was astonished! I read the preface and the last chapter, and several bits here and there; I saw that it was great, greater and greatest, with prodigious possibilities in its power. I was determined to master the book and set to work. I was very ignorant. I had no knowledge of mathematical analysis (having learned only school algebra and trigonometry, which I had largely forgotten), and

> thus my work was laid out for me. It took me several years before I could understand as much as I possibly could. Then I set Maxwell aside and followed my own course. And I progressed much more quickly.

In his efforts he was, indeed, to be successful beyond all possible expectation. But first there was the matter of the electric telegraph.

Tech Note: Where Is the Fault?

The sort of mathematical analyses that the young Heaviside was doing in the early years was typically the application of simple electrical theory to telegraph cables in the steady state. One notebook entry (NB 1A:94), for example, dated January 16, 1871, shows how he calculated the location of a fault (a leakage path to ground caused by a defect in the cable sheathing) in an underwater cable between Newbiggin-by-the-Sea and Sondervig. The analysis was strictly algebraic, but clever. He denoted the cable resistance from one end to the other (assuming no fault was present) by a, the cable resistance from the English end to the fault by x, and the resistance of the fault itself by y. The sketch he made of all this is shown in Fig. 2.1.

Heaviside easily calculated a, writing "Cable is about 360 knots [nautical miles, a nonstandard usage of *knot* as a distance], say 6 ohms per knot, therefore $a = 2160$, roughly." He then defined $x + y = b$, which could be measured by open-circuiting the right end of the cable, applying a battery of known voltage to the left end, and measuring the current. He defined the resistance as seen from the left end *when the right end was grounded* as c, which could be measured by the same method as used for b. Applying Ohm's law (and noting the fault and the $a - x$ section of the cable are in parallel), Oliver wrote

$$c = x + \frac{y(a-x)}{y+a-x}$$

from which he solved for x (a nice exercise for the reader!) as

$$x = c - \sqrt{(a-c)(b-c)}.$$

This method for locating faults from one end only was not invented by Heaviside, but is known as the *Blavier* method after the French telegraph engineer Edouard Ernest Blavier (1826–1887). It was generally considered not to be a very reliable technique because "the further end of the cable is to earth during one stage of the test, and free during another, a change in the conditions which makes it very difficult to maintain a constant current flow, and causes the fault to become very restive."—quoted from *The Electrician,* vol. 47, pp. 953–954, May 3, 1901. For his particular cable, Heaviside had the values $b = 1040$ and $c = 970$, giving $x = 682$ ohms or, at 6 ohms per knot, he concluded the fault was located 113 and 2/3 knots out at sea from the English end. In this same analysis are the happy words "All over. Dined roast beef,

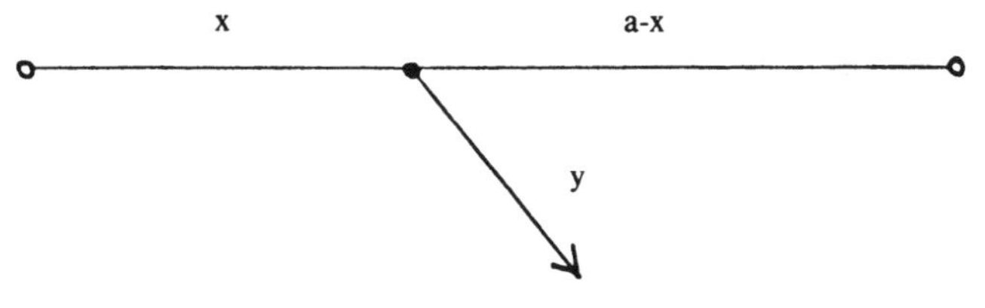

Fig. 2.1.

apple tart, and rabbit pie, with Claret, and enjoyed ourselves." Clearly, for Heaviside, finding cable faults could have its pleasant moments!

Notes and References

1. Sir Edmund Whittaker, "Oliver Heaviside," *The Bulletin of the Calcutta Mathematical Society,* vol. 20, pp. 199–220, 1928-29, facsimile reprinted as the foreword to D. H. Moore's *Heaviside Operational Calculus: An Elementary Foundation,* New York, NY: American Elsevier, 1971.
2. This was quite common, and had a sound economic basis. Until the repeal in 1851 of a law dating back to the late 17th century, every window in a dwelling (beyond a certain number) was subject to taxation.
3. C. Booth, *Life and Labour of the People in London,* vol. 5 (First Series: Poverty), London: Macmillan, 1902; in particular see the "Descriptive Map of London Poverty 1889," North-Western Sheet (block E-3).
4. The letter, dated June 2, 1897, was written by Heaviside to his dear friend, the Irish physicist G. F. FitzGerald. Heaviside carried on a twelve-year correspondence with FitzGerald, which both men greatly valued, until FitzGerald's death in 1901, and yet they met only twice, for a total of eight hours, according to Heaviside. See F. E. Hackett's "FitzGerald as revealed by his letters to Heaviside," *Scientific Proceedings of the Royal Dublin Society,* vol. 26, no. 1, pp. 3–7, 1952/54.
5. EP 2, p. 348.
6. R. Appleyard, *Pioneers of Electrical Communication,* London: Macmillan, 1930, p. 217.
7. Heaviside's oldest brother, Herbert, was born in 1842, and his other two brothers, Arthur and Charles, were born in 1844 and 1846, respectively. Oliver would, later in life, collaborate professionally with Arthur, and live with Charles and his family. He apparently had little to do with Herbert who left the family home on King Street after an argument with his father while Oliver was still quite young.
8. R. Taft, *Photography and the American Scene: A Social History,* New York, NY: Dover, 1964 (originally published in 1938).
9. One brother, John, was sufficiently talented to find his way into *Bryan's Dictionary of Painters and Engravers,* vol. III (H–M), London: Bell, 1927, p. 25.
10. Sir George Lee, "Oliver Heaviside—The man," in *The Heaviside Centenary Volume,* London: The Institution of Electrical Engineers, 1950, p. 11.
11. There is at least one recorded reference by Oliver to his early interest in science. In a 1918 letter referring to his pioneer work (inductive loading) in telegraphy he recalled "... my experiments with knotted clothes lines ... done in the backyard at the age of 12 to 13. They sank in, tho' I had no notion of any application then." What Heaviside was probably getting at are observations on how regularly spaced knots (mass concentrations) on a line affect the propagation of oscillations as one end of the line is wiggled. This is a mechanical analog of electrically exciting a telegraph cable. The ages would place Heaviside still at the King Street house. See K. Maynard's "Oliver Heaviside as seen in his books and letters," *Technology Review,* vol. 35, no. 6, p. 214 and p. 234, March 1933.
12. "The teaching of mathematics," *Nature,* vol. 62, pp. 548–549, October 4, 1900. In an essay on the same subject prepared a year later for a meeting of the British Association for the Advancement of Science (often called the B.A., which in turn Oliver would unkindly reinterpret as the "British Asses" as did Maxwell before him), Heaviside was even more to the point: "Boys are usually exceedingly stupid in anything requiring concentrated reasoning. It is not in the nature of their soft brains that they should take kindly to Euclid and other stuff of that logic-chopping kind."
13. The College of Preceptors, an institution originally created to raise the professional standing of teachers (in 1872 the College approached the Privy Council seeking the establishment of chairs of education in English universities), began administering uniform tests to students in 1850. This was part of a mid-19th century reform movement to reduce the importance of nepotism and patronage in gaining admission to the Civil Service and military academies.
14. EMT 1, p. 12.
15. EMT 1, pp. 404–405.
16. The definitive biography of Wheatstone is by B. Bowers, *Sir Charles Wheatstone,* London: Science Museum, 1975. This book will probably remain the standard on Wheatstone for a very long time, and is packed with illustrations and clear descriptions of all the prolific inventor's gadgets. It was Bowers who discovered the curious fact that Wheatstone's first child was born less than three months after its parents married, showing Wheatstone must have had a certain flair for passion and persuasion. Of course, this might for all we know be better said, instead, of Emma.
17. The single exception to this statement is Wheatstone's discovery of the principle of the stereoscope, which he publicly displayed to the Royal Society in 1838 (by then he had had a working stereoscope for six years), *before* the invention of photography. Wheatstone, of course, is also usually credited with inventing the electric telegraph (along with William Fothergill Cooke), but this invention was clearly also the work of other men. Wheatstone was involved with the development of the self-excited dynamo, but again others were, too. The stereoscope, however, is solely Wheatstone's.

18. R. D. Altick, *The Shows of London: A Panoramic History of Exhibitions, 1600-1862,* Cambridge, MA: Belknap Press, 1978, p. 360. Wheatstone's musical interests were shared by his good friend at the Royal Institution, Michael Faraday.
19. O. Heaviside, "Pianoforte touch," *Nature,* vol. 91, p. 397, June 19, 1913. Heaviside once modestly related to a friend, "I have no technical knowledge (of music) nor am I a pianist, though I once taught myself B.'s Opus 90. I liked it better than anything else."
20. Sir George Lee, *Oliver Heaviside and the Mathematical Theory of Electrical Communications,* New York, NY: Longmans, Green and Co., 1947, p. 1.
21. G. F. C. Searle, "Oliver Heaviside: A personal sketch," in *The Heaviside Centenary Volume,* London: The Institution of Electrical Engineers, 1950, p. 93.
22. Heaviside revealed his feelings toward foreign languages when he wrote (EMT 3, p. 52), "Though sad, it is a fact that few Britons have any linguistic talent. This is not due to laziness, but mainly to a real mental incapacity, combined with the feeling that one language is quite enough." As was his style, this observation is stuck right in the middle of a detailed, theoretical analysis. The complaint was motivated by Heaviside's irritation at having to read Lorentz in the original German (which, despite his grumbling, he *could* do).
23. Heaviside had remarkably little to say about the personal aspects of life, except for occasional remarks scattered about in letters and his research notebooks (which now and then served double-duty as his diary). He did, once, allude to a proposed autobiography ("When I started this work [volume 3 of *Electromagnetic Theory*] I had the idea of four volumes ... . Circumstances have prevented this. They will be described not here, but in my Biography in the Chapter entitled 'Wicked People I Have Known' "). This biography, if ever actually written, has not been found. See the paper by B. R. Gossick, "Where is Heaviside's manuscript for volume 4 of his *Electromagnetic Theory*?", *Annals of Science*, vol. 34, pp. 601–606, 1977.
24. As Heaviside was occasionally inclined to do, a parable in one of his books surely tells us something of his youth: "More than a third part of a century ago, in the library of an ancient town, a youth might have been seen tasting the sweets of knowledge to see how he liked them. He was of somewhat unprepossessing appearance, carrying on his brow a heavy scowl that the 'mostly fools' consider to mark a scoundrel. In his father's house were not many books, so it was like a journey into strange lands to go book-tasting. Some books were poison; theology and metaphysics in particular; they were shut up with a bang. But scientific works were better; there was some sense in seeking the laws of God by observation and experiment, and by reasoning founded thereon. Some very big books bearing stupendous names, such as Newton, Laplace, and so on, attracted his attention. On examination he concluded that he could understand them if he tried, though the limited capacity of his head made their study undesirable." This passage is from EMT 3, p. 135, and is dated March 21, 1902. The opening words would seem to date the "youth" during Oliver's two-year period of self-study.
25. This description is from an obituary written by Sir Oliver Lodge which appeared in *The Journal of the IEE* (Institution of Electrical Engineers), vol. 63, pp. 1152–1155, 1925.
26. Young Brown's tragic death was, of course, just one of many such similar cases in the "Great War." The most infamous example of this sort of thing is the case of Henry Gwyn-Jeffreys Moseley, who used x-rays to study atomic structure (the concept of *atomic number* is due to him), who died in 1915 at Gallipoli (where Brown was invalided with dysentery). In his letter to the Browns, Heaviside addressed this issue when he wrote "I do not think the Military Authorities should have accepted him as a fighting soldier. Ruffians are wanted for that. And I think the Military Authorities were very wrong in not overcoming W.G.B.'s refusal [not to accept a less dangerous post] by the simple process of compulsorily *promoting* him to one of their numerous scientific departments in which higher mathematics would have been more useful than in the trenches."
27. NB 3A:4.
28. J. C. Maxwell, *A Treatise on Electricity and Magnetism*, vol. 1, New York, NY: Dover, 1954, p. 482 (article 350).
29. Although he was now no longer an active professional, and seems never to have attended a Society meeting, Oliver did keep up with the technical literature. As proof of this (as well as an example of his need to speak his mind bluntly on unrelated matters, right in the middle of a technical discussion) there is a letter he wrote to *The Electrician*, vol. 16, p. 271, February 12, 1886. This letter begins with observations on a topic dear to Heaviside, the art of technical nomenclature, but it quickly detours wildly. The technical portion is reprinted in EP 2, p. 28, but the following words were cut from the book for reasons I think obvious: "I may perhaps also be allowed to mention, whilst on the subject of nomenclature, a difficulty thrown in the way of learning by the S.T.E. and E. [Society of Telegraph Engineers and Electricians], which some time ago had some trouble with its own nomenclature. Having changed name three times, it might be called 'The Society of Variable Nomenclature and of the Redundant Conjunction'. I should like it to have been called simply 'The Electrical Society', following the example of so many other societies who are satisfied with simple and comprehensive names. For instance, the Physical Society does not call itself the Society of Physicists and of Natural Philosophers, nor does the Mathematical Society call itself the Society of Mathematicians and of Arithmeticians and Geometricians. But that is not the point to which I would refer, which is, that the S.T.E. and E. has its journals sewed and wired in such a manner that a real difficulty is placed in the way of getting at their contents. They won't lie open, for one thing, which has been called the unpardonable sin of [a] book; they must therefore be held open; more than that, they must be strongly stretched before the words nearest the wires become readable, as they are nearly hidden by reason of the curvature. It is as exasperating as a tight boot."

30. There is a curious story behind how Heaviside came to join the Society of Telegraph Engineers (S.T.E.). It was told by Heaviside himself in a 1922 letter to the president of the Institution of Electrical Engineers: "I got into it (S.T.E.) nearly 50 years ago, and had trouble with it all along. I was then a scientific enthusiast, filled with a strong sense of duty to impart my knowledge to others and help them. But I found them very stupid usually. They wouldn't come out of their old grooves. A.W.H. [his brother, Arthur—the W. stood for West, their mother's maiden name] said I should join the new Society. But there were snobs in those days. There are not so many now, I think. On inquiry he [Arthur] was told they didn't want telegraph clerks! What would Edison say to that, if he were here now? I was riled. So I went to Professor W. Thomson [William Thomson, at that time president of the S.T.E.] and asked him to propose me. He was a real gentleman, and agreed at once. So I got in, in spite of the snobs." Obviously, Heaviside had no lack of nerve in his youth, approaching a famous scientist like Sir William Thomson (the professor had been knighted in 1866) to intervene in his behalf. It didn't hurt, of course, that Thomson knew Oliver's uncle, Sir Charles Wheatstone, quite well—they had, for example, served together on a consulting committee to the Atlantic Telegraph Company prior to the laying of the 1865 Atlantic Cable. In any case, by 1885 Heaviside had fallen so far behind on his dues that he was dropped from the membership list. In 1908, however, by which time the S.T.E. had evolved into the IEE, he was elected an Honorary Member in recognition of his then appreciated accomplishments. See R. Appleyard's entertaining *The History of the Institution of Electrical Engineers (1871-1931),* London: IEE, 1939, p. 94.
31. R. Appleyard, "A link with Oliver Heaviside," *Electrical Communication,* vol. 10, pp. 53–59, October 1931. Appleyard further recorded Heaviside was remembered by his co-workers as "lonely and silent, but always ready to explain the construction of the [telegraph] apparatus." As to being "unruly," perhaps the following shows how: "One day at Newcastle, he refused to work as a pasting-down clerk, remarking 'I am not a book-binder.' "
32. "The academic side of technical training," an address delivered at University College, London, by Dr. A. B. W. Kennedy, Emeritus Professor of the Faculty of Science, *The Electrician,* vol. 55, pp. 435–437, June 30, 1905. At the time of his original appointment (1874) Kennedy also observed, "The fact of the matter was that at that time it was exceedingly difficult to find anyone who would take up the position of an engineering professor." One reason for this, no doubt, was lack of money. As one of Kennedy's younger colleagues (J. A. Fleming) put it, "When I was appointed a Professor of Electrical Engineering [in 1884], all that could be done at first was to provide me with a piece of chalk and a blackboard." See *Memories of a Scientific Life,* London: Marshall, Morgan and Scott, 1934, p. 99.
33. *The Conflict of Studies,* London: Macmillan, 1873, p. 17.
34. I make this assertion despite Heaviside's own apparent claim to the contrary (EMT 2, pp. 10–12), because he used it mostly as a springboard to launch some of his most sarcastic comments at Cambridge mathematicians. *Why* he felt this way we'll see in Chapter 10. He labeled the work of some of them "distressing and soul-destroying," and accused them of being unfair. His particular phrasing is interesting, however, telling us, I believe, of his own assessment of his early youth: "I regret exceedingly not to have had a Cambridge education myself, instead of wasting several years of my life in mere drudgery, or little more."
35. Quoted from the obituary notice by J. Bethenode, "Oliver Heaviside," *Annales des Postes Telegraphs,* vol. 14, pp. 521-538, 1925. This letter was written in English by Heaviside, translated by Bethenode into French, and then retranslated back to English for inclusion here. To soothe fears that something awful may have happened during this back-and-forth process, Bethenode states in a footnote that he made the original translation "as literally as possible in order not to change the meaning." The second translation was so done, too.

3
The First Theory of the Electric Telegraph

Does it seem all but incredible to you that intelligence should travel for two thousand miles, along those slender copper lines, far down in the all but fathomless Atlantic; never before penetrated ... save when some foundering vessel has plunged with her hapless company to the eternal silence and darkness of the abyss? Does it seem ... but a miracle ... that the thoughts of living men ... should burn over the cold, green bones of men and women, whose hearts, once as warm as ours, burst as the eternal gulfs closed and roared over them centuries ago?

— A somewhat gruesome 19th century tribute to the Atlantic cable

Sir W. Thomson's theory of the submarine cable is a splendid thing.

— Oliver Heaviside, 1887

...with the exception of the work of Oliver Heaviside, the theory of the submarine telegraph cable remains very much where it was left by Sir William Thomson in 1855.

— *The Electrician*, 1912

Thomson and Stokes

When Heaviside began his studies of just what goes on inside a telegraph cable the existing theory was Lord Kelvin's. He had worked this out in late 1854, while still just plain Professor William Thomson (at Glasgow University), as the result of a correspondence with Professor George Gabriel Stokes,[1] under the stimulus of what would culminate in the 2000-mile Atlantic Cable Project. To many, the laying of such a gigantic submarine cable seemed an absurdity, and Professor George Biddell Airy (the Astronomer Royal) declared it to be nothing less than a mathematical impossibility. For his participation in this effort, one of the most astounding feats of 19th century engineering, Thomson would be knighted (at least in part, on the recommendation of Charles Wheatstone) in 1866. It is instructive to look at Thomson's analysis,[2] as it was the starting point for Heaviside's extension two decades later. This will also give us a look at Thomson (a man with enormous personal impact on Heaviside's career) at an early stage of his own professional life.

Thomson began by modeling a submerged telegraph line as a very long wire conductor along the axis of a cylinder of perfect electrical insulation. This arrangement, today, is called a *coaxial cable*, formed by two concentric, conducting cylinders, separated by the perfect (zero conductivity) insulator, as shown in Fig. 3.1. The inner conductor is obviously the telegraph line, and the outer conductor is formed by the insulator/seawater interface.

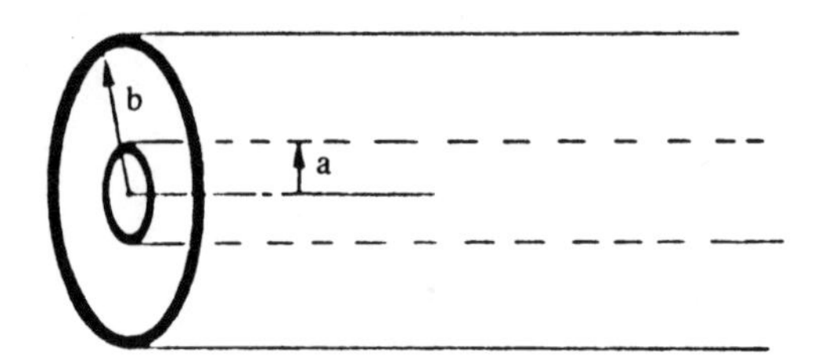

Fig. 3.1. The geometry of Thomson's submarine cable.

With ϵ denoting the electric permittivity of the insulator, Thomson next introduced the "electrostatical capacity" *per unit length*; in modern notation

$$C=\frac{2\pi\epsilon}{\ln(b/a)}, \qquad \epsilon=n\epsilon_0,\ n=\text{relative permittivity}, \qquad \epsilon_0=\frac{1}{36\pi}\times 10^{-9}\ \text{farad per meter.}$$

(This basic result, first worked out by Thomson,[3] is now a classic textbook example in modern college physics and electrical engineering classes.) Using gutta-percha for insulation (an Asian vegetable gum with insulating properties superior to those of rubber), with a relative permittivity of $n = 3.1$, a cable with a b/a ratio of 10 had a capacitance of 0.12 microfarad per mile. This was quite large compared to that of overhead telegraph lines.[4]

Thomson next introduced the resistance *per unit length* of the wire, and called it K. The mathematical description of this cable can then be found by applying Kirchhoff's current law (the conservation of electric charge) and Ohm's law to an infinitesimal length of cable. If $v(x, t)$ is the voltage at time t at distance x from one end of the cable, the result is a partial differential equation (*partial*, as there are two independent variables, distance and time):

$$\frac{\partial^2 v}{\partial x^2}=KC\frac{\partial v}{\partial t}.$$

Thomson called this result "...the equation of electric excitation in a submarine telegraph wire, perfectly insulated... ." And then he informed his readers that this is a well-known equation[5]: "This equation agrees with the well-known equation of the linear motion of heat in a solid conductor; and various forms of solution which Fourier has given are perfectly adapted for answering practical questions regarding the use of the telegraph-wire." It isn't surprising that Thomson would be well aware of Jean Baptiste Joseph Fourier's work, as he had studied it intensively as an undergraduate at Cambridge, and so impressed was he by it that he called Fourier's 1822 seminal work *Theorie analytique de la chaleur* a "mathematical poem."[6] Thomson would, in 1862, use this same equation to calculate the age of the Earth; his result threw both the geologists and the evolutionists into turmoil.[7] Later in this book we shall see how, again, Heaviside played a subsequent role in modifying Thomson's pioneering work (this time in "theoretical" geology).

With the cable equation now established, Thomson could (and did, in a letter to Stokes dated October 1854) attack and solve many interesting special cases. It was actually Professor Stokes, however, who later (in a letter to Thomson in November) formally solved the most general case considered by the two men. Stokes wrote[2]:

> In working out for myself various forms of the solution of the equation $dv/dt = d^2v/dx^2$ under the conditions $v = 0$ when $t = 0$ from $x = 0$ to $x = \infty$ [that is, Stokes was assuming

Inscribed on this photograph, in Kelvin's own hand and dated Nov. 23, 1892: "(W Thomson) reading a letter or letters from Fleeming Jenkin, about experiments on submarine cables probably about March 1859." The photo was taken by the Reverend Dr. David King who was married to Thomson's eldest sister, Elizabeth.

an initially uncharged, infinitely long cable with $KC = 1$]; $v = f(t)$ when $x = 0$ from $t = 0$ to $t = \infty$ [that is, Stokes was allowing an *arbitrary* input signal, $f(t)$], I found that the solution ... was...

$$v(x, t) = \frac{x}{2\sqrt{\pi}} \int_0^t (t - t')^{-3/2} e^{-x^2/(4(t-t'))} f(t')\, dt'.$$

This result by Stokes is highly useful for understanding Thomson's most important results. For example, suppose $f(t)$ represents the sudden application of constant voltage of unity amplitude (this particular $f(t)$ was, for a time, in the early decades of the 20th century, given the special name of the *Heaviside step* since Oliver used it so much in his work). Then Stokes' integral (generalized for arbitrary values of K and C) reduces to

$$v(x, t) = \frac{2}{\sqrt{\pi}} \int_{x(1/2)(KC/t)^{1/2}}^{\infty} e^{-u^2}\, du, \qquad \begin{matrix} t \geq 0 \\ x \geq 0. \end{matrix}$$

George Gabriel Stokes (1819–1903)

Stokes was one of William Thomson's dearest, lifelong friends. Thomson tried hard to get Stokes to join him at Glasgow University, but, because this required taking an oath to be a practicing member of the Church of Scotland, Stokes refused. He remained in England till the day he died. And when the Cambridge professor did die (since 1848 Stokes had been Lucasian Professor at Pembroke College), the Glasgow professor was remembered as standing at the grave "almost overcome by emotion, saying in a low voice: 'Stokes is gone and I shall never return to Cambridge again.' " Quoted from A. Schuster, *Biographical Fragments,* London: Macmillan, 1932, p. 242. (From an 1875 sketch in *The Popular Science Monthly.*)

It can also be shown[8] that the current in the cable is given by

$$i(x,\ t) = \left(\frac{C}{\pi K t}\right)^{1/2} e^{-x^2 KC/4t}, \qquad \begin{array}{l} t \geq 0 \\ x \geq 0. \end{array}$$

Fig. 3.2 shows the behavior of the voltage and the current at any point along the cable at any time after the application of the step (note that these are "universal arrival" curves, as early electricians called them, and that the horizontal axis is an *amalgamation* of both independent variables, x and t). This figure graphically displays one of the most important and surprising conclusions to come out of Thomson's analysis.

The Law of Squares

As both curves in Fig. 3.2 show, the voltage (or current) reaches a given amplitude after a time lapse that increases as x *squared*, and not linearly with x as one might intuitively expect.

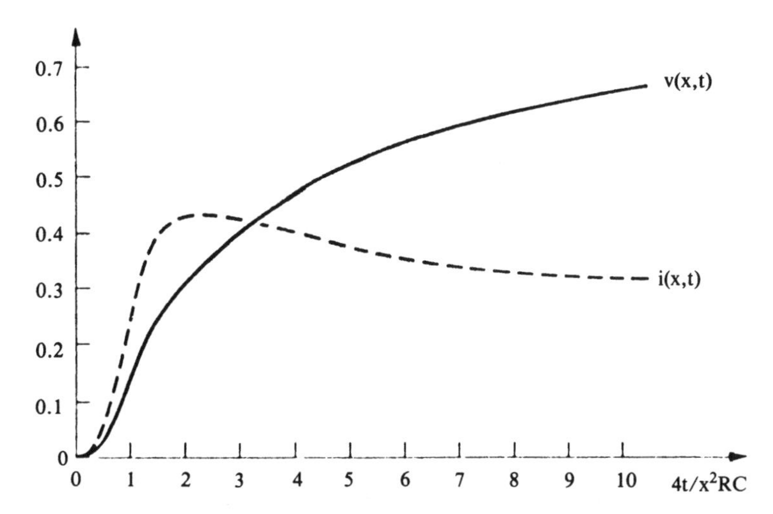

Fig. 3.2. Voltage and current behavior on Thomson's cable due to a Heaviside step.

The voltage curve rises monotonically to a fixed, steady value as the cable charges, but the case is slightly more complicated for the current. For it, at any given distance x, there will be an *overshoot*, i.e., a *maximum* current greater than the steady value will flow. Therefore, if there is an electromagnetic instrument (e.g., a meter) connected to the cable that responds to the current, then there will be an instant of time, t_{max}, at which the instrument response will be maximum. We can calculate this time by setting the time derivative of the current equal to zero and, in this case, we obtain

$$t_{max} = \frac{1}{2} KCx^2.$$

That the retardation of the maximum signal current depends on *x squared* was a surprising result,[9] and indeed not everyone believed it. This curious result is not peculiar to the special case of a Heaviside step input. For example, a far more realistic telegraph signal is not the step (which represents closing a telegraph key and keeping it closed forever), but rather a quick pulse of voltage (which is a good model for briefly closing a telegraph key and then opening it). Thomson considered this particular input signal in some detail, but his analysis is extremely cumbersome by today's standards. This is, however, a perfect example of the value of reading scientific papers in the original for in so doing we cannot help but admire how Thomson's genius overcame his lack of today's more powerful mathematical tools.

Thomson first worked out the solution of the cable equation for a pulsed input signal of finite duration T, and then calculated the limiting form of this solution as T tends toward zero (but in such a way that a fixed value, Q, of charge is always injected into the cable end at $x = 0$). To arrive quickly at Thomson's result we could just let $f(t)$ in Stokes' integral solution be the unit impulse function (which is, in a certain sense, the time derivative[10] of the Heaviside step). Alternatively, we could apply a powerful theorem from linear systems theory that states if we know the response of a system to a step, then the response to the derivative input is the derivative of the original response. Either approach quickly gives us the unit impulse voltage response of Thomson's cable. This is precisely what Thomson derived and it leads immediately to his result "...the time at which the maximum electrodynamic effect of connecting the battery for an instant...,'' which he found to be

$$t_{max} = \tfrac{1}{6}KCx^2$$

which again, with a minor change in the multiplicative factor, is a squared-distance relationship. This simple result would be misinterpreted, however, and consequently have a profound *retarding* effect on the practice of telegraphy and long-distance telephony during the last two decades of the 19th century. We'll take this up in Chapter 8.

The Atlantic Cable

This fundamental result had tremendous implications for the very long Atlantic cable. An acceptable delay in a 100-mile cable would not just be 20 times longer in a 2000-mile cable, but 400 times longer! Not everyone, however, accepted Thomson's theoretical results. One who objected was Edward Orange Wildman Whitehouse, chief electrician for the Atlantic Telegraph Company. In 1856, at a meeting of the British Association, he strenuously and dramatically rejected Thomson's ''Law of Squares'' by stating[11]:

> In all honesty, I am bound to answer, that I believe nature knows no such application of that law; and I can only regard it as a fiction of the schools, a forced and violent adaptation of a principle in Physics, good and true under other circumstances, but misapplied here.

This statement by Dr. Whitehouse (he was a physician before declaring himself an electrical expert) about ''a fiction of the schools'' was not an uncommon belief among the ''practical men'' of the day. Mathematical analysis was considered second-rate (and a poor second, at that) to educated judgments by ''men of experience.'' This philosophy was bluntly stated in an editorial[12] published by an important electrical trade journal:

> We may remark that of late years the (experimental) facilities for obtaining definite ideas on the subject of electrical phenomena have greatly increased. Sciences generally become simplified as they advance, albeit in some branches they may attenuate into the abstruse. In electricity there is seldom any need of mathematical or other abstractions, and although the use of formulae may in some instances be a convenience, they may for all practical purposes be dispensed with.

And as late as 1887 the important weekly journal *The Electrician* printed a letter from a practicing electrical engineer who wrote

> ...to enter a protest against the growing tendency to drag mathematics into everything. I have as high an admiration of mathematics as any one, but I like them as servants, not as masters; and I like to take the shortest, simplest, and surest road to my object, which, to my mind, mathematics do not always furnish... . Mathematics and theory are dangerous tools; you can never be certain you are using them aright until you have clearly proved them by practice.

Whitehouse, however, did little to help the cause of the ''practical man'' when he ruined the first successfully laid transatlantic cable of 1858. Convinced (by his practical experience, presumably) that only a high voltage could send signals through thousands of miles of cable, he constructed enormous, five-foot-long induction coils generating nearly 2000 volts. This absurdly high voltage quickly broke down the cable insulation and the cable failed after only a few weeks. As one writer much later put it,[13] ''He [Whitehouse] sent a stroke of lightning over the cable, which required only a spark.'' After this fiasco (and after Thomson successfully refuted Whitehouse in the technical literature), Whitehouse was dismissed by the Atlantic Telegraph Company.

Loading the 1865 cable (2233 miles long, weighing 5000 tons) into the *Great Eastern*, five times larger than any other ship then afloat.

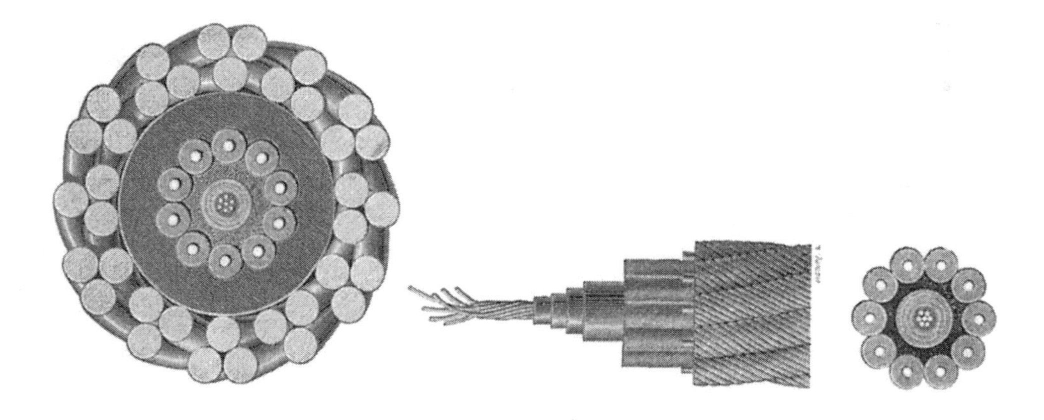

The cables of 1865 and 1866. The cross-sectional view on the left is of the extra-strong shore end, and the middle and right views are of the main ocean section. The seven-strand copper core was covered by four layers of gutta-percha, wrapped in tarred hemp and protected by ten steel wires, each wrapped in impregnated hemp.

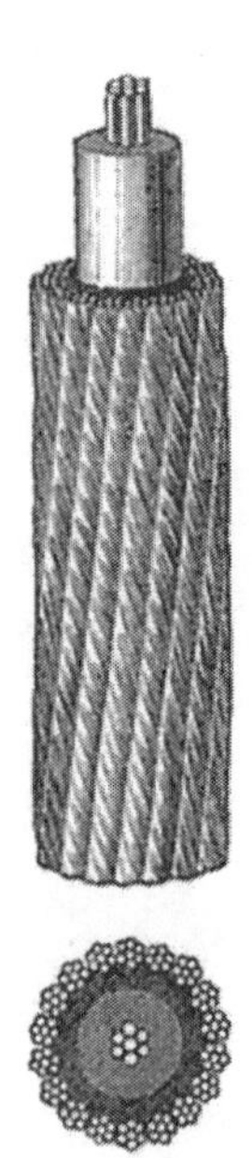

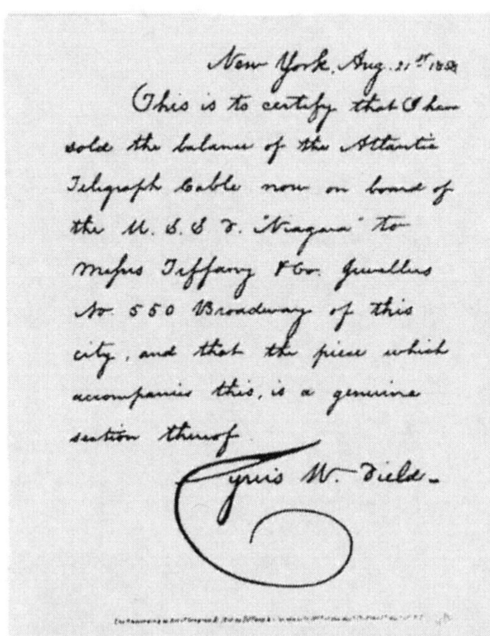

New York, Aug. 21st 1858

This is to certify that I have sold the balance of the Atlantic Telegraph Cable now on board of the U. S. S. F. "Niagara" to Messrs Tiffany & Co. Jewellers No. 550 Broadway of this city, and that the piece which accompanies this, is a genuine section thereof.

Cyrus W. Field.

Burndy Library

A capitalist in the best meaning of the word, C. W. Field made money even from leftover cable.

The departure of Whitehouse, who denied the law of squares had any validity at all, was followed by the rise of a widespread belief among electricians that Thomson's result was practically a general law of nature—the pendulum swung from one extreme to the other. It would take some time to put matters right, and Heaviside played a role in the affair.

The Speed of the Current

Heaviside called Thomson's law of squares ''rather neat,''[14] and so it is as long as the limitations inherent in its derivation are kept in mind (no inductance or leakage in the cable). Others, however, forgot to do that (or never knew they should) and tried to read far too much into the law, such as when they argued the ''speed of electricity'' in a cable depends on the length of the cable. This gave Heaviside an opportunity to poke some fun, and in 1887 he ''proved'' the variable-speed advocates wrong in his usual undiplomatic way. After first informing[15] his readers that ''recent researches'' had ''proven'' the speed of the current varies ''not quite so fast as the square of the length,'' nor ''quite so slow as the inverse square of the length,'' he continued with

> Is it possible to conceive that the current, when it first sets out to go, say, to Edinburgh, *knows* where it's going, how long a journey it has to make, and where it has to stop, so that it can adjust its speed accordingly? Of course not; it is infinitely more probable that the current has no choice at all in this matter, that it goes just as fast as the laws of Nature, preordained from time immemorial, will let it; and if the circuit be so constrained that the conditions prevailing are constant, there is every reason to expect the speed will be constant, whether the line be long or short. QED.

Earlier, in 1885, he had been less sure of things, writing[16]: "We know nothing about the velocity of electricity; it may be an inch in a year or a million miles in a second. Following this up, it may be nothing at all." A quarter of a century before, Thomson himself had written an essay ("Velocity of electricity") addressing this matter.[17] After listing the widely varying experimental values obtained by a number of people (including Wheatstone's 1834 speed of 288,000 miles per second) he observed, "Now it is obvious, from the results which have been quoted, that the supposed 'velocity' of transmission of electric signals is not a definite constant … ." And then, at the end of his essay, he may well have planted the seed that grew into the belief Heaviside later ridiculed, "retardations [are] proportional to the squares of the distances travelled over by the impulse… . In other words, the 'velocity' of propagation might be said to be inversely proportional to the distance travelled."

The use of the word *retardations* shows Thomson knew what was what, but others could easily have interpreted this as a statement about electricity itself.

Phase Distortion

The second basic result that came out of Thomson's analysis, besides the law of squares, had to do with the *phase distortion* produced by the cable (this is a modern term, one not used by Thomson). The arbitrary input signal assumed by Stokes, $f(t)$, can be represented either as a sum (a *Fourier series* with possibly an infinite number of harmonically related sinusoidal functions), or an integral (a *Fourier integral*, again involving sine functions, only now with *all* frequencies continuously present). Thomson showed that each of these frequency components would diffuse into the cable at a *different speed*. This results in a spreading of the signal, and what might start off as a short-duration, crisp telegraph pulse at the transmitter key, could arrive at its destination as a weak, smeared-out dribble.

Two such pulses, separate and distinct at the start of their journey ($x = 0$; $t = 0$; and $x = 0$, $t = T$), could easily end up overlapping each other at the receiver (as the faster moving frequencies of the second pulse overtake the slower moving frequencies of the first pulse) unless T is made sufficiently large. To prevent such confusion (Heaviside referred to this as "mixing-up the signal"—it is called *intersymbol interference* today), each new pulse would have to be keyed into the cable only after having let the previous pulse "get clear," so to speak. The end result is a sharp reduction in the signaling rate, as compared to shorter overland telegraph systems.

It is easy to derive this second important result from Thomson's analysis; the details are given in the Tech Note at the end of this chapter.

That analysis tells us, first, that the input signal at $x = 0$ suffers exponential attenuation in amplitude as it diffuses into the cable and, second, that the diffusion *speed* of a signal component of angular frequency w is given by $(2w/KC)^{1/2}$, which is obviously frequency dependent. Since such a cable pulls apart (or *disperses*) the input signal, Thomson's cable is an example of a *dispersive* (to use the modern term) system. This is a conclusion of enormous importance, because it tells us that it isn't enough just to apply a huge voltage (as did Whitehouse) to the cable to overcome the attenuation. The cable also distorts the original signal by literally tearing it apart. This is the effect, in fact, displayed earlier in the "arrival curve" for a step input.

To see the magnitude of these difficulties it is helpful to do a little arithmetic. From Rayleigh[18] we learn that a typical KC product for an Atlantic cable was, in cgs units, 5×10^{-17}. Therefore, for the exponential amplitude factor to decay by a factor of e, the signal must

Lord Rayleigh (1842–1919), who was the second (after Maxwell) Cavendish Professor of Physics at Cambridge University until his resignation in 1884, once remarked "Professorships would be nice, if it were not for the students." (From a steel engraving published in *The Electrician*, 1893.)

be diffused into the cable a distance of

$$x=\left(\frac{2}{wKC}\right)^{1/2}=\frac{497}{\sqrt{f}}\text{ miles.}$$

Thus, a 100-Hz frequency component would decay by a factor of e every 49.7 miles. In a 2000-mile cable there would be more than a forty-fold effect, and a 1-volt sine wave would emerge at the receiving end as just 3.34×10^{-18} volt! A 10-Hz frequency component, however, would have a decay distance (by a factor of e) of 157 miles, and a 1-volt input signal would arrive as 2.97×10^{-6} volt (a *trillion* times larger!), which while small is measurable (signals were sent on the Atlantic cable by ordinary telegraph key, but received by Thomson's incredibly sensitive mirror galvanometer which, in effect, used a light beam as a meter needle). It is the very fact (now obvious from these numbers) that only low frequencies are of any concern on very long cables that justifies ignoring inductive effects.[19]

The diffusion speed is given by $(2w/KC)^{1/2}$ which works out to be

$$3107\sqrt{f} \text{ miles per second.}$$

Therefore the 100-Hz frequency component travels at 31,000 miles per second, while the 10-Hz component loafs along at "only" 9800 miles per second,[20] and this is the effect that pulls apart and stretches out the original signal.

The problem of dispersive phase distortion would remain the ultimate block to high-speed telegraphy (and long-distance voice transmission over telephone lines) until Heaviside's theoretical work in the 1880s.

Tech Note: How Thomson Thought Electricity "Soaks" into an Infinitely Long Cable

We begin[21] by letting the input signal be the real part of a complex[22] sinusoidal signal

$$v(0,\ t) = \text{Re}\ \{e^{jwt}\} = \cos\ (wt), \qquad j = \sqrt{-1}$$

where w is the radian frequency of some particular signal component (w is related to the frequency, f, in Hertz—what is still often called "cycles per second"—by $w = 2\pi f$). We assume the time variation of $v(x,\ t)$ is everywhere along the cable of the form e^{jwt}, and that the space variation can be separated out, i.e.,

$$v(x,\ t) = \text{Re}\ \{A(x)e^{jwt}\}, \qquad A(0) = 1.$$

Direct substitution of this equation into the fundamental cable equation leads to

$$\frac{d^2A}{dx^2} = jwKCA, \qquad A(0) = 1.$$

This can now be solved by assuming a solution of the form $A(x) = e^{px}$, where p is independent of both time and distance, which upon substitution leads to

$$p = \pm(jwKC)^{1/2} = \pm[(\tfrac{1}{2}wKC)^{1/2} + j(\tfrac{1}{2}wKC)^{1/2}].$$

Only the negative root makes physical sense *on an infinite cable* (the root with the positive real part gives rise to an unbounded $v(x,\ t)$), and thus $v(x,\ t)$ is the real part of

$$e^{-((1/2)wKC)^{1/2}x}e^{j[wt-((1/2)wKC)^{1/2}x]}$$

or, as Thomson concluded,

$$v(x,\ t)e^{-((1/2)wKC)^{1/2}x}\cos\ [wt - (\tfrac{1}{2}\ wKC)^{1/2}x], \qquad \begin{matrix} t \geq 0 \\ x \geq 0. \end{matrix}$$

From this we see that the cable voltage is a time-varying sinusoidal signal, but with an amplitude that decays exponentially with the distance from the input end. The higher the frequency, the more rapid the decay.

To understand what is meant by the speed of diffusion, imagine a surfer "riding the voltage

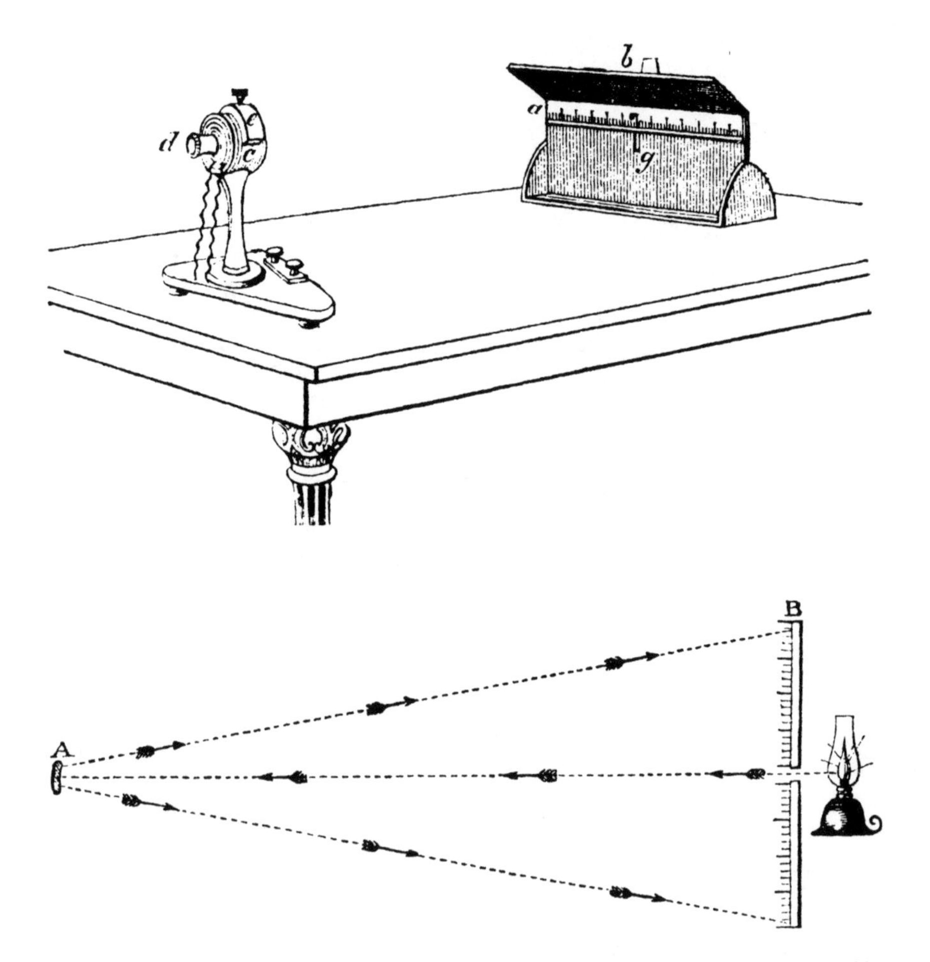

Thomson's sensitive marine galvanometer for receiving telegraph signals over the Atlantic cable.

signal into the cable'' at a speed just such that he observes only the exponential amplitude decay, but not the sinusoidal time behavior. This is what a real surfer would observe as he rides a water crest, seeing only a decline in water height but no change in *shape*. This condition requires the value of the cosine factor to remain constant or, in what amounts to the same thing, that the cosine argument (phase angle) remain constant. This defines a diffusion velocity of

$$\frac{dx}{dt}=\left(\frac{2w}{KC}\right)^{1/2}.$$

The higher the frequency, the greater the diffusion speed. In general, then, high-frequency signals ''live a fast life but die quickly''—no doubt a moral lesson is in here, somewhere, for all of us.

This double effect of signaling through a dispersive medium is the bugaboo that prevents direct radio (voice) communication with submerged submarines at all but the lowest frequencies.[23]

Notes and References

1. In his 1889 Inaugural Address as the new President of the Institution of Electrical Engineers (IEE) Lord Kelvin recalled how Stokes (who was in the audience) had become involved. "I was hurriedly leaving the meeting of the British Association (Liverpool, 1854), when a son of Sir William Hamilton, of Dublin, was introduced to me with an electrical question. I was obliged to run away to get a steamer by which I was bound to leave for Glasgow, and I introduced him to Professor Stokes, who took up the subject with a power which is inevitable when a scientific question is submitted to him." See the *Journal of the IEE*, vol. 18, p. 10, 1889.
2. W. Thomson, "On the theory of the electric telegraph," *Proceedings of the Royal Society,* May 1855, and reprinted in *Mathematical and Physical Papers,* vol. 2, Cambridge: Cambridge University Press, 1884, pp. 61–76.
3. "On the electro-statical capacity of a Leyden phial and of a telegraph wire insulated in the axis of a cylindrical conducting sheath," published in 1855 in *Philosophical Magazine,* and in *Reprints of Papers on Electrostatics and Magnetism,* London: Macmillan, 1884, pp. 38–41.
4. See, for example, Heaviside's 1880 paper "On the electrostatic capacity of suspended wires" (EP 1, pp. 42–46). There he showed that the capacitance per unit length of an overhead telegraph wire of diameter d, strung at height h (above a perfectly conducting earth, which is a good approximation), is $2\pi\epsilon_0/\ln(4h/d)$, a result he attributed to Thomson (NB 1:148); "Thomson first gave the solution of this, but I have seen no proof anywhere." This was no doubt the motivation for Heaviside's paper, to provide a proof in the literature, although the result itself had already been published (Heaviside himself referred his readers to Professor Fleeming Jenkin's *Electricity and Magnetism,* which gives the formula without proof or attribution). In a letter to *The Electrician* (vol. 15, p. 375, September 25, 1885) not included in his books, he said he did the analysis because of Jenkin's remark that theory and experiment were in disagreement by a factor of 2. Using Heaviside's numerical example (which he took from Jenkin's book) of $h = 3$ meters and $d = 4$ millimeters, we get a capacitance of 0.01 microfarad per mile, much less than that of the Atlantic cable.
5. Today's electrical engineers refer to Thomson's model of the electric cable as a *leakage-free* (because of the perfect insulator), *noninductive* transmission line. Thomson did eventually relax the leakage-free restriction, but it was Heaviside who, in 1876, removed the noninductive constraint (and who really studied, in depth, the implications of both leakage and inductance). Gustav Robert Kirchhoff had earlier included inductive efforts in his studies of 1857, but he based his work on pre-Maxwellian, action-at-a-distance concepts. Kirchhoff's work was translated into English and published in the *Philosophical Magazine* ("On the motion of electricity in wires," series 4, vol. 13, pp. 395–412, June 1857). In this prescient paper Kirchhoff derived the wave equation, spoke of electric wave propagation and of the possibility of reflected waves on finite-length wires (as a function of the termination conditions), and even included very modern-looking drawings of traveling wave fronts! In his irascible book *Electromagnetics* (New York, NY: Longmans, Green and Co., 1938, p. 533) Alfred O'Rahilly—who thought little of Heaviside—quoted the French scientist Marcel Brillouin as declaring Kirchhoff's work to be based on "adventurous simplifications." Heaviside wrote in 1886 (EP 2, pp. 81–82) that his work of 1876 (discussed in the next chapter) was done "in ignorance of Kirchhoff's investigation," and when he did learn of it he could "make neither head nor tail of it." The cable equation derived by Thomson is today just called the *diffusion* equation, and it is still very important in electrical physics.
6. S. P. Thompson, *The Life of William Thomson, Baron Kelvin of Largs,* London: Macmillan, 1910, p. 1139. Fourier had an enormous impact on Thomson, with one writer asserting "there are good indications that Fourier's work as a whole was the major influence on Thomson's career as a mathematical physicist." See J. Herival, "The influence of Fourier on British mathematics," *Centaurus,* vol. 17, pp. 40–57, 1972. Heaviside shared Thomson's high opinion, writing: "No one admires Fourier more than I do. It is the only entertaining mathematical work I ever saw. Its lucidity has always been admired. But it was more than lucid. It was luminous. Its light showed a crowd of followers the way to a heap of new physical problems." See EMT 2, p. 32.
7. But not everyone would use this strong a word—see, for example, S. G. Brush, "Nineteenth-century debates about the inside of the Earth: Solid, liquid or gas?", *Annals of Science,* vol. 36, pp. 225–254, May 1979, and in particular Footnote 2, p. 226.
8. D. K. Cheng, *Analysis of Linear Systems,* Reading, MA: Addison-Wesley, 1961, p. 368. Cheng solved this problem by performing the inversion of a Laplace transform that is transcendental in the transform variable, a technique unknown to Thomson because it hadn't been invented yet (it was destined to appear a half-century later as an offshoot, incredibly, of yet *another* aspect of Heaviside's work, the operational calculus). But still, Thomson arrived at the correct solution, and this is instructive as to how a brilliant mind is never at a loss merely for the lack of routine methods. Herival (Note 6) rightfully calls Thomson's solution "ingenious, and in fact brilliant."
9. The delay, itself, was no surprise, having already been observed in real cables. The surprise was the dependency of the delay on x^2.
10. "But the outstanding feature of Heaviside's creative imagination was the calm assurance with which he differentiated the function

$$f(t) = \begin{matrix} 1 & \text{if } t \geq 0 \\ 0 & \text{if } t < 0 \end{matrix}$$

to give the impulse function $\delta(t)$ which is zero if $t > 0$ or $t < 0$ and infinite if $t = 0$." Quoted from G. Temple's *100 Years of Mathematics: A Personal Viewpoint,* Berlin: Springer-Verlag, 1981, p. 159. The impulse function will be discussed in more detail later, as part of Heaviside's operational calculus.

11. H. Sharlin, *Lord Kelvin: The Dynamic Victorian,* University Park, PA: The Pennsylvania State University Press, 1979, p. 133. As an article on Kelvin in *Electrical World* (vol. 19, p. 36, January 16, 1892) put it, "... when the possible success of the Atlantic Cable came up Thomson's prediction of the retardation that would be experienced struck practical electricians aghast. They were very much inclined not to believe in this mathematical newcomer... ."
12. Quoted from *The Heaviside Centenary Volume,* London: IEE, 1950, p. 53.
13. P. B. McDonald, *Saga of the Seas, The Story of Cyrus W. Field and the Laying of the First Atlantic Cable,* New York, NY: Wilson-Erickson, 1937, p. 83. McDonald also related how very far from the mark Whitehouse was with his monster induction coils: "... after the 1866 cable was laid, an experiment was made in which a message was sent successfully from Newfoundland to Ireland by using ... a small copper percussion-cup and a tiny strip of zinc, activated by a drop of acidified water—a truly pigmy battery."
14. EMT 1, p. 415.
15. EP 2, pp. 128–129.
16. EP 1, p. 435. In 1888 (EP 2, p. 394) he changed the "million miles in a second" to "immensely great."
17. *Mathematical and Physical Papers* (Note 2), pp. 131–137.
18. Lord Rayleigh (Professor John William Strutt), *The Theory of Sound,* vol. 1, London: Macmillan, 1894, p. 466. Strutt, a Victorian genius (he received the 1904 Nobel prize in physics), every bit as talented as his good friend Thomson, was well aware of Heaviside's abilities and occasionally mentioned him favorably in his papers.
19. There is a note (dated July 2, 1852) in one of Thomson's research notebooks that shows he *did* initially consider induction, but eventually concluded it too unimportant to even mention in his analysis of a long submarine cable. The same note also shows he *did indeed* realize induction might play a dominant role in other situations. See D. W. Jordan, "The adoption of self-induction by telephony, 1886–1889," *Annals of Science,* vol. 39, pp. 433–461, September 1982.
20. The diffusion speed expression indicates that if $f > 3590$ Hz, then the speed exceeds that of light. This is not possible, of course, and merely demonstrates the partial differential equation of diffusion isn't really correct. Heaviside was well aware of this, writing (EMT 2, p. 73): "All diffusion formulae (as in heat conduction) show instantaneous action to an infinite distance of a source, though only to an infinitesimal extent. It is a general mathematical property, but should be taken with salt in making applications to real physics. To make the theory of heat diffusion be rational as well as practical, some modification of the equations is needed [induction, as we'll see in the next chapter] to remove the instantaneity, however little difference it may make quantitatively, in general." Heaviside claimed, in fact, that by 1875 he knew this. In his words (written in 1897; EMT 2, p. 396), as part of an analysis concerning the propagation of electromagnetic waves along a transmission line, "I set myself the above problems 22 years ago, when I first realized as a consequence of Maxwell's theory of self-induction, combined with W. Thomson's theory of the electric telegraph, that all disturbances travelled at finite speed... ." Faster-than-light (FTL) or hyperlight velocities didn't bother anyone until the very end of the 19th century because at that time nobody believed the speed of light (or any other speed) was an upper bound. Heaviside, in fact, often wrote about what he thought might be possible FTL electromagnetic effects; these will be discussed in Chapter 7.
21. The approach given is not the one used by Thomson, but rather the much more direct (and modern) one put forth by Lord Rayleigh—see Note 18.
22. When dealing with trigonometric functions in differential equations it is almost always useful to express them in complex exponential form, as exponentials are easier to differentiate. We merely do all the analysis with these complex exponentials and then, as the observed physical variables (e.g., voltage and currents) must have no imaginary parts, we keep only the real part of the resulting complex answer. The whole idea depends on Euler's relation, $e^{jx} = \cos(x) + j\sin(x)$, where $j = \sqrt{-1}$.
23. See, for example, the special issue on extremely low frequency (ELF) communication in the *IEEE Transactions on Communications,* vol. 22, April 1974.

4
Heaviside's Early Telegraphy Work

He was a mathematician at one moment, and a physicist at another, but first and last, and all the time, he was a telegraphist.

— W. E. Sumpner, in his 1932 Kelvin Lecture
to the IEE, "The work of Oliver Heaviside"

A Full-Time Student

After Heaviside resigned from his telegraph operator's job in Newcastle in 1874, he returned to London to live with (and off) his parents in their home at 117 Camden Street, Camden Town. Soon after (June 1874) *Philosophical Magazine* published his paper "On telegraphic signalling with condensers" in which he analyzed the transient behavior of a cable excited by a step input at one end, and terminated in a capacitance (*not* an earth ground). The approach is all in the spirit of Sir William Thomson (Heaviside treated Thomson's cable equation, but with an additional term to account for leakage through a nonperfect, i.e., realistic, cable insulation). While actually a routine extension of Thomson's paper of nearly twenty years earlier (a fact he admitted, writing[1] in the introduction "...the only way [to solve the problem]...is to follow the method given by Sir William Thomson in 1855..."), it is mathematically quite sophisticated and displays a skilled knowledge of Fourier series, partial differential equations, and the use of boundary conditions. From the publication date Oliver must have actually written it while still employed in Newcastle. This paper is theoretical in nature, but it is quite clear that the motivation behind it was rooted in practical issues, and Heaviside made conclusions in this paper concerning an actual, in-use telegraph system.

In 1876 his parents moved to 3 St. Augustine's Road, St. Pancras, and it was here that Oliver really began to work in earnest. He quickly settled into a routine which most people certainly must have found just a bit peculiar[2]:

> His habit was to retire to his room at about 10 o'clock at night and to work there until the early hours of the morning. He closed his door and window, lighted his oil lamp, and allowed the air to become hot and stifling. He worked also during the day, in seclusion. In order that he might not be disturbed, his food was placed outside his door, and there it remained until he was disposed to take it. Thus he transgressed most of the rules that modern conventions prescribe for health. On the other hand, he enjoyed walking, and occasionally he had more vigorous exercise, for he was a good gymnast.

This description of Heaviside's penchant for strange study conditions is similarly related by one[3] who knew Heaviside later in life, and who recalled how Oliver liked to work in a room "hotter than hell," and that he kept the windows of this room shut in spite of the smoke. At

Heaviside's house from 1876 to 1889, at 3 St. Augustine's Road, where he did a major part of his research. This photograph was taken in 1950. Rollo Appleyard's book *Pioneers of Electrical Communication* has a picture circa late 1920s, from a different angle, showing the house in better times.

some time in his life Heaviside picked up the habit of smoking a pipe; it is easy to imagine him bent over his desk puffing away above his pen, papers, and electrical apparatus (for some years after his early retirement, until at least 1887, Heaviside conducted actual experiments with hardware in his room, and on at least one occasion—which left his father unamused—filled the entire house with the odor of battery acid[4]).

The Telegraph Papers

The first five years after his retirement were among the most productive of his life, during which he wrote three remarkable papers on the theory of the electric telegraph. The trilogy of papers was an impressive display of their author's enormous practical knowledge of real telegraph systems, as well as his mathematical skills. They were jam-packed to the point of overflowing with results of immediate, practical importance to the then rapidly developing telegraph industry. They certainly brought Heaviside to the attention of individuals within the Post Office—and yet he had no impact. Why this was so makes a story of intrigue, vendetta, and ignorance. But first the theory.

August 1876 saw the appearance in *Philosophical Magazine* of the first of these three papers which extended the mathematical understanding of telegraphy far beyond William Thomson's submarine cable theory. The initial paper in the trilogy,[5] with the intriguing title "On the extra current," would have been an impressive piece of work even if it had been written by the pen of a distinguished Cambridge don, and a close study of it makes it clear that Heaviside (a 26-year-old unemployed nobody) was already a brilliant talent.

In this paper Heaviside treated the *charging* and *discharging* of a *finite* length of cable with the effects of induction included. He quickly, seemingly without effort (which leaves open the question of why it hadn't been done before), obtained the differential equation for the voltage, $v(x, t)$, along the cable. Assuming a uniform resistance, capacitance, and inductance per unit length, k, c, and s, respectively, he arrived[6] at

$$\frac{\partial^2 v}{\partial x^2} = kc\frac{\partial v}{\partial t} + sc\frac{\partial^2 v}{\partial t^2}.$$

This equation is now much more than the diffusion equation of Thomson's submarine cable (containing it as the special case for $s = 0$). The equation now allows finite-velocity *wave* propagation (although Heaviside wouldn't fully appreciate the implications of this for several more years); it would be the key to the solution of the phase-distortion problem bedeviling and impeding the widespread use of the telephone.

Heaviside did, however, appreciate the interdependence of his two problems, as he first solved the mathematically easier discharge problem (easier because it has homogeneous boundary conditions), and then used that solution to write, *by inspection*, the solution to the charging problem. Specifically, if a battery with voltage V has been applied across a cable of length ℓ for a very long time, then the steady-state current and voltage are

$$\left.\begin{aligned} \text{current} &= \frac{V}{k\ell}, \\ \text{voltage} &= V\left(1 - \frac{x}{\ell}\right), \end{aligned}\right\} \quad 0 \le x \le \ell.$$

Heaviside then shorted the ends of the cable (he was imagining each end of the cable now connected to a perfect earth ground) and found that the effect of the new induction term in the cable equation was to create oscillations in the voltage as it decayed during the discharge. He presented Fourier series expressions for both the discharge current and the voltage, $i(x, t)$ and $v(x, t)$.

He next argued that if you start with a discharged length of cable and then suddenly apply a voltage V to the end at $x = 0$ (the end at $x = \ell$ is grounded) then the voltage and current will go through their discharge behavior in reverse, back toward the steady-state values of the charged cable. That is, if $v'(x, t)$ and $i'(x, t)$ are the charging voltage and current, then

$$v'(x, t) = V\left(1 - \frac{x}{\ell}\right) - v(x, t)$$

$$i'(x, t) = \frac{V}{k\ell} - i(x, t).$$

It is here that we finally learn what Heaviside meant by his enigmatic title—he calls the decaying, oscillatory $i(x, t)$ "the extra current" in the charging problem.[7]

Obviously, then, v' and i' will oscillate, too. As Heaviside wrote,[8]

> ...we may be sure that, in virtue of the property of the electric current which Professor Maxwell terms its "electromagnetic momentum", whenever any sudden change of current or of charge takes place in a circuit possessing an appreciable amount of self-induction, the new stage of equilibrium is arrived at through a *series of oscillations* [my emphasis] in the strength of the current, which may be noticeable under certain circumstances.

He then gave some examples of such oscillations, thereby explaining earlier, puzzling observations, e.g., anomalous behavior of telegraph receivers, and the discharge of coiled (highly inductive) submarine cables.[9] This interest in the oscillatory behavior of inductive circuits was the motivation for the paper, and Heaviside wrapped it up by deriving an interesting formula for a *lumped* parameter circuit (as compared to the *distributed-line* parameters of the first part of the paper). If L, R, and C are the *total* inductance, resistance, and capacitance of a series circuit, and if C is charged and allowed to discharge, then the discharge will be oscillatory if

$$L > \frac{R^2C}{4}$$

and furthermore, if the current oscillations have n zero-crossings per second (Heaviside used the terminology "current reversing"), then[10]

$$L = (2C\pi^2n^2)^{-1}[1 + (1 - \pi^2R^2C^2n^2)^{1/2}].$$

The most interesting aspect of this result is the idea of expressing L in terms of an observed frequency (the value of n). This is a very natural thing to do today with inexpensive oscilloscopes (commonly available in high schools where teenagers make them from kits) able to observe signals up to 10 MHz and beyond. In 1876 (the oscilloscope wasn't invented[11] until 1897) this concept was far ahead of its time.

"On the extra current" is a paper of high technical merit, but with the exception of people of the caliber of Maxwell or Thomson, hardly anyone involved in electrical matters could have understood much of it (certainly no one at the Post Office, even if they had tried, which it seems they didn't). The blame for that lies mostly with Heaviside himself. He refused to expend the extra time and effort required to make this and subsequent papers not just correct, but also *readable*, to those who could make practical use of his discoveries. In this he was foolish and obstinate, and complaints[12] about the extreme difficulty involved in reading his papers were always curtly dismissed by Heaviside, with no apologies. For that he would pay dearly almost all his life.

Indeed, he defended his lack of aid for the reader in the Introduction to the first volume of

Electromagnetic Theory with words that must have seemed arrogant: "Fault has been found with these articles that they are hard to read. They were harder, perhaps, to write." No doubt true, but certainly no excuse for seeming to thumb his nose at those less gifted. Later in the book[13] Heaviside again complained bitterly about this: "...I find that there has been too much of an idea that I have merely given some very complicated formula which no fellow can understand, and have made some dogmatic and paradoxical statements about the consequences." Heaviside did try to justify his position by claiming he *had* to eliminate explanatory passages to achieve sufficient compactness to be published in space-cramped journals. This was perhaps reasonable to some extent for the original appearance in print, but it was *not* true for book publication. His five books (*Electrical Papers* and *Electromagnetic Theory*) contain almost exact reprints of his journal articles.

In spite of direct appeals to sit down and write a fresh, new book from the beginning, Heaviside refused to do so. On February 24, 1894, for example, George Francis FitzGerald wrote to Heaviside "Perry was suggesting the other day that if you would write a book that was easy to be understood it might pay." The reference was to John Perry, at that time a professor of mechanical engineering at Finsbury Technical College, London, who was one of Heaviside's strongest supporters. We'll meet Perry again, during the age-of-the-Earth controversy with Lord Kelvin which involved Heaviside and his controversial mathematics. On February 26 (mail service evidently being very swift in those days) Heaviside replied to FitzGerald in a long, rambling letter of over 2500 words. His words were a flat rejection. "I do not write for the masses... . It demands such a heap of trouble in expanding, and trying to make readable, and to produce connections... . In fact, the work I like best is not writing at all, but investigation... . It is no use for Perry or anyone else to complain that I am hard to read. He must make the best of me that he can... . I cannot favour Perry's notion of a joint book. I am sure he could do it better by himself, if he is aiming at popularity... ." Not even the lure of possibly making some badly needed money was able to sway Heaviside.

The Problem of Signal Rate Asymmetry

The second paper[14] in the trilogy ("On the speed of signalling through heterogeneous telegraph circuits") appeared in *Philosophical Magazine* in March 1877. His personal observations, to which he specifically refers, of the dependence of the maximum usable information transmission rate on the *direction* of transmission over the 1868 Anglo-Danish cable (see Chapter 2) were the motivation for this paper. His explanation of this puzzling effect was brilliant in its simplicity.

Heaviside began by stating that *the asymmetry could not be due to the underwater cable or its associated land-lines*. This really was an astonishing claim because it seemed to go directly against intuition. It had, after all, been frequently observed that if the underwater cable was in the electrical center of the total circuit (that is, with equal resistance land-lines at each end), then the maximum transmission rate was the same in both directions. If, however, the terminating land-lines were of unequal resistance, then the station connected to the cable via the "electrically short" land-line could transmit faster than could the station at the other end. It therefore seemed obvious that the geometry of the cable and its land-lines must cause the transmission rate asymmetry.

Heaviside flatly declared this was not true, writing "...when the light of theory is thrown upon this view of the matter it is at once found to be untenable." He showed this by invoking what is known today as the *reciprocity theorem*. As he wrote,

> It is easily shown [but he *didn't* show it] that if condensers be distributed in any arbitrary

George Francis FitzGerald (1851–1901)

He tried to explain some of the social realities of the world to Heaviside—and failed. He could often find startlingly refreshing ways to express himself. In an 1892 letter to *Nature*, for example, he wrote in defense of Academe: "If Universities do not study useless subjects, who will?"

> manner along a line which is to earth at each end, dividing it into sections having any resistances, and the condensers be all initially discharged, the introduction of an electromotive force in the first section will cause the current to rise in the last section, in the same manner as the same electromotive force in the last section will cause the current to rise in the first section. Furthermore, it may be shown [but he didn't] that if leaks be introduced on the line in any arbitrary manner, the same property will hold good [see Note 15]. ...Now every telegraph-line, however irregular it may be in its resistance, capacity, and insulation in different places, may be considered as such a system of condensers and leaks, infinite in number if necessary; whence it follows that on *any line* [my emphasis] there is absolutely no difference in the retardation in either direction... . Therefore, to account for the facts, which cannot be gainsaid, we must look *outside the line* [my emphasis] and fix our attention on the sending and receiving apparatus.

Heaviside modeled a marine telegraph cable with land-based send/receive stations as a cable with total resistance c and total capacitance S, connected to land-lines with resistances a and b. The receiving instrument has resistance g, while the sending device (battery and key) has resistance f, as shown in the Tech Note at the end of this chapter. In his analysis Heaviside ignored induction.

Letting T and T' denote signal retardation times, Heaviside's results were

$$T=\frac{S}{R}\left(\frac{c}{2}+a+f\right)\left(\frac{c}{2}+b+g\right) \qquad \text{when sending left to right}$$

$$T'=\frac{S}{R}\left(\frac{c}{2}+a+g\right)\left(\frac{c}{2}+b+f\right) \qquad \text{when sending right to left.}$$

From these expressions for the retardations one could, as Heaviside did, observe that

> ...if $a = b$, $T = T'$; also if $f = g$, $T = T'$; but if $a < b$...$T < T'$ if $f < g$, and $T > T'$ if $f > g$. Or, in plain English, the retardation is the same in both directions if the land-lines have equal resistances, whatever may be the resistances of the battery and receiver; it is also the same in both directions if the battery and receiver have equal resistances, whatever may be the resistances of the land-lines; but if the resistances of the land-lines are unequal, the retardation is greatest when the station nearest the cable is receiving, if at the same time the battery is less than the receiver resistance, and least in the contrary case.

Thus, with amazing ease, Heaviside accounted for *all* the previously puzzling transmission rate effects on marine telegraph circuits.

A "Mathematical Monster"

No one would apply the description of "amazing ease" to the final paper in the trilogy, "On the theory of faults," which appeared in *Philosophical Magazine* in the summer of 1879. This paper is *very* advanced mathematically, and (in my opinion) nearly impossible to read. It would be interesting to be able to go back in time, as a fly on the wall, to observe the reaction of subscribers when they read, with no preliminary analysis *at all* (the following is the *first* equation in the paper):

> The maximum strength Γ of the received-current waves is

$$\Gamma=\frac{E}{K\ell}\frac{8n\sqrt{2}}{\pi}(e^{n}+e^{-n}-2\cos n)^{-1/2}\{e^{n}+e^{-n}+2\cos n + \frac{1}{2nz}(e^{n}-e^{-n}+2\sin n)+\frac{1}{8n^{2}z^{2}}(e^{n}+e^{-n}-2\cos n)\}^{-1/2}.$$

It is easy to sympathize with those readers (probably nearly *all* the readers) who found this about as meaningful as they would a line of hieroglyphics. It must have appeared, as one writer has put it, as "a superior form of gibberish,"[16] and the Institution of Electrical Engineers (IEE) in London has a letter written by Heaviside in which he acknowledged that some people thought of him as a "mathematical monster." This is not to say there are no important results in the paper—there are, and that makes it twice as unfortunate that Heaviside chose to present his discoveries in an obscure, intimidating manner. To gain a hint of what he was up to, the following is about as clear as he got:

> When a natural fault, or local defect in the insulation is developed in a cable, it tends to get worse—a phenomenon, it may be observed, not confined to cable-faults. Under the action of the current the fault is increased in size and reduced in resistance, and if it not be removed in time, ends by stopping the communication entirely. Hence the directors and officials of

> submarine-cable companies do not look upon faults [see Note 17] with favor, and a sharp look-out is kept by the fault-finders for their detection and subsequent removal. But an *artificial fault* [my emphasis], or connection by means of a coil of fine wire between the conductor and sheathing, would not have the objectionable features of a natural fault.

Next comes the most interesting (historically) part of the paper, with Heaviside using his theory to suggest (in vain) improvements to existing technology:

> If properly constructed it [the artificial fault] would ...considerably accelerate the speed of working [of transmission of information]. The best position for a single fault would be the centre of the line; and perhaps 1/32 of the line's resistance would not be too low for the fault.

In this suggestion he was decades ahead of everyone else. It would be nearly twenty years before Silvanus Thompson would independently suggest the same thing, and people would belatedly recall Heaviside's analysis. But of course the overly complex presentation of his work was a major reason for this delay.

Even Heaviside must have had some internal uneasiness over the mathematical mountains he had erected in this paper for his readers to climb (if they could), as evidenced by the following two lines:

> All mathematical investigations of physical questions are approximative; and being such, impossible results arise in extreme cases [see Note 18].

and

> ...the completion of a solution is of far greater importance than any proof that the solution is possible.

This last assertion is a concise statement of a point of view Heaviside retained all his life. It got him into hot water time and time again with many mathematicians who might otherwise have viewed his work more sympathetically.

Heaviside wrote a very practical (and for him almost tutorial) paper[19] the year before the last paper in the telegraph trilogy ("On electromagnets, etc.," in the *Journal of the Society of Telegraph Engineers*). It is an important part of his early work, however, because it introduced the expressions for the *a–c* impedances of resistors, inductors, and capacitors now routinely used by all electrical engineers. This alone, of course, ensures the historical value of the paper, but there is one passage in it that also provides the basis for some amusing speculation. After discussing the high-frequency dominance of inductive impedance in a series resistor–inductor circuit (e.g., a telephone[20]), he wrote the interesting line:

> As...the telephone is sensible to very much more rapid reversals than 1000 per second, the enormous speeds possible on short [telephone] lines is easily conceivable, would the action be sufficiently magnified and recorded, so as to *appeal to the eye instead of the ear* [my emphasis].

What a nice description of cable TV (in 1878!).[21]

These four papers show he was thinking in terms of voltage and current, which he took to be the fundamental physical quantities of interest. Later, as he applied Maxwell's theory to the same and other problems, the electric and magnetic field vectors would take center stage. Heaviside himself remarked on this transition by writing[22] a footnote in another paper which referenced his earlier work: "The present article will be found to be a sort of missing link between the earlier articles on propagation and the later ones, in which the subject is discussed on the basis of Maxwell's theory of the ether as a dielectric."

Arithmetic Drudgery

A characteristic feature of Heaviside's work was already clearly evident in these early papers. His solutions almost always were in the form of complicated formulas and/or infinite series. To calculate numerical values for graphs and special cases of interest must have been just awful, mind-dulling work. Today's researchers take for granted the availability of inexpensive electronic calculators that can calculate inverse trigonometric functions and logarithms to any base, and raise any number to any power, all to ten digits of accuracy in less than a second. And for the really complicated stuff, programmable desk-top personal computers can chew it all up in no time at all (as well as create plots directly on paper). Heaviside, of course, had none of this, but only his pen, paper, a *lot* of ink, a book of logarithms, and a slide rule[23] (a device now as extinct as the pterodactyl). He explained his use of series by writing,[24] "Orthodox mathematicians, when they cannot find the solution of a problem in a plain algebraical form, are apt to take refuge in a definite integral, and call that the solution."

It is interesting to note how evolving technology alters our feelings about what constitutes a "good" mathematical solution. Heaviside, as noted, did not consider definite integrals to be the end to a problem; "...a definite integral is ...of no use until it is evaluated..."[25] "Perhaps, on the whole, it is as well to keep away from the definite integrals..."[26] and "My usual practice is to avoid definite integrals... ."[27] Solutions in the form of series expansions, however, are in just the right state to allow numerical work and can be easily converted to computer code. As Heaviside wrote[28] after deriving a solution in the form of an infinite sum of Bessel functions, "The ...result is ornamental. Putting it in a power series, its meaning can be seen, and is easily calculable." Yet John Carson of AT&T (a great admirer of Heaviside) wrote[29] in 1926, "Its [a power series] chief defect, and a very serious defect indeed, is that except where the power series can be recognized and summed, it is usually practically useless for computation... . This disadvantage is inherent and attaches to all power series solutions. For this reason I think Heaviside overestimated the value of power series... ." Carson suggested the use "of a planimeter... or ...numerical integration" to evaluate definite integrals. When Carson wrote his opinion (which stated a perfectly reasonable position in his day) he could have had no inkling of the computer revolution that lay a half-century in the future and which would make Heaviside's series solutions so natural (although even Heaviside was once moved to call[30] a series "horrid").

How Heaviside kept from giving in to despair[31] as he plodded through his arithmetic (his notebooks are full of calculations carried out to seven decimal places and in one place he went out to *sixteen* places!) is a mystery to me. He did make it quite plain, however, that he didn't enjoy this aspect of his studies at all, writing at one point[32]

> Let us now dig something out of the above formulae. This arithmetical digging is dreadful work, only suited for very robust intellects.

In 1898, clearly fatigued from too many fractions, Heaviside commented again on the onerous nature of arithmetic when, after arguing his way *by words* through a wave propagation problem, he wrote[33]

> Perhaps some electrical student who possesses the patient laboriousness sometimes found associated with early manhood may find it worth his while to calculate the waves thoroughly and give tables of results, and several curves in every case. It should be a labour of love, of course... ."

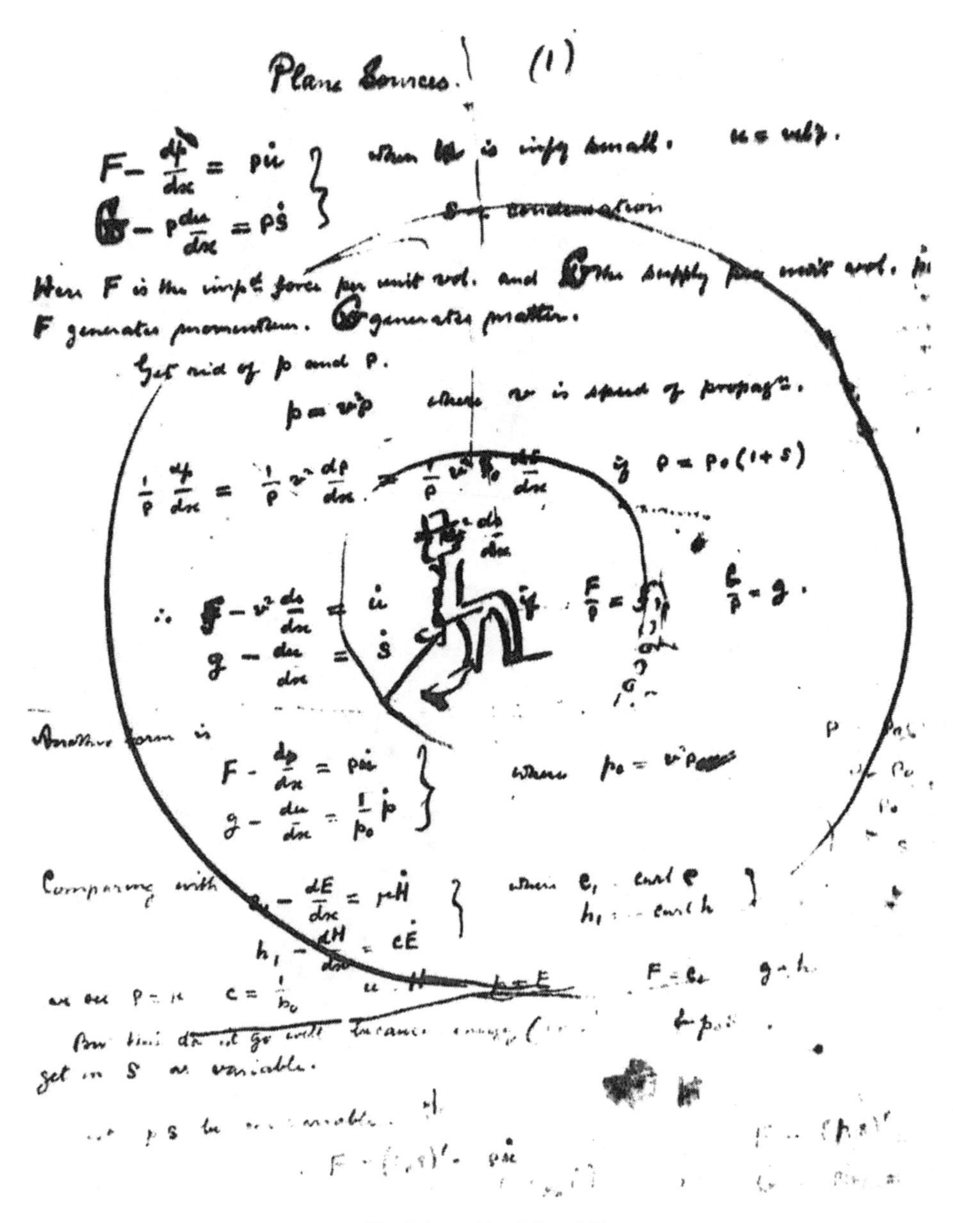

The lighter side of Heaviside.

Heaviside had more than arithmetic to block his way, however. It was during this early period of his career that he made an enemy for life. The man he faced as an adversary was a powerful one and, while not nearly as smart as Oliver in electrical matters, he could run circles around the "crank from the London slum" in every other way. The rest of Oliver's life was to take the form of a David and Goliath confrontation with this person, even after the man died. True to the myth, Heaviside would finally win, but it would be a bloody, bitter, Pyrrhic victory for him.

i.e. pot. of uniformly charged sphere is 140 times that of a single electron due to itself, when $b = 100$, and $1{\cdot}73 \times 10^8$ volts is pot. of uniform, etc.

$$U_b = \frac{\frac{1}{2}(4\pi b^2)^2 (n^2 \times 10^{-20})^2}{b} \times 9 \times 10^{20} \quad \text{if } n \text{ per linear cm.}$$

$$= \frac{720\, n^4 b^3}{10^{20}} = 7{\cdot}2 \times 10^{12} \quad \text{when } n = 10^6 \quad b = 10^2$$

U_b varies as 4th power of n, and cube of b.

$$\Sigma U_0 = U_0 \times 4\pi b^2 n^2 \quad \therefore \quad \frac{[illegible]}{\Sigma U_0} = \frac{4{\cdot}1}{10^7} \times 4\pi b^2 n^2$$

$= {\cdot}000140$ when $n = 10^8$, $b = 10^2$

$$\frac{U_b}{\Sigma U_0} = \frac{720\, n^2 b}{10^{20} \cdot 4\pi \cdot 4{\cdot}1} \times 10^7 = \frac{14 n^2 b}{10^{15}} = 140 \text{ when } n = 10^6,\ b = 10^2$$

So it is easy to make U_b less than ΣU_0; by reducing n or b. as above.

Of course the mutual energy is important in this case of a sphere. Perhaps we may say safely that pot. at any electron due to the rest is about same value as pot. in case of uniform distribution, so that the actual electric energy is increased from U_b (uniform) to $U_b + \Sigma U_0$ by condensing the surface electrification into electrons. This argument should apply when ΣU_0 is small compared with U_b. But how about a small charge given to sphere, say $n = 1$. Then

$$\frac{U_b}{\Sigma U_0} = \frac{14 b}{10^{15}} \quad \text{very small unless } b \text{ is large.}$$

This would need a considerable revision of formulae of electrostatics, (capacity, etc).

$$V' = \frac{4\pi b^2 n^2 10^{-20}}{b} \times 9 \times 10^{20} = 113\, b n^2 \quad (\text{uniform})$$

Now if $n = 1$, this would be $113 b$, which is quite easily measurable.

$V' = \frac{9}{b}$ = potential of one electron spread over sphere rad. b. This might be measurable, as against $8{\cdot}2 \times 10^{13}$ pot of electron itself.

Evidently, a lot of humouring is wanted. There is action of the surface electrons inwardly as well as outwardly.

Shrinkage of Earth under gravitation.

If G is displ. [illegible] eq. in elastic solid

$$\rho \ddot{G} = (\lambda + \mu)\nabla \cdot \nabla G + \mu \nabla^2 G - \rho \nabla E.$$

If there is no impressed force $-\rho \nabla E$. Take divergence (s = dilatation)

$$\rho \ddot{s} = (\lambda + 2\mu)\nabla^2 s - \rho \nabla^2 E.$$

If the impressed force is that of gravitation, $+\nabla^2 E = \gamma\rho$ (to make force $= -\rho \nabla E$)

so $\rho \ddot{s} = (\lambda + 2\mu)\nabla^2 s \mp \gamma \rho^2$.

In equilm. $+\nabla^2 s = \frac{\gamma \rho^2}{\lambda + 2\mu} = +\frac{\rho}{\lambda + 2\mu}\nabla^2 E$

or $s = \frac{\rho}{\lambda + 2\mu}(E - E_a)$; where $-E = \Sigma \frac{\gamma\rho}{4\pi R} = \gamma$ pot ρ

$$= \frac{\gamma\rho}{4\pi}\left(\frac{\frac{4\pi r^3}{3}}{r} + \int_r^a \frac{4\pi r^2 dr}{r}\right) = \gamma\rho\left(\frac{r^2}{3} + \frac{a^2}{2} - \frac{r^2}{2}\right) = \gamma\rho\left(\frac{a^2}{2} - \frac{r^2}{6}\right)$$

$E_0 - E_a = \gamma\rho \frac{a^2}{3}$ and $-s = \frac{\rho}{\lambda + 2\mu}\frac{a^2 - r^2}{6}\gamma\rho$ and $-s_0 = \frac{\gamma\rho^2 a^2}{6(\lambda + 2\mu)}$.

The above is rational, so that $\frac{\rho\rho\gamma}{4\pi R^2}$ is attraction of ρ and ρ. Put $\frac{\rho^2}{4\pi} = \rho'^2$

$\therefore -s_0 = \frac{4\pi\gamma\rho'^2 a^2}{6(\lambda + 2\mu)}$ = compression at centre.

Also $-s = \frac{d}{dr}(rG) = [illegible]$ $\therefore \frac{d}{dr}(r^2 G) = \frac{\rho\gamma}{\lambda + 2\mu}\frac{a^2 r^2 - r^4}{6}$

$r^2 G = \ldots \frac{\frac{a^2 r^3}{3} - \frac{r^5}{5}}{6}$ $G = \ldots \frac{\frac{a^2 r}{3} - \frac{r^3}{5}}{6}$ $\therefore G_a = \frac{4\pi\rho'^2\gamma}{\lambda + 2\mu}\frac{a^3}{45}$ at surface.

So the central condensation for water, as earth in size, is

$$-s_0 = \frac{4\pi \cdot 6{\cdot}65 \cdot 10^{-8}(6{\cdot}37)^2 \cdot 10^{16}}{12 \cdot 10^{10}} = 2{\cdot}82$$

Much too big for theory. Let $a = \frac{1}{10}$ of earth's radius, $\therefore -s_0 = {\cdot}0282$.

This is reasonable, and the shrinkage at surface is

$$G_a = -s_0 \times \frac{6a}{45} = \frac{28}{1000}\frac{2}{15}a = \frac{56}{15000}a = \frac{11}{3000}a.$$

So we cannot use ord. [illegible] of elasticity for the real earth, the effect of gravitation is to produce condensation [illegible] far in excess of

Typical pages from one of Heaviside's surviving notebooks.

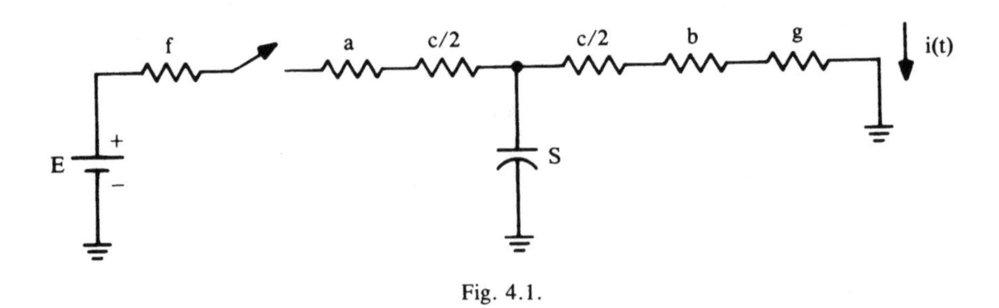

Fig. 4.1.

Tech Note: Why a Cable Is Slower in One Direction than in the Other

Before the sending key on the left side of Fig. 4.1 is closed the receiver current, $i(t)$, is zero. Heaviside wrote the current once the key is closed (at time $t = 0$) as an exponential decay asymptotically approaching a final, steady-state value; it was already well known that that is the generic behavior of all circuits containing only resistances and capacitances. The steady-state current (reached when S is fully charged and therefore of no further influence) is obviously E/R, where R is the total resistance as "seen" from the viewpoint of the key. Therefore, by inspection,

$$i(t) = \frac{E}{R}\left[1 - e^{-t/T}\right], \qquad t \geq 0$$

$$R = f + a + c + b + g.$$

In all such resistance/capacitance circuits the value of T is the product of a resistance and a capacitance. Heaviside's (tricky) contribution was the observation that the value of T is *not* the same in left-to-right as in right-to-left transmission. Heaviside called T the *retardation* (not too much later the modern term *time constant* came into common use among electrical scientists); obviously, the larger T is, the slower $i(t)$ will increase and thus the slower the maximum possible transmission rate will be. Unfortunately, Heaviside provided no clue as to how he arrived at his expressions for T (they are, of course, correct), and this is just one more example of how a few extra lines of explanation would have significantly increased the number of those able to understand him. To appreciate how little more he had to do to aid his readers, redraw the circuit of Fig. 4.1 as shown in Fig. 4.2: $K_1 = c/2 + a + f$, $K_2 = c/2 + b + g$, $R = K_1 + K_2$.

The time constant for the received current, $i(t)$, when S is charging is the same as when it discharges *if* the battery is replaced with its equivalent resistance (zero, or a short-circuit, if the battery is ideal). When S discharges it does so through K_1 and K_2 *in parallel*, i.e., the discharge resistance is $K_1K_2/(K_1 + K_2) = (c/2 + a + f)(c/2 + b + g)/R$. Heaviside's expression for

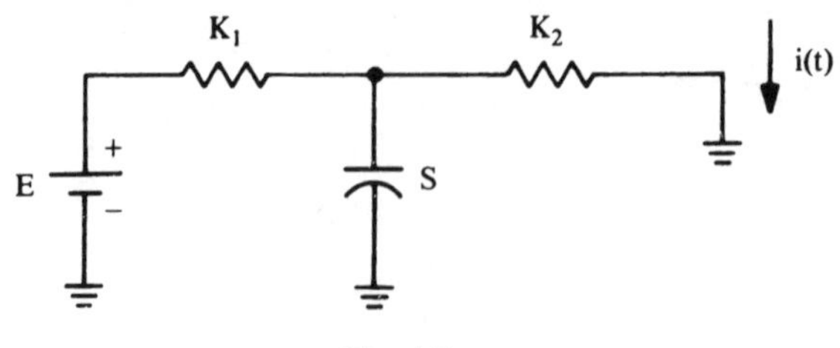

Fig. 4.2.

T is just this effective resistance multiplied by S. For right-to-left transmission (i.e., f and g are interchanged), then $K_1 = c/2 + a + g$ and $K_2 = c/2 + b + f$ and the expression for T' results. This is all Heaviside had to write; unfortunately he didn't, and thus no doubt left many of his readers mystified.

Notes and References

1. EP 1, p. 48.
2. R. Appleyard, *Pioneers of Electrical Communication*, London: Macmillan, 1930, pp. 215–216.
3. Dr. Ludwik Silberstein (1872–1948), as related in the *Heaviside Centenary Volume*, London: IEE, 1950, p. 14. There is a line in the first volume of *Electromagnetic Theory* (p. 5) that may explain Heaviside's fondness for the late hours of the night to do his work: "How is it possible to be a natural philosopher when a Salvation Army band is performing outside: joyously, it may be, but not most melodiously?"
4. Appleyard (Note 2), p. 222. In Heaviside's early papers there are several detailed descriptions of his experiments with carbon contacts, microphones, and telephones (used as acoustic indicators, which would seem to indicate that deafness was not *always* a problem). See, for example, EP 1, pp. 181–190 and 314. His hardware interests included non-telegraphic applications, too. One of Heaviside's notebooks shows, for example, that he was interested in the experimental design of a metal detector to locate bullets lodged within a human body. That was in the days before x-rays, of course, and was a problem of great interest (see, for example, A. G. Bell's "Probing by electricity," *Nature*, vol. 25, p. 40, November 10, 1881, in which a gadget supposedly inspired "on the occasion of the sad attempt upon the life of President Garfield" is described).
5. EP 1, pp. 53–61.
6. Heaviside provided a terse explanation for how he arrived at his equation. More formally, it can be derived by a nearly trivial extension of the method shown in the previous chapter for obtaining Thomson's submarine cable equation. The implications of the extended equation (known today as the *Telegrapher's equation*) are, however, far from trivial, as Heaviside went on to show.
7. Heaviside's use of the term "extra current" varied in his papers. For example, in an 1878 paper "On electromagnets, etc." (EP 1, p. 96) he said it is "the [current] that flows in the circuit after the electromotive force that produced the current in the first place is removed... ." In general, Heaviside's extra current is what we call today the *natural* or *relaxed* response of a circuit, i.e., the transient response due to the dissipation of *internally stored energy*.
8. EP 1, p. 59.
9. Writing ten years later (EP 2, p. 83), Heaviside provided a little more background to his interest in this problem. He said his brother Arthur "...called my attention to certain effects observed on telegraph lines which could be explained by the combined action of the electrostatic and electromagnetic induction, causing *electrical oscillations* [my emphasis]... ." I must be careful, however, not to claim too much for Heaviside in the study of circuit oscillations. It was William Thomson in 1853, in fact, who first studied the series discharge of a capacitor through resistance and inductance (see his "On transient electric currents" in *Mathematical and Physical Papers*, vol. 1, Cambridge: Cambridge University Press, 1882, pp. 534–553). And it was James Clerk Maxwell who was the first to publish (in 1868) an analysis of a resistor, capacitor, and inductor in series with an *alternating* voltage source, as well as to study mathematically the phenomenon of resonance.
10. Heaviside derived this formula at the tail end of a detailed analysis of the series R, L, C circuit with an initial charge in C. To derive it quickly and directly in the modern way, set the generalized circuit impedance (a word, by the way, coined by Heaviside), $Z(s) = R + sL + 1/sC$, equal to zero (because if $I = E/Z$ is to exist with $E = 0$, then $Z = 0$, too). Solve for the time constants (there are two) of the natural response current as C discharges. For this current to be oscillatory these time constants must be complex (see Note 22 in Chapter 3), i.e., the expression under the resulting square root must be negative. This condition gives the relationship among R, L, and C for oscillations to exist, as well as the frequency of the oscillations, and leads immediately to Heaviside's formula. At this point I can't resist quoting the remark of a "practical man": "...You cannot generate electricity out of the square root of minus one." It seems, however, that that is where it mostly comes from! See H. J. Ettlinger's "Four sparkling personalities," *Scripta Mathematica*, vol. 8, pp. 237–250, 1941. In addition to Heaviside, the "sparkling personalities" belong to Charles Steinmetz (the genius of General Electric, whom Heaviside greatly admired), Michael Pupin (who will appear later in Heaviside's story), and Vladimir Karapetoff (to whom the "practical man" directed his misinformed comment). Karapetoff was a professor of electrical engineering at Cornell University and a great admirer of Heaviside. In particular, he invented an ingenious gadget (called the "Heavisidion") for mechanically solving the differential equations of long transmission lines (see his paper for the construction details and a photograph of the Heavisidion in *Transactions of the American Institute of Electrical Engineers*, vol. 42, pp. 42–53, February 1923). Heaviside knew of Karapetoff, and in a letter dated September 23, 1922 (written to John S. Highfield, then President of the IEE) he gave this reaction: "Last winter one named Karapetoff in the U.S. proposed to name a machine he was making to imitate loaded transmission lines, 'the Heavisider'! This does not provoke me to commit Heavicide."
11. It was invented by the German Ferdinand Braun, who would later share the 1909 Nobel prize in physics with the

Italian Guglielmo Marconi (both men were honored for their work in wireless telegraphy). As late as 1902 even one of Braun's own doctoral students, the Russian Leonid Isakovich Mandelstam, found he couldn't use his professor's new cathode-ray oscilloscope at the high frequencies he encountered during his research. His dissertation "Determination of the period of oscillation of an oscillatory condenser discharge" sounds close to the spirit of Heaviside's analysis. See F. Kurylo and C. Susskind's *Ferdinand Braun*, Cambridge, MA: MIT Press, 1981, pp. 145–146.

12. Lord Kelvin came right to the point when he wrote (in an 1888 letter to J. J. Thomson, discoverer of the electron): "I think O.H. is... unintelligible to anyone who has not read all O.H.'s papers, and it and everything else would be unintelligible to anyone who had. No brains would be left." See Lord Rayleigh's *The Life of Sir J. J. Thomson*, Cambridge: Cambridge University Press, 1943, p. 33. A more personal comment was made by George Frederick Charles Searle at the 1950 IEE Centenary Celebration of Heaviside's birth: "On one occasion... he asked me to verify paragraph 535 of vol. 3 of *Electromagnetic Theory* [on p. 499], relating to the motion of an electrified straight line in any direction relative to its length. After an effort, I succeeded. I begged for an insertion of a few words of guidance, which would have reduced the labour into merely 'turning the handle', but all he did was to put '(by work)' after his original 'It follows'." See *Heaviside Centenary Volume* (Note 3), p. 9.
13. EMT 1, p. 417.
14. EP 1, pp. 61–70.
15. Heaviside then gave a clue at this point as to why the reciprocity property is a reasonable one, writing "The differential equation of the current, which is linear and of the same degree as the number of condensers, is the same for the first and last sections; and the conditions to determine the arbitrary constants are the same." As late as 1909 Heaviside was corresponding about the reciprocity theorem with Searle (Note 12).
16. D. W. Jordan, "The adoption of self-induction by telephony, 1886–1889," *Annals of Science*, vol. 39, pp. 443–461, September 1982.
17. One very curious cause of cable faults was early on found to be fish bites! As one writer reported (*The Electrician*, vol. 7, pp. 186–187, August 6, 1881), a cable laid in 1874 was soon after found to have suffered "at least four indubitable fish-bite faults... where the iron sheathing had been forcibly crushed up and distorted from the core as if by the powerful jaws of some marine animal." The writer presented convincing evidence pointing to the *Plagyodus ferox* ("one of the most formidable of deep-sea fishes") as the culprit.
18. Heaviside often liked to push his mathematical results to the limit, just to see what would happen. For example, he occasionally got involved with *negative* resistance (at best, an alien concept in those pre-electronic days). In 1882 (EP 1, p. 148) he declared this to be "physically impossible." Today, electrical engineers wouldn't miss a step over that (e.g., an *oscillator* is an *amplifier* with an input impedance having a negative real part), considering it to represent a *source* (supplier) of energy, as opposed to an energy *sink* (dissipater). Indeed, that is precisely (EMT 2, pp. 163–170) what Heaviside finally (in 1895) came to say: "A source of energy is involved, of course... ." A little later he went on to explain his interest in extreme situations: "After so much detail concerning this simple abnormal case, others of a similar character may be very briefly treated. As for why they are considered at all, an anecdote about Dr. Elliotson comes in useful. One of his students said he did not see the use of studying morbid physiology; it was so unnatural. The doctor told him he was a blockhead, adding, 'It is only by studying the morbid that the true conditions of health can be ascertained.' " The reference is probably to Dr. John Elliotson (1791–1868), a London surgeon, highly acclaimed as a clinical teacher. Later in his career he took up mesmerism and phrenology, which got him in trouble with the medical establishment. He eventually lost his practice and died in poverty. The similarity of their careers must have made him a sympathetic character to Heaviside.
19. EP 1, pp. 95–112.
20. Heaviside called the Bell telephone "this most sensational application of electricity... ."
21. No, I do *not* really believe Heaviside was thinking about transmitting pictures over a telegraph line when he wrote this, although it was precisely during this time that others *were*—see J. D. Ryder and D. G. Fink's *Engineers & Electrons*, New York, NY: IEEE PRESS, 1984, pp. 149–151, and the editorial "Seeing by electricity," *The Electrician*, vol. 24, pp. 448–450, March 7, 1890, which concluded, "There is more hope of seeing through the proverbial brick wall than of seeing through a copper wire." More optimistic was the correspondent who proposed to send *people* by telegraph from London to New York after dissolving them into a "human solution"! See *The Electrician*, vol. 6, p. 263, April 9, 1881 (and the humorous reply on May 7, p. 328, signed by a Mr. B. A. M. Boozle).
22. EP 1, p. 141.
23. In 1901 the B.A. held its annual meeting in Glasgow, during which John Perry led a discussion on the subject of teaching mathematics. This discussion, along with written remarks from non-attendees (including Heaviside), was later published as the book *Teaching Mathematics*, London: Macmillan, 1901. As part of his presentation, Perry asserted (p. 25) the value of teaching students the "principle underlying the construction and method of using the common slide rule," and "the use of a slide rule in making calculations." To this Heaviside replied (p. 66), "I may not think so much of the slide-rule as you do."
24. EMT 2, p. 11.
25. EMT 2, p. 314.

26. EMT 2, p. 383.
27. EMT 3, p. 234.
28. EMT 3, p. 301.
29. *Electric Circuit Theory and the Operational Calculus*, New York, NY: McGraw-Hill, 1926, pp. 31–32 and 95.
30. EMT 2, p. 416.
31. Some years ago I wrote a short science-fiction story (''Newton's gift,'' *Omni*, August 1980) based on my guess as to Newton's reaction to a time-traveler presenting him with a modern hand-calculator to do *his* heavy numerical work. I imagined that Newton, a deeply religious man, would have thought such a mysterious gadget the work of the Devil. I think Heaviside, on the other hand, would have been beside himself with joy, no matter *where* he suspected it might have come from!
32. EP 2, p. 73.
33. EMT 2, p. 433.

5
The Scienticulist

Even this address may be quoted as an illustration of the ignorance of the great Victorian period.

— W. H. Preece, from his inauguration speech upon becoming the new President of the Institution of Electrical Engineers, 1893

True theory does not require the abstruse language of mathematics to make it clear and to render it acceptable. ...All that is solid and substantial in science and usefully applied in practice, have been made clear by relegating mathematical symbols to their proper store place—the study.

— From Preece's 1893 speech, showing the previous quote to be a prophetic one

A few men may confound mathematics with metaphysics (strange delusion!), and vent their scorn upon the former—sour grapes to them.

— Oliver Heaviside

Heaviside's Nemesis

William Henry Preece was born[1] in Carnarvon, Wales in 1834, where his father became mayor when William was ten. By the time he died in 1913, also in Carnarvon, he had enjoyed a life rich with public acclaim and financial rewards; his was a life where yachting and shooting were regular recreations. Starting at age nineteen, he rose steadily through the ranks of the new communications business to a level light-years beyond that of Oliver's telegraph operator's job, eventually becoming Engineer-in-Chief of the British General Post Office (the GPO) in 1892.

Preece learned his trade in the school of hard-knocks, and had little analytical insight into the nature of electricity. This was during the formative years of the electrical industry when the now familiar concept of electrical conduction in metals due to a "sea" of negative charge carriers (electrons) drifting under the influence of an applied electric field was still far in the future.[2] Preece's idea of electricity was analogous to the behavior of an incompressible fluid in a pipe (Oliver Lodge derisively called this the "drain-pipe theory"[3]), and the mathematical complexities of Maxwell's field theory (or even just those of a-c circuit theory) were *always* to be beyond him. In this he was not alone, of course, but Preece *always* refused to admit the simple concepts of his youth would no longer do, even in the 1890s when it was perfectly clear he was wrong, and he was quite bitter in his scorn of theory.

William Henry Preece (1834–1913)

Electrician and Engineer-in-Chief of the British General Post Office, he was Heaviside's principal (and most powerful) adversary.

It was, therefore, possibly more of a provocation than we can fully appreciate today when Oliver Lodge publicly declared[4] "The phenomena connected with alternating currents are peculiar and surprising, and I believe there is no person now alive who has anything like the intimate familiarity with their idiosyncrasies than Mr. Oliver Heaviside has attained to." Heaviside, with his highly mathematical papers, was a natural target for Preece's wrath at upstart theoreticians who dared to challenge his position as a leading authority in all matters electrical.[5]

He was, indeed, an influential man. In 1881 Preece was elected a Fellow of the Royal Society and, in 1899, he was knighted and thereby became Sir William. During the previous year he had served as President of the Institution of Civil Engineers. Earlier, in 1880, Preece had been President of the Society of Telegraph Engineers, and he served as President again in 1893 after the Society had become the Institution of Electrical Engineers. During the 1893 inauguration ceremonies he was introduced by his predecessor (Professor W. E. Ayrton) as "...undoubtedly the most popular man in the whole electrical engineering profession."[6]

Possibly so, but he would not have gotten Oliver Heaviside's vote. It was with this august personality that Heaviside became bitterly entangled. Their decades-long conflict[7] over who had the true insight into the role of induction in telegraph/telephone circuits would, at first, be a one-sided affair. Heaviside had nothing but an acid tongue with which to fight all the power of the "establishment of practical men" that Preece so ably commanded.

In 1850, when Heaviside was just a baby a few months old, the seventeen-year-old William enrolled at King's College, London, intending to eventually acquire a commission in the army. Due to family financial difficulties, however, he withdrew in 1852 and never returned.[8]

The GPO on St. Martin's-le-Grand, as it looked in 1896. The Greek-style building was built during the years 1824–1829, and was demolished in 1910.

Sometime later he obtained a job that would influence his entire life, becoming an assistant on the Engineer's staff of the Electric & International Telegraph Company (Edwin Clark was the Engineer and his brother, Josiah Latimer Clark[9] was his chief assistant—Josiah would also become Preece's brother-in-law, marrying his sister Margaret in 1854).

It was during this period of employment that Preece participated in some experiments (along with the Astronomer Royal G. B. Airy[10] and Michael Faraday) on electric signaling over cables. This work, in 1853, led to the experimental discovery of signal retardation (recall, as discussed in Chapter 3, that Thomson and Stokes did not perform their theoretical studies until late 1854). This early, personal interaction with Faraday had an understandably profound impact on the young Preece which he took to be almost a laying-on of hands, and he went to his grave proudly relating, to all who could stand to hear the story just one more time, how he had "sat at the feet of Faraday."[11]

These experiments with long, capacity-dominated cables, and Thomson's "Law of Squares" a year later, no doubt combined in Preece's mind, resulting in his later stubborn adherence to the infamous "*KR*-law." This "law," announced by Preece in 1887, was supposed to yield the distance over which telephony could be satisfactorily conducted as a function of certain electrical parameters of the telephone cable. It was mostly over this issue (to be discussed in Chapter 8), in fact, that Heaviside and Preece squabbled for years, although by 1887 they had already been taking shots at each other for nearly a decade.

The cable experiments with Faraday were not, in fact, the first time Preece had interacted with the great man. Earlier, before joining the Electric Telegraph Company, William had

Preece in his office at the GPO. This photograph accompanied an essay-interview published in the December 1, 1892 issue of *Lightning* (later *Electrical Times*).

assisted Faraday in his public lectures at the Royal Institution (of which Faraday was Director). It was also about this time that Preece first met Professor William Thomson (Lord Kelvin) with whom he would become quite friendly.[12] It was during this still early stage in his life that Preece gave evidence of his lifelong negative attitude toward theoretical analysis. During a chance encounter at the Royal Institution with Thomson (who was visiting Faraday) the professor declared (as he examined a new book on mathematics), "This is the only proper language of engineers." To which Preece replied with all the confidence of youth, "I cannot worry myself, when I can get it done for thirty shillings a week.'' This sort of unthinking remark must have been like turpentine on a cat's back to math whiz Thomson, and the professor promptly set him straight. Preece later[13] claimed to have been suitably enlightened, but as will become evident the lesson apparently didn't stick. Sir William would become one of the leaders of the ''practical school of electricians,'' as opposed to the new breed of mathematical electrical engineers of which Heaviside was an outspoken member.

Preece's electrical horizons were always sharply limited by not much more than Ohm's law, while Heaviside clearly saw things with a far broader view: ''There are wheels within wheels, and Ohm's law is merely the crust of the pie.''[14] As a man thoroughly at ease with mathematics, Heaviside could hope to make something of the way the wheels turned, while the only fate that awaited Preece and his fellow ''practicians'' was to get ground up in them.

Subdividing the Electric Light

One such example of this fate has become a minor classic in the history of electrical engineering, and gives a good illustration of Preece's style. In the late 1870s the only sources of lighting other than candles and oil lamps were gas lights and the electric arc. The arc is a low-resistance/high-current device, and emits a brilliant, intense light (as well as copious noxious fumes). It was totally unsuitable for interior home lighting, and found its primary use during the 19th century as a form of street lighting; it survives today in the searchlights used for nighttime antiaircraft raids and Hollywood-style movie premieres. The invention of the incandescent lamp promised big changes, however, and indeed the value of gas company shares plummeted in the resulting scare immediately after the announcement of the Edison light. It was Preece who sought to calm frightened widows and orphans (as well as the capitalists) who had their money invested in gas. He did this with some impressive calculations which he claimed ''proved'' it is impossible to operate many incandescent lamps from a central energy source or, as he put it,[15] ''the extensive subdivision of the light must be ranked with perpetual motion, squaring the circle, and the transmutation of metals.'' Reassuringly, he declared ''electricity cannot supplant gas for domestic purposes.''

To all who today pay a monthly electric utility bill, of course, it is evident that somewhere in his mathematics Preece must have gone astray (see the Tech Note at the end of this chapter for a discussion of Preece's analysis). Preece was in good company, however, as Lord Kelvin himself for a time held the same negative opinion as to the viability of the electric light.

The Age of the ''Practical Man''

Not being a mathematician was not Preece's sin (or at least not a necessarily fatal one), of course, as even the genius Faraday[16] used no mathematics beyond simple arithmetic. Unlike Faraday, however, Preece took public pleasure in announcing his pride at being a nonmathematical ''practical man,'' and seemingly gloried in his analytical ignorance. What else could anyone conclude, after reading the following typical comments?

> The advance of our knowledge in this branch of electrical development has been very much retarded by the phantasies of visionary mathematicians who monopolize the columns of our technical literature and fill the mind of the student with false conclusions. I have no sympathy with the pure mathematician who scorns the practical man, scoffs at his experience, directs the Universe from his couch, and invents laws to suit his fads. [Note 17]

and

> ...I cannot recall to mind one single instance where I have derived any benefit from pure theory. [Note 18]

and finally, possibly the best summation of Preece's attitude toward theory, he boldly declared[19] that he had

> ...made mathematics his slave...

and that

> ...there were those amongst mathematicians who certainly allowed mathematics to make them its slaves... .

This last assault on mathematicians was delivered as part of an address Preece made at the British Association (B.A.) meeting at Bath (1888). To support his assertion that mathematicians were generally an unreliable bunch that practical men would do well to be wary of, Preece recited a litany of historical examples of erroneous predictions (one of which was Airy's claim of the impossibility of an Atlantic cable). Such remarks created a bit of a stir, naturally, and later at the same meeting Lord Rayleigh was sufficiently moved to observe[20] that "during Mr. Preece's address one felt that 'mathematician' was becoming almost a term of abuse." Moments later Sir William Thomson rose to add his voice to Rayleigh's.

To counter Preece's anecdotal "proof" of the superiority of the engineer over the mathematician, Rayleigh recalled how "His thoughts went back to a well-known book, *The Life of Stephenson*, by Smiles, in which a great authority of the day is said to have offered to eat a steam engine wheel for breakfast if the locomotive ever attained a speed of twelve miles an hour. That, he [Rayleigh] believed, was a great engineer, and not a mathematician."[21]

Preece responded to Rayleigh and Thomson's criticism in the most sycophantic, obsequious manner imaginable, declaring "It would be folly and absurdity...to call into question for one moment the conclusions of...masters in mathematics." He went on to say they "...were all students, who lay at the feet of their Gamaliel [the teacher of the apostle Paul—see The Acts 22:3] at the end of the room, Lord Rayleigh. Everything Lord Rayleigh wrote, everything he said, and everything he did, was studied with the very greatest care..." and further, so Preece said, he "...would be very grieved if anybody left [the] room and thought that [he] had called into question anything that masters like Lord Rayleigh and Sir William Thomson...had ever said." Preece then stated his concern was actually with "...some of those young fellows coming out [of technical institutes] with a smattering of mathematics; they wrote Papers for the technical journals, and they thrust upon the electrical world conditions and conclusions...with a coolness and effrontery that was simply appalling."

These amusing exchanges (always delivered, from the reportorial tone of *The Electrician*, with the genteel politeness that is now the stereotyped hallmark of civilized Victorian gentlemen) were only the tip of the iceberg. The conflict between the old and the new had been years in the making, simmering at a low heat. Some years before Bath, for example, we can find evidence of the conflict even in a book review[22]: "...We have noticed a tendency now and then in the technical journals, on the part of men of practice, evidently ignorant of the history of the science they apply, to depreciate unduly the services of their theoretical brethren."

A Public Debate

The public brawl over experience versus theory between Preece and men such as Lodge and Heaviside, finally brought the issue to a boil and dragged it across the pages of the important trade weekly *The Electrician* ("It has the largest circulation of any English Electrical paper, and circulates all over the World,'' the masthead modestly announced in the third person). Before the B.A. meeting at Bath was completed things had gotten sufficiently hot that Lodge was reduced[23] at one point to poetic doggerel:

> Some talk of Isaac Newton, of Euler and Clairaut
> Of Kepler and Copernicus, and old Galileo;
> But of all the mathematicians
> There's none who can compare
> With the row dow dow de dow dow dow
> Of the British Engineer.

Lodge then went on to quote (in a way that displayed his distain) from a recent editorial in the then oldest engineering journal published (*Engineer*) which flatly stated ''...the world owes next to nothing to the man of pure science,'' and that ''...the engineer, and the engineer alone, is the great civiliser. The man of science follows in his train.'' Clearly then, Preece was *not* an oddball exception in his rejection of theory and his astonishing (by today's standards) position was *not* unusual a hundred years ago.

That this intellectual dichotomy was much more than a mere quibbling over personal styles was made clear by the following incident. Lodge gave labels to the issue when he wrote [24] ''The opposite camps may be styled the Practical *versus* [Lodge's emphasis] the Theoretical.'' Lodge then attempted to summarize the technical positions taken by each of the ''camps'' and while doing so made the assertion ''Electric charges splash about in a struck [by lightning] mass of metal... .'' This brought a quick response[25] from Preece who, just a week later, struck back with ''When Professor Lodge wrote the following sentence, 'Electric charges splash about in a struck mass of metal, as does the sea during an earthquake or when a mountain top drops into it,' I wish he had added 'and as does a theorist when he takes a header into a bath of cold practice.' ''

Things were evidently now beginning to take a turn to the nasty, and the editors of *The Electrician* tried to play a moderating, soothing role by publishing what they hoped would be taken as a reasoned[26] balancing of the pros and cons of each ''camp.'' It didn't work. The very next week the journal printed a reply from Heaviside in which he stated[27] ''...It is the duty of the theorist to try to keep the engineer who has to make the practical applications straight, if the engineer should plainly show that he is behind the age, and has got shunted on to a siding.'' If Preece, by some remote chance, was too dense to recognize himself in this, Heaviside then proceeded to draw arrows to him by naming him no less than *eight* times in the next five paragraphs, including ''...Mr. Preece, in the presence of some distinguished mathematicians, recently boasted that *he* made mathematics his *slave*, yet it is not wholly improbable that he is a very striking and remarkable example of the opposite procedure... .'' Other insults (all founded in truth) on electrical issues followed, and Preece must surely have been nothing less than outraged.

Heaviside's public words were matched in anger in his private journals, with one entry from this episode stating ''Preece made a violent attack upon mathematical physicists with a great laudation of himself... . Also made attacks upon me in particular, without mentioning my name... .''

While Preece was no doubt still smarting over Heaviside's public letter, another attack arrived in the form of an unsigned essay in *Nature*.[28] It, too, was obviously aimed at Preece, and the word *absurd* (with all its variants) was sprinkled liberally throughout the text. It could not have been a happy time for William Henry Preece and, indeed, in retrospect it marked the turning of the tide against the antitheoreticians.

Why Preece Prevailed (for a While)

The tide, however, didn't change overnight. It had, in fact, been running long and strong in favor of Preece. He and others had had decades before Bath to make solid and secure positions for themselves that no mere verbal broadsides from Heaviside and Lodge could easily damage.

More than two years after Bath, in fact, Heaviside and Lodge, along with Silvanus P. Thompson, were again involved in a debate on the pages of *The Electrician*, this time with John T. Sprague. Sprague was clearly an intelligent man superior to Preece in his knowledge of electricity, but still he was a member of the "practical" camp and, indeed, acknowledged[29] he belonged to the class of "half-educated electricians" who "have not surrendered...the belief that Ohm's law represents the actual facts of nature." This remark was prompted by the inertia of old-time electricians who couldn't understand a-c phenomena, including the idea that the resistance of a circuit is, in general, frequency dependent. Sprague's position is perhaps best shown by the following two quotes from *The Electrician*: "There are people so constituted that they flourish on a mental diet of $\frac{1}{2} VD(L_1N^2) + \sqrt{\mu} - R^2$, and so on, to whom scalars and vector potential act like...champagne. I prefer to form concrete pictures of what I suppose really to occur in nature, ..." and "I need scarcely say I entirely differ from his [Heaviside's] opinion as to the superior value of abstract theory. I take it that our only mode of attaining truth, is to find out what Nature does, not to invent imaginary schemes about her works."

Preece made his career a success by the Horatio Alger route of long, faithful and, to his credit, not entirely unproductive service to his employers. These were virtues especially valued in Victorian times (ones conspicuously absent for the most part, I might add, from Heaviside's career). For many, Preece's inability to master the mathematically delicate, abstract subtleties of the "new electricity" wasn't the awful evil that Heaviside would have them believe.

In 1855 he made his first independent engineering effort and received a patent for a duplex telegraph system. Duplex was an approach that interested those looking for an economical solution to the increasing message congestion on existing wires, and the idea of being able to handle traffic simultaneously in both directions was attractive. The Electric Telegraph Company (ETC) did not actually incorporate Preece's work into their system, but obviously still thought highly enough of him that by 1856 he was promoted to the position of Superintendent of the Southern District with headquarters in Southampton. Within a few years of taking that job Preece's career also became tied to the fast developing British railroad system.

Railroads, naturally enough, found the telegraph assuming an ever increasing role in their operations (at long last one could communicate from one end of a train to the other without having to physically run the length of the thing, for example). In 1860 the ETC system became so important in the day-to-day activity of the London and South-Western Railway that its directors asked Preece to design a system exclusively for their use. ETC, to promote goodwill with one of its best customers, agreed to let him take on this additional job.

Preece was now doing quite well financially and he married in 1863. By 1866 he was earning a respectable 350 pounds per year and from then on his annual income increased steadily to the 1000 pounds he earned in 1892 as Engineer-in-Chief of the GPO with headquarters at St. Martin's-le-Grand (he had been appointed the GPO's Electrician in 1877). He eventually became the proud owner of *two* grand homes, one at Wimbledon and the other at a prestigious Queen Anne's Gate address. Heaviside, of course, continued to mostly mooch—it wouldn't be until 1896, after the deaths of his parents, that Heaviside would finally be on his own. Heaviside's income was always *much* less than Preece's.[30]

Preece's interest in applying electricity to railroad operations, including lighting trains and sending messages while in-route, was intense. Between the years 1862 and 1882, for example, he took out a total of seven patents with direct railroad application. In 1900, when looking back over his life, Preece was moved to comment,[31] "I have always looked on my railway career as the best time of my life."

Perhaps he believed this since it was a period relatively free from contention at a personal level. With the takeover of the private telegraph companies (like the old ETC) in 1870 by the Post Office, however, Preece's life took a new direction, one that as we've seen would involve him in public, occasionally wild, acrimonious debates. His adversaries were not always limited to good-natured professors (Lodge) and sharp-tongued eccentrics (Heaviside). In 1877, for example, the year he became the GPO's Electrician, Preece was officially dispatched to America to see what all the fuss was about concerning the Bell telephone (announced the previous year). He met with Bell, who gave him a set of telephones (more on this later), and later with Edison at his Menlo Park, New Jersey laboratory. It was during this meeting that Preece learned of Edison's work toward developing a sensitive audio transducer, or *microphone*.

The next year, long after Preece's return to England, his good friend, the highly successful inventor David Hughes,[32] presented a paper to the Royal Society. The topic was his latest invention—the microphone. Immediately an international uproar developed (a decade later Heaviside would also become involved in an occasionally unpleasant debate with Hughes) with Edison claiming Preece had "committed a gross infringement of the confidence obtained under the guise of friendship." Edison soon escalated the intensity of the squabble by accusing Preece of "piracy," "plagiarism," and "abuse of confidence."

Preece and Hughes denied any wrongdoing,[33] and the affair gradually faded away. Historians now generally credit Hughes with being the true inventor of the microphone, and dismiss the incident as just one more of the many rash episodes that spotted Edison's occasionally less than admirable career. But still, it had to be a most embarrassing time for Preece. It would not be the last.

A Clash of Personalities

With his new job with the State in 1870 Preece became Engineer in charge of the Southern Division, with Oliver's brother (Arthur) reporting to him as a District Superintendent. No doubt it was at this time that Preece first became aware of Heaviside's existence. If not, he soon did. In June 1873 Heaviside published a paper[34] in *Philosophical Magazine* on duplex telegraphy circuits, and it began with a brief history of the subject. Nowhere is Preece's 1855 effort mentioned, and this evidently greatly distressed the man. Shortly after Heaviside's paper appeared Preece wrote a letter in which he angrily declared[35] "Oliver Heaviside has written a most pretentious and impudent paper in the *Philosophical Magazine* for June. He claims to have done everything,... . He must be met somehow." To which came the reply[36] "O. Heaviside shows what is to be done by cheek. ...We will try to pot Oliver, somehow."

Preece was apparently upset mostly by Heaviside's claim[37] to have actually gone beyond duplex to achieve *quadruplex* (without actually making one). A few years later, however, even Preece admitted[38] the truth in Heaviside's 1873 quadruplex claim. Still, historians have traditionally credited Edison with the invention of quadruplex, and there seems to be no reason (in spite of Heaviside's claim, which was surely legitimate) to change that view.[39] One thing is clear—by 1873 Preece not only knew of Heaviside but had a distinct dislike of him as well.

In turn, Heaviside's eventual public distain for Preece's abilities was certainly developing rapidly during the 1870s. For example, Preece's joint authorship (he was senior author) in 1876 of a book on telegraphy could only have convinced him that Preece was a borderline technical incompetent who quite often just didn't know what he was talking about. This book was reviewed[40] in *Nature* and the review was devastating. The opening line gave warning of worse yet to come, "It is with feelings of great disappointment that we lay down the latest book on Telegraphy." Later, the reviewer stated "Everything scientific is in fact left out." Even though Preece and his coauthor stated they couldn't be expected to discuss *everything*, they somehow found space enough to explain how to preserve timber for telegraph poles, but not enough to give even Ohm's law! In the end, the reviewer was reduced to sarcasm at the technical absurdity that abounded in the book:

> Interesting information is given as to rates of working. Some of these show very wonderful results of practice. For example, we learn that an experienced operator usually punches forty-five words per minute [Note 41]. Now a word contains 4.5 letters, and if we take it that an average letter contains, including the space that divides it from the next letter, four dots, we find that at this rate of punching 13.5 dots per second are made. If three more dots could be made per second, the strokes would nearly cease to be heard...and a deep musical note four octaves below middle C on the piano-forte would be the result. We wonder whether this could be done were the operator to punch a few times over some sentence that he knows by heart.

As an experienced telegrapher who had sent a few messages in his time, this must have made Heaviside laugh aloud (and to dismiss Preece as a fool, at best). Indeed, he soon got into the habit of using words and phrases of disrespect when writing about Preece. Never at a loss for words,[42] Heaviside had several mocking ways of referring to Preece, such as "the man of brass" and "Mr. Prigs." It isn't at all obvious why he used these particular phrases, but perhaps one explanation might be found somewhere in his familiarity with the more unsavory characters in Dickens. For example, Betsey Prig in *Martin Chuzzlewit* was an ill-tempered, bearded nurse with a voice "more like a man's." "Man of brass" might just be a play on the British use of *brass* as slang for money, as well as its more conventional use for an individual displaying (possibly without cause) blatant self-assurance. But Heaviside would have loved best of all the *triple* play on the word, recalling from *The Old Curiosity Shop* that Sampson Brass was "an attorney of no very good repute," often involved in shady business.

The most obviously insulting Heaviside name for Preece, however, was "The eminent scienticulist," which first appeared on the pages of *The Electrician* in 1887. The occasion was prompted by a new boast from Preece of his early work on signal retardation on submarine cables: "Mr. Preece is much to be congratulated upon having assisted at the experiments upon which (so he tells us) Sir W. Thomson based his theory; he should therefore have an unusually complete knowledge of it. But the theory of the eminent scientist does not resemble very closely that of the eminent scienticulist." Later, when reprinted in *Electrical Papers*,[43] "scienticulist" was softened to "practician." On other occasions when even Heaviside feared to go too far toward libel, he would refer to Preece as "the Nameless One."

Heaviside, in fact, had been poking fun at Preece on the pages of *The Electrician* for years. Four years earlier, for example, in December 1883, Preece wrote a letter to *The Times* (of London) putting forth his reasons for believing electricity to be the cause of the recent dramatic sunsets. A few months before, in late August, the volcano on the island of Krakatoa in the Sunda Strait between Java and Sumatra had exploded, sending five cubic miles of rock ash and lava into the atmosphere (in addition it created far-ranging sea waves over 100 feet high that killed 36,000 people, as well as what has been called "the loudest noise heard in the history of man"). Preece believed the material blasted into the air to be negatively charged[44] and, once suspended, the similarly charged particles of debris had repelled each other and so had spread out around the entire world (high-altitude winds are the modern theoretical explanation). A month after this letter, in an article[45] in *The Electrician*, Heaviside made fun of Preece's ideas without actually mentioning his name. But when he wrote of the "unscientific speculator" letting his "imagination run riot," it is difficult to believe readers, and Preece, didn't know whom Heaviside had in mind.

Preece's Ability

Surely one of the more fascinating questions about Preece must be how such a technically weak man could rise to the very high position of Chief Engineer of the GPO? And do so in spite of awful blunders that sometimes brought ridicule down on his head from establishment critics more immediately credible than Heaviside (who might be dismissed by many as just a crank). One answer is that Preece's flaws were not unique to him, and were generously shared by many other Victorian electricians who also couldn't assimilate the new discoveries that seemed to be made almost daily. Preece was perhaps slower than most, even if we limit our study to those of the "Practical Camp," but he was also something else. He was, in fact, *ahead* of his time as a big-business, big-time bureaucratic executive operating within a complex, structured hierarchy.

Today, an Oliver Heaviside, unable to function in the modern high-pressure, team approach to serious science (as well as being a queer duck, to boot), wouldn't stand a chance. Heaviside wasn't disciplined enough to be even an academician, and might well be dismissed as a "loony" with his regular tirades against the establishment. Even people who admired and respected Heaviside's intellectual abilities could grow tired of his continual complaints, and be irritated by his blunt writing style. During the 1891 exchanges in *The Electrician* with Sprague, for example, Heaviside, as usual, took some shots at nonmathematical electricians. Sprague, naturally, took objection. In reply (and in support) Oliver Lodge wrote[46]: "...If he [Sprague] can divert Mr. Heaviside from tirades against 'anti-mathematicians' and get him to give us either some more solid investigation or else some vivid explanation, he will be doing good service."

From those who didn't appreciate Heaviside's talent in spite of his flaws, the reaction could be far more unpleasant than Lodge's mild rebuke. Consider, for example, the following exchange from *The Electrician* in 1887. On June 24 Heaviside wrote a letter suggesting *Mac* for the name of the unit of inductance "...in honor of the man [Maxwell] who knew something about self-induction, and whose ideas on the subject are still to be fully appreciated. This was very much his own fault." This remark on the difficulty of Maxwell's *Treatise* (and anybody who has tried to read it can hardly deny Heaviside had a point) tremendously offended one reader, sufficiently so to move him to reply on July 1:

> Mr. Heaviside...asserts (as I think somewhat impertinently) that "it was very much his own fault" if Maxwell is "still to be fully appreciated"—which appears to be Heavisidean English for "not yet quite understood." Well, if Maxwell's expositions are anything approaching Heaviside's in obscurity, no wonder.

The writer (who signed himself "Amicus") then worked up into something just short of a frenzy in his conclusion:

> Mr. Heaviside tells us that the term "mac" is quite euphonious and unobjectionable. If that be granted, so, too, is smack; and for any scientist, who takes unwarrantable liberties with

> his readers, or who drags his personal feelings into public print, or makes...perversions of the English language, smack is really the only absolutely proper thing to use.

Heaviside replied on July 8 and declared he was applying a *scienticulometer*

> to the investigation of a quite new phenomenon,...the learned grammaticulist. Up to the present time, however, I have failed to extract anything but froth, with a parting smack of vulgarity. Froth is excellent in its way, but is rather too insubstantial. As for the other, it is certainly not the proper thing to use, especially for so accomplished a literary critic, who should, I think, endeavor to copy the refined style of...polite writers.

In contrast to these sorts of antics, the polished Preece, if alive today, would almost certainly become, at the least, a vice-president for engineering development and research at some Fortune 500 company. And as an ambitious man with an undeniably dynamic personality, he would be *good* at it! Preece, in fact, was neither a scientist nor an engineer, but rather an accomplished manager of skilled labor. Of all men to cross swords with, Heaviside could not have found a more formidable foe.

Preece made it quite plain where his talents and interests actually converged when he declared:

> The monopoly of the Post Office is...not a mere commercial industry, maintained for the benefit of the few who have risked their capital in speculation.

And, after citing the impressive technical advances in telegraphy made following the transfer in 1870 of the private telegraphs to the State (and after naming a few of the responsible individuals) he stated:

> All of these things have been done without recourse to the Patent Office. Dozens of improvements have been effected which, in ordinary commercial life would have been patented and published, and received recognition.

These sentiments from his 1893 IEE speech[6] were not spur of the moment utterances. He had consistently over the years preached this point of view. As far back as 1878, for example, in an address to the B.A., he said[47]:

> It has been publicly stated by very high authorities that, since the transfer of the telegraphs to the State, invention in that art has left the shores of the United Kingdom and flown to those of America. Moreover, it has been intimated that the monopoly in telegraphy possessed by the State has checked improvement. Such statements are made in ignorance of the facts. Indeed, improvement in telegraphy was never more active in England than it has been since the Government has managed the business.

In 1882, while describing to the B.A. the state of telephony (which was then entering a stage of open competition among various private companies), he said[48]:

> The free traffic in patents, however, leads to jobbery and speculation of the worst type. The public have wildly rushed into ill-matured schemes that have swollen the purses of gambling promoters, have turned the heads of inventors, have retarded the true progress of the beneficial application of this new science to the wants of man, and have thrown away millions upon imperfect schemes. Much has been said against the monopoly of the Post Office in telegraphic business, but at any rate it has the merit that it has checked the rapacity of company promoters and patent-mongers in that branch of the practical application of electricity, while no one can assert that it has checked the progress of telegraphy.

In 1887 (the jubilee year of telegraphy) he attacked the impact of patent law on the very *morality* of society when, in yet another address[49] to the B.A., he first listed the technical advances since 1870 and then said:

> Two reasons exist why these great advances have not received notice; they were not patented, and they emanated from Government service. A patent has certainly one great use, it fixes a date and it defines an invention, but it also attracts attention to novelties and improvements, and if the patent be lucrative it incites the immoral [!] to try and do the same thing in another way, which tends to litigation, though often resulting in still further improvements. There is a very ridiculous conception abroad and widely circulated, that commercial enterprise alone is competent to excite inventive skill.

and

> Had the telegraph service been conducted under the severe competition of private enterprise, each form of transmitter and of receiver, the relay, the high-speed repeater, the shunted condenser, and every other improvement would have been patented, and would have been found worthy of substantial reward.

In other words, Preece would, for the purposes of public image, pay lip-service to individuals but his actual feelings were that of the high-placed civil servant whose goals are the homogenization of invention, the total control of creativity, and the forcing of the individual to submit to the dominance of the monopolistic organization. No free-enterprise man was Preece, but one who would feel right at home in a State-regulated economy.

The Telephone Affair

In 1901 J. A. Fleming published[50] an historical essay on the detrimental impact of the state monopoly in electrical communication. To be sure his position would be obvious from the start he entitled it "Official obstruction of electrical progress," and Preece is one of the few individuals named. As Fleming pointed out for his readers,

> When the telephone made its appearance in 1877, and telephone exchanges began to be devised in 1879, the question arose whether a telephone was a telegraph within the meaning of the Act [passed by Parliament transferring the operation of telegraphs to the Government]. The Government telegraph officials feared that their monopoly was threatened; hence with the assistance of the Crown lawyers they proceeded to stake out a big claim, and to obtain an interpretation of the Telegraph Acts, passed to legalise the purchase of the old electric telegraph companies, which was equivalent to an authoritative statement that the Post Office possessed the sole right to transmit intelligence by electrical means in return for payment, not merely as the art was then known, but by all and every method which the wit of man could or might throughout everlasting ages devise.

In the end, the private providers of telephone service had to pay the GPO a ten percent tax on their gross receipts (for which the GPO did nothing); by the end of 1900 this tax burden on the new technology totaled more than a million pounds. This was particularly outrageous because, as Fleming said, "When the telephone first made its appearance the technical experts of the Post Office [in particular, Preece[51]] laughed at it as a toy."

When Fleming considered the result of the monopoly, he concluded:

> It is a pure waste of time for an inventor to spend days and nights over a telegraphic invention, or invest capital in patenting it, unless he can get it tried, and, if it succeeds, market his invention to a purchaser. He is not generally a philanthropist, but is spurred to work by the hope of reward. But in electric telegraphy he can try nothing and market nothing

"I say, Harold, those ghastly people, the Dudd-Robinsons, have asked us to one of their filthy dinners on Tuesday. Tell me what excuse to make while I keep my hand over the telephone."

As this 1922 cartoon from *Punch* shows, the telephone remained, for many, a strange and hazardous gadget even a half-century after its invention.

unless he first persuades or pleases the permanent officials of the State Telegraph Department. He has to overcome their inertia, opposition, or it may be ill will, before he can even get a trial of his telegraphic apparatus, and when at last he demonstrates an important advance, he is entirely at their mercy whether it shall be adopted or not, and, if so, what price he shall receive for it.

Heaviside Refuses to Be Shackled

There is, in fact, some evidence that Preece quite early in the game tried to gain this sort of stranglehold control over Heaviside. In 1881 Arthur (who saw Preece regularly as part of his Post Office duties) wrote[52] to Oliver "Preece states that the Western Union of America are about to adopt the Wheatstone [telegraphic equipment invented by their uncle], having ordered 24, and that he has the nomination of about six clerks to manage them, with salaries of about 250 pounds, *and then he asked after you and I told him you were a student still* [my emphasis]—obvious—should you apply I believe he would nominate you."

Heaviside declined to apply, thereby no doubt disappointing his family (who by now must have been near despair over whether or not Oliver would ever return to the real world of employment for wages and adult responsibility), and probably also Preece (who would have found it all the easier to "pot Oliver," if necessary, with such an arrangement).

From Oliver's point of view, of course, he was insensitive to the mundane financial concerns of his family, or to the possible Machiavellian plots of Preece. *He* was hot on the trail of all the exciting discoveries he knew lay waiting to be found along the path outlined by Maxwell's new theory of the electromagnetic field. He definitely was not about to become just a telegraph clerk again, not with his powers at their maximum.

With Clerk Maxwell's ideas to spur him on, Heaviside was off to blaze his own trail.

TECH NOTE: PREECE'S ANALYSIS OF THE ELECTRIC LIGHT

A few words, first, on just what Preece meant by the "subdivision of light." Imagine an energy source (a battery) with a known internal resistance. Preece and other writers almost always began their analyses by first determining the external load resistance (e.g., that of a lamp filament) that would maximize the delivered power. It was known that this occurs when the external resistance equals the internal resistance.[53] These writers next worried about how to distribute or *divide* up that maximum possible power among many lamp filaments. However, the actual power consumed by any practical array of lamps is *far less* than this maximum.

Preece considered[54] a d-c source of internal resistance ρ and voltage E feeding electricity to n lamps, either all in series or all in parallel ("multiple arc," as he called it), each lamp having resistance l. The connecting wire was taken as having resistance r. Figs. 5.1 and 5.2 show the circuit form.

He then calculated the total heat energy dissipated per unit time in the lamps. Calling this quantity H, he arrived at the correct results:

$$H_s = E^2 \frac{nl}{(\rho + r + nl)^2}, \qquad H_p = E^2 \frac{l/n}{(\rho + r + l/n)^2}.$$

Preece then assumed n to be a number sufficiently large that $\rho + r \ll nl$ or $\rho + r \gg l/n$. With the large-n assumption,

$$H_s = E^2 \frac{1}{nl}, \qquad H_p = E^2 \frac{l}{n(\rho + r)^2}.$$

In both cases the total heat energy dissipated by the n lamps decreases with the first power of n, and thus the heat energy in any *individual* lamp decreases with the *second* power of n. From this Preece concluded that the lamps would grow dim at a rate faster than their increase in numbers, and so the subdivision of the electric light was reduced to "a possibility which this demonstration shows to be hopeless." Why was Preece wrong?

Consider the parallel case, which is the way electric lamps are actually wired in a house. Preece went wrong with his assumption of $\rho + r \gg l/n$. In Edison's original d-c utility (the Pearl Street facility in New York City) $\rho + r$ was a fraction of an ohm, while l for the Edison lamp was on the order of 200 ohms. For Preece's inequality to have held, n would have to have been on the order of a *thousand* lamps or more. The proper assumption was, in fact, exactly

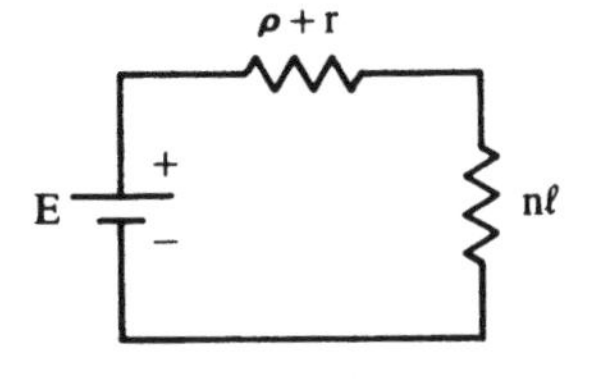

Fig. 5.1. Lamps in series.

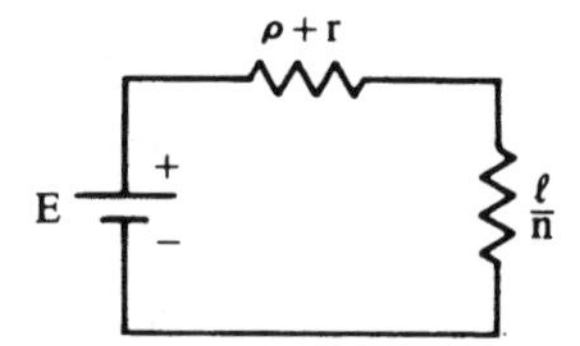

Fig. 5.2. Lamps in parallel.

the reverse of Preece's, i.e., $l/n \gg \rho + r$. Then,

$$H_p = E^2 \frac{n}{l} .$$

This is just what is required, with total lamp energy *increasing* with the first power of n, and individual lamp energy remaining *fixed* at E^2/l.

Less than three years later Preece admitted to having made a grievous error. In a talk[55] to the Society of Arts describing his recent visit to the 1881 International Exhibition of Electricity in Paris he said of Edison's lighting display,

> The system is self-regulating, if the electromotive force is kept constant, and the resistance of the lamps be uniform. We have the dynamo-machine at one end of the circuit, and a lamp at the other. The circuit is complete; a small current flows, which is determined by the resistance of the lamp alone, if the main conductors are made sufficiently large to neglect their resistance.

He then declared the key to the success of Edison's system, contrary to his earlier analysis, to be

> ...the value of high resistance in the lamps, and low resistance in the armature of the dynamo-machine.

He at last stated that

> Many unkind things have been said of Mr. Edison and his promises; perhaps no one has been more severe in this direction than myself. It is some gratification for me to be able to announce my belief that he has at last solved the problem [the subdivision of the electric light] that he set himself to solve.

It is almost impossible today to appreciate why the "subdivision of light" was so difficult to understand a mere century ago. Perhaps a century hence somebody will write the same about our present confusion over time-travel! Still, if the electrical engineers of the 19th century were bewildered by the electric light, that was nothing compared to the state of those who were not technically trained. At the end of Preece's talk, for example, one of Edison's representatives in the audience commented on the plans then afoot to allow the electric lamp "to be used by uneducated or unscientific people," and "to bring the lamps within the care of ordinary house servants, no matter how ignorant they might be."

Those were the days, after all, when it was still thought by "common folk" that electrical wires were conduits for "fluid fire"![56]

Notes and References

1. The definitive source on the life of Preece is the interesting (but slightly hagiographic) biography by E. C. Baker, *Sir William Preece, F.R.S.: Victorian Engineer Extraordinary*, London: Hutchinson, 1976. Other sources of

information are the several lengthy obituaries that appeared (for example, *Engineering*, vol. 96, pp. 661–663, November 14, 1913; *The Electrician*, vol. 72, pp. 253–255, November 14, 1913; and *Nature*, vol. 92, pp. 322–324, November 13, 1913).

2. This is most often called the "free electron theory of metals," but for semiconductors it cannot explain observed behavior. It then becomes necessary to consider the environment of the charge carriers (the rigid atomic lattice of the semiconductor crystal) and to introduce quantum mechanics. This leads to the energy-band theory of electrical conduction in solids, one of the truly impressive successes of quantum theory.
3. "Mr. Preece on lightning protection," *Nature*, vol. 47, p. 536, April 6, 1893. Heaviside's feelings about electrical fluids changed with time. In 1884 (EP 1, p. 338) he wrote, "Electricity, in conductors, is subject to the same law of continuity as an incompressible liquid." (Just two pages earlier, however, he had made it clear he thought the fluids were not actually *real*.) In 1886 (EP 2, p. 182) Heaviside was still referring his readers to earlier work (EP 1, pp. 378–384) where he had, himself, used the "waterpipe analogy" to study inductive effects. By 1888 (EP 2, p. 486), however, when writing about Lodge's work on lightning, he dismissed the concept totally with "The fluids are played out; they are fast evaporating into nothingness. The whole field of electrostatics must be studied from the electromagnetic point of view to obtain an adequately comprehensive notion of the facts...."
4. *The Electrician*, vol. 21, p. 303, July 13, 1888. Lodge will appear in this book frequently from now on as his career often intersected those of Preece and Heaviside. Heaviside, in his later years, called Lodge "the other Oliver."
5. In fairness to Preece I should admit that a-c circuit analysis provided some interesting surprises even for a genius like Lord Rayleigh. In volume 1 of his *Theory of Sound* (London: Macmillan, 1894, pp. 442–443), for example, he discussed the curious ability of an alternating current in a main circuit to divide into two parallel branch circuits in such a way that each branch current, *individually*, is numerically *greater* than the main current!
6. *Journal of the IEE*, vol. 22, p. 35, 1893. In an ironic twist, Preece had to wait a few minutes for Oliver's brother Arthur (who worked for Preece) to receive an IEE award before assuming office.
7. Heaviside was not the only one to bash heads with Preece. Sir William also had a monumental public brawl with Oliver Lodge over the subject of lightning rods (in which Heaviside's ideas played a role). This dispute was a personal, bitter one, too, and Preece made Lodge pay dearly for his impertinence, later, in the affair with Marconi during the earliest days of wireless telegraphy (but that is discussed later in this book).
8. Thereafter all references to Preece's education were in the form of "he completed his education at King's College," or "he passed through King's College," but in fact he never graduated from college. However, in view of his worldly success and influence, he was elected a Fellow of King's College in 1885. He did not, in the end, go to his grave unanointed, eventually receiving an honorary Doctor of Science from the University of Wales.
9. It was Latimer Clark, in fact, who conceived the "pigmy battery" Atlantic cable experiment described in Chapter 3, Note 13.
10. Airy was interested in the telegraph as a means of "instantaneous" communication of time signals between widely separated astronomical observatories. Preece described in some detail the signal retardation (which was fatal to Airy's time-keeping goal) experiments he and Latimer Clark did with Faraday in *The Electrician*, vol. 22, pp. 101–102, November 30, 1888.
11. In his 1893 IEE Presidential Address, for example, Preece recalled his 1880 Presidential Address and said "...I took [then] the opportunity to formulate the theoretical views of electricity that I had acquired at the feet of Faraday." This coat-tailing on Faraday's immense reputation obviously irritated Heaviside and his reply can be found in the following passage from EMT 1, p. 337, dated March 10, 1893 (Preece delivered his address on the evening of January 26, 1893): "If you have got anything new...you need not expect anything but hindrance from the old practitioner even though he sat at the feet of Faraday. Beetles could do that... . But when the new views have become fashionably current, he may find it worth his while to adopt them, though, perhaps, in a somewhat sneaking manner, not unmixed with bluster, and make believe he knew all about it when he was a little boy!"
12. After Kelvin's death in 1907 his widow commissioned a marble bust of him to be placed in the Institution of Electrical Engineers. She asked Sir William to make the presentation and he did so in February 1912. This solemn event, however, gave Heaviside yet a new opportunity to insult Preece. During his remarks Preece declared his intention to present the IEE with a bust of Faraday to match that of Kelvin (which was finally done in 1914 by Preece's son). In a bitter parody of that promise, Heaviside wrote (in August 1912) in the Preface to the third and final volume of *Electromagnetic Theory*, "If my life is spared, I hope to be able to present a bust of the eminent electrician who invented everything worth mentioning to the Institution over which he once ruled, to be placed under that of Faraday." Heaviside, to the day he died, never budged an inch from his mean feelings toward Preece, and it is easy to imagine him chuckling at the thought of the unspeakable acts passing dogs would perform upon Preece's bust sitting on the floor at Faraday's feet. Today, the busts of Faraday and Kelvin gaze at each other across the marble lobby of the IEE, but nowhere is there a bust (or even a portrait) of Preece. There is however, a "Heaviside Room" at the IEE, in which hangs a color portrait of O.H.
13. Baker (Note 1), p. 54. This little episode, supposedly recalled by Preece many years after the fact, has in my opinion all the flavor of an apocryphal tale, at least as far as Preece's response to Thomson. For a young man, who had not yet hired anyone, to say "I can get it done..." seems a bit too precocious. I suspect it was just a way

for Preece to state his true feelings about theoretical analysis, yet maintain the image that he, of course, was no Philistine.

14. EMT 1, p. 41.
15. *The Electrician*, vol. 2, p. 167, February 22, 1879. In this same essay he also made his famous remark that "the sub-division of the light is an absolute *ignis fatuus*." See also Preece's essay "Gas versus electricity," *Nature*, vol. 19, pp. 261–262, January 16, 1879, in which he wrote, "There can be no doubt that the use of electricity for the production of light is a very wasteful as well as costly process...." Preece was in good company, it must be emphasized—a few months earlier the editors of *Nature* had written (vol. 18, pp. 609–610, October 10, 1878), "Let the directors of gas companies do all they can to improve their gas. They may be certain that it will never cease to be required; a considerable splitting up of the electric current is impossible...."
16. We can find Heaviside's views on Faraday's lack of mathematics scattered about in his writing. For example, in EP 1 (p. 415) "I do not mention Faraday in this connection [those having correct ideas about retardation in submarine cables], for that great genius had all sorts of original notions, wrong as well as right, and not being a mathematician, could not effectively discriminate, especially as he had so little practical experience with cables," and in EMT 3 (p. 437) "...I do not see how the true results could be got without mathematics, not even by Faraday." And yet, somewhat paradoxically, Heaviside also wrote (EP 1, p. 195), "But earnest students, if they will not or cannot learn the mathematical methods, need not therefore be discouraged, for the name of Faraday will shine forth to the end of time as a beacon of hope and encouragement to them. He was no mathematician, yet achieved results apparently only attainable by such methods."
17. From Preece's 1893 Address (Note 6), p. 67.
18. From a letter to Oliver Lodge, quoted in B. J. Hunt's " 'Practice vs. theory': The British electrical debate, 1888–1891," *Isis*, vol. 74, pp. 341–355, September 1983. As Hunt observes, this was no doubt true.
19. *The Electrician*, vol. 21, p. 645, September 21, 1888.
20. *The Electrician*, vol. 21, p. 674, September 28, 1888.
21. Ibid. Rayleigh was referring to S. Smiles' biography of the pioneer railroad man George Stephenson (Boston, MA: Ticknor and Fields, 1858, p. 254). Rayleigh's memory failed him slightly, in that the "great authority" thought not twelve miles an hour absurd, but just ten! Further, Smiles reported not a "steam" engine wheel, but rather a "stewed" one.
22. *Nature*, vol. 25, p. 238, January 12, 1882.
23. *The Electrician*, vol. 21, p. 622, September 21, 1888. All the mathematicians mentioned are no doubt familiar to the reader, with the possible exception of Alexis Claude Clairaut (1713–1765). He was a French geometer now best remembered for his work in differential equations.
24. *The Electrician,* vol. 21, p. 663, September 28, 1888. Some years earlier, in a different dispute, Lodge had coined the phrase "*non-scholastic* school of electricians" for what he later called the "practical camp" (*The Electrician,* vol. 2, pp. 296–297, May 10, 1879).
25. *The Electrician*, vol. 21, p. 712, October 5, 1888.
26. "Practice versus theory," *The Electrician*, vol. 21, p. 730, October 12, 1888. An earlier editorial on the same topic appeared in vol. 10, pp. 252–253, January 27, 1883. The journal's final opinion of the debate was that it was "a somewhat unseemly and wholly profitless controversy" (vol. 22, p. 654, April 12, 1889).
27. *The Electrician*, vol. 21, p. 772, October 19, 1888.
28. "Empiricism versus science," *Nature*, vol. 18, pp. 609–611, October 25, 1888. There is evidence, in addition to the details in the text itself, that this essay, quite bitter in tone, was written by Lodge—see Hunt (Note 18), p. 352. In one of his notebooks (NB 3A:48) Heaviside remarked on this, writing "L. also wrote a stinging leader in *Nature*, Empiricism v. Science." The October 19th issue of *The Electrician* (p. 749) carried an anonymous note that Hunt also believed was due to Lodge (citing another entry in one of Heaviside's notebooks for support). The author of the note was referred to by the journal only as "a valued correspondent," and he said (among other things) "...the only cure for the 'practical' generation is to die off...." It isn't entirely clear if Lodge really was the mysterious author, however, as in the October 26th issue (p. 800) Lodge, himself, wrote in reply "The correspondent...is guilty of a serious exaggeration...." Lodge may, of course, also be guilty of a bit of dissembling.
29. *The Electrician*, vol. 26, p. 340, January 16, 1891. Sprague was not alone in being left behind (Thompson warned Sprague that he was "dropping down from the position of pioneer to that of pessimist" and was in danger of becoming "a fossil"—see *The Electrician*, vol. 26, p. 375, January 23, 1891). So many other "practical men" were in the same danger, in fact, that *The Electrician* ran an editorial ("Alternating currents and mathematics," vol. 26, pp. 298–299, January 9, 1891) in an attempt to make plain that progress demanded changes in old ways of thinking. Sprague's comments are in *The Electrician*, vol. 26, p. 543 and pp. 671–672, 1891.
30. Heaviside's earnings were either from his technical writing (on the order of 40 pounds a year—see Hunt, Note 18) or from a Civil List Pension he began to receive in 1896 (120 pounds per year) and which was increased to 220 pounds per year in 1914. In 1896 Preece was earning 1150 pounds per year, and even Arthur had a GPO salary of 600 pounds per year. Heaviside's correspondence shows he often received loans from friends.
31. Baker (Note 1), p. 82. Details of Preece's work can be found in "Electric inter-communication in railway trains," *The Electrician*, vol. 42, pp. 540–542, February 10, 1899.

32. J. O. Marsh and R. G. Roberts, "David Edward Hughes: Inventor, engineer and scientist," *Proceedings of the IEE*, vol. 126, pp. 929–935, September 1979. One of Hughes' inventions was the induction balance used in Bell's bullet detector (Chapter 4, Note 4). Preece claimed (*The Electrician*, vol. 7, p. 299, September 24, 1881) to have served as Bell's contact with Hughes for information on this use of the balance, and in 1886 he read a paper at the B.A. meeting in Birmingham describing how Hughes had used the balance to locate a broken sewing needle embedded in the hand of Preece's daughter (*The Electrician*, vol. 17, p. 363, September 10, 1886).
33. Hughes defended himself in a letter to *Nature* (vol. 18, pp. 277–278, July 11, 1878), claiming his work to be "a discovery too great and of too wide bearing for any one to be justified in holding it by patent [a curious position, indeed!], and claiming as his own, that which belongs to the world's domain." William Thomson soon after (*Nature,* vol. 18, pp. 355–356, August 1, 1878) wrote in strong support of Preece and Hughes, ending his letter with "I cannot but think that Mr. Edison will see that he has let himself be hurried into an injustice and that he will therefore not rest until he retracts his accusations of bad faith publicly and amply as he made them."
34. EP 1, pp. 18–24 (a second mathematical paper followed in 1876, pp. 24–34). Heaviside's 1873 duplex paper did not pass unnoticed, as its ideas were utilized by the Indian Government Telegraph Department (see the letter to *The Electrician*, vol. 2, p. 262, April 19, 1879, from that Department's "Lately Officiating Electrician," who specifically credited Heaviside). Heaviside's brand of humor was operating in grand style in that paper. He wrote, for example, "Prior to 1853, it is said to have been the current belief of those best qualified to judge, that to send two messages in opposite directions at the same time on a single line was an impossibility; for it was argued that the two messages, meeting, would get mixed up and neutralize each other more or less, leaving only a few stray dots and dashes as survivors (after the manner of the Kilkenny cats [Heaviside was referring here to the *Mother Goose* rhyme], who devoured one another and left only their tails behind). However, Dr. Gintl effectually silenced this powerful argument by going and doing it." The reference is to the Viennese Wilhelm Gintl, who in 1855 published a book on his pioneering 1853 duplex experiments.
35. Baker (Note 1), p. 109.
36. Baker (Note 1), p. 110.
37. EP 1, p. 24. "...From experiments I have made, I find it is not at all a difficult matter to carry on *four* correspondences at the same time...." In December 1874 Edison demonstrated a *working* quadruplex.
38. *The Electrician*, vol. 1, p. 165, August 24, 1878.
39. Appleyard reported (*Pioneers of Electrical Communication*, London: Macmillan, 1930, p. 221) a Heaviside notebook entry stating "I was credited in America with having described quadruplex first, or suggesting it." The fact remains, however, that it was Edison who first *built* a quadruplex (R. W. Clark, *Edison*, New York, NY: G. P. Putnam's Sons, 1977, pp. 50–55). Heaviside took no back seat to Edison in ingenuity, however. In 1875 Heaviside published an extremely clever paper in the *Telegraphic Journal* entitled "Notes on Mr. Edison's electrical problem" (EP 1, pp. 34–38). It was a reply to Edison's publication (a few months before in the same journal) of a difficult challenge problem in telegraphic circuitry. Heaviside wrote, tongue-in-cheek, "None of the readers of this journal have as yet come forward with any solution. Why is this?...I can only suppose that an excess of modesty has prevented many of the readers of this journal from sending a solution for publication." Heaviside then proceeded to analyze and solve Edison's problem in brilliant style. We can be sure Edison was aware of Heaviside's existence once this appeared, and particularly so after reading the last sentence, "Mr. Edison's own solutions would also be very acceptable." There is no record of any reply (with solutions) from Edison.
40. "Preece's telegraphy," *Nature*, vol. 13, pp. 441–442, April 6, 1876. This review was unsigned, but the editorial archives of the journal show it was written by James Thomson Bottomley, professor of physics at Glasgow University, nephew, colleague, and collaborator of William Thomson.
41. This is incredible, certainly for routine sending. The modern radio "ham" holding the highest level of license (Amateur Extra Class) is tested at just 20 wpm *receiving*—and transmission during the test is by prepunched tape run through a code machine. Preece's fascination for high-speed telegraphy was shared by many others, particularly in America. See "The American telegraphers' fast-sending tournament," *The Electrical Engineer* (New York), vol. 9, pp. 266–267, April 23, 1890, for an example of just how silly things could get.
42. One individual, clearly on the edge of despair, began a letter to *The Electrician* (vol. 26, p. 554, March 6, 1891), with the plea "Cannot something be done to check Mr. Oliver Heaviside's prolific invention of new words?"
43. EP 2, p. 119.
44. Preece had written in support of an earlier letter to *The Times* by Norman Lockyer, the editor of *Nature*, who had speculated that electricity might be involved in some (unspecified) way.
45. EP 1, p. 332.
46. *The Electrician*, vol. 26, p. 375, January 23, 1891.
47. *The Electrician*, vol. 1, p. 164, August 24, 1878. The "high authorities" who rejected Preece's point of view included the President of the Society of Telegraph Engineers, and Alexander Graham Bell.
48. *The Electrician*, vol. 9, p. 389, September 9, 1882. *Jobbery* is English slang for the use of unfair or politically corrupt means to obtain one's ends.
49. *The Electrician*, vol. 19, pp. 423–426, September 23, 1887. In this address, by the way, Preece displayed his occasionally puzzling misunderstanding of electrical matters as, for example, when he declared "Every electromagnet wound to a resistance R has its own coefficient of self-induction L, which determines the rate at which a

current rises or falls, and the time constant is expressed by the ratio L/R. It is the time the current takes to rise from zero to *full value* [my emphasis]." To rise to full value would take *infinite* time (the current increasing as a decaying exponential). The time constant in this case is the time to reach $1 - e^{-1}$ (about 63%) of the final value.

50. *The Nineteenth Century and After*, vol. 49, pp. 348–363, February 1901.
51. Prior to leaving for America to examine Bell's invention, Preece had declared he would unmask it as a fraud. As Silvanus Thompson put it in 1887, "We all know how perfectly sure—in fact cocksure—Mr Preece is when he forms opinions or makes statements. He was cocksure that the early forms of telephones were only toys; he went to America cocksure that he would in a quarter hour expose Graham Bell." (Quoted from R. Appleyard, *The History of the Institution of Electrical Engineers (1871–1931)*, London: IEE, 1939, p. 102.) Later, after he had returned from America, Preece continued to deny the telephone had any *practical* value, saying in 1879 "I have one in my office, but more for show, as I do not use it because I do not want it. If I want to send a message to another room, I use a sounder or employ a boy to take it." (Quoted from J. K. Kingsbury, *The Telephone and Telephone Exchanges*, New York, NY: Longmans, Green, and Co., 1915, p. 209.) By 1893 Preece apparently had forgotten his own words and Thompson's sarcasm, and that year he wrote "I had the good fortune in 1877 to bring back to England the first pair of practical telephones. They had been given to me in New York by Graham Bell himself.... Who at that time could have imagined that the instruments which were then but toys would, within sixteen years, have become a necessity of commercial and almost of domestic life?" (Quoted from the GPO's house organ *St. Martin's-le-Grand*, vol. 3, p. 140, 1893.)
52. Appleyard (Note 39), p. 222.
53. This is a special case of the *maximum power transfer theorem*. More generally, if the energy source *impedance* is $R + jX$, then the load impedance that results in the maximum dissipated power in the load is $R - jX$, the complex conjugate impedance. This is the proper criterion to use, for example, when coupling tiny amounts of radio energy from one power amplifier stage to the next, but it is *not* appropriate for high energy transmission by a utility. A utility wants to maximize the *efficiency* of transmission, and the maximum power transmission criterion results in just 50% efficiency, i.e., the source consumes as much energy as the load, which can result in a hot power plant!
54. *The Electrician*, vol. 2, p. 84, January 4, 1879; and pp. 94–95, January 11, 1879.
55. *Journal of the Society of Arts*, vol. 30, pp. 98–107, December 16, 1881.
56. In this respect, the "common folk" were not that far behind the scientists. In the late 18th century electricity and fire were commonly associated by scientists. Preece, himself, liked to present a somewhat more generous view of the common folk. In 1885, for example, he spoke at the B.A. meeting in Aberdeen on the lighting in his house in Wimbledon (the lamps were powered by batteries kept charged by a gas engine). *Nature* (vol. 32, p. 537, October 1, 1885) reported on his comments, "It was said that he, as an expert, could make things go which would fail in ordinary hands; but he mentioned several cases where coachmen, butlers, gardeners, and grooms had been found perfectly competent and intelligent enough to attend to everything."

6
Maxwell's Electricity

Ten thousand years from now there can be little doubt that the most significant event of the 19th century will be judged as Maxwell's discovery of the laws of electrodynamics. The American Civil War will pale into provincial insignificance in comparison.

— Richard P. Feynman, *Lectures on Physics*

From here on this book is absolutely unreadable.

— The opinion of an unknown 19th century college lecturer in mathematical physics, as found written partway through his copy of James Clerk Maxwell's *Treatise on Electricity and Magnetism*

It was great, greater and greatest.

— Oliver Heaviside's opinion of the *Treatise*

He is a genius, but one has to check his calculations... .

— Kirchhoff on Maxwell

To anyone who is motivated by anything beyond the most narrowly practical, it is worth while to understand Maxwell's equations simply for the good of the soul.

— John R. Pierce, *Electrons and Waves*

The most fascinating subject at the time I was a student was Maxwell's theory.

— Albert Einstein, *Autobiographical Notes*

Introduction

To understand the scientific content of Heaviside's work in electrical physics after 1880, it is necessary to understand the state of affairs concerning electricity as Maxwell left them at the time of his tragically early death from cancer in 1879 (at age forty-eight). This could be done most quickly by merely listing the famous equations, and in a certain sense that would be an intellectually defensible approach. To quote no less an expert than Hertz,[1]

> To the question, "What is Maxwell's theory?" I know of no shorter or more definitive answer than the following:—Maxwell's theory is Maxwell's system of equations.

This is, in fact, a favorite approach of the theoretically most extreme mathematicians who

James Clerk Maxwell (1831–1879)

One of the great theoretical physicists of all time, and Heaviside's hero (he called him "heaven sent"). This photograph first appeared in the *Telegraphic Journal & Electrical Review* the year after Maxwell's death.

want to derive everything as logical deductions from a given set of axioms (e.g., the Maxwell equations). But such an approach is essentially sterile and implicitly denies the element of *human toil.* It avoids the essential issue of the *origin* of the Maxwell equations. Maxwell, after all, did not just wake up one morning and write down his equations after a cup of coffee!

Heaviside's friend at Cambridge, G. F. C. Searle, made exactly this point in his book review[2] of the second volume of Oliver's *Electromagnetic Theory*: "These descriptions are not the work of a mere mathematician driving blindly some analytical machine, but of a man possessing a singular insight into the physical processes involved in the propagation of electromagnetic waves."

The Men before Maxwell

Maxwell, of course, did not work in a vacuum. In the century before his time isolated fragments of electrical knowledge had been slowly and painfully uncovered by such men as

Franklin, Coulomb, Priestly, Cavendish, Volta, Poisson, Oersted, and Andre Marie Ampere (1775–1836), professor of mathematics at the Ecole Polytechnique. It is from Ampere that we have the wonderful insight that magnetism is an effect due to circulating electrical currents; it was Ampere who mathematically formulated the magnetic interaction of two parallel, current-carrying wires. The link between Ampere and Maxwell was, of course, Faraday's magnificent, brilliant concept of the *field*.

Even though Ampere's particular formulation of magnetic force interactions has long since been discarded, his influence on electrical science has been enormous. Certainly he deeply impressed Maxwell, who wrote[3]

> The experimental investigation by which Ampere established the laws of the mechanical action between electric currents is one of the most brilliant achievements in science. The whole, theory and experiment, seems as if it had leaped, full grown and full armed, from the brain of the "Newton of electricity". It is perfect in form, and unassailable in accuracy, and it is summed up in a formula from which all the phenomena may be deduced, and which must always remain the cardinal formula of electro-dynamics.

Great praise, indeed, but still a bit puzzling to understand in light of the oblivion to which the "cardinal formula" has been consigned. Some years later, in 1888, Heaviside put forth his own opinion, in his usual forceful way[4]:

> It has been stated, on no less authority than that of the great Maxwell, that Ampere's law of force between a pair of current-elements is the cardinal formula of electrodynamics. If so, should we not be always using it? Do we *ever* use it? Did Maxwell, in his treatise? Surely there is some mistake. I do not in the least mean to rob Ampere of the credit of being the father of electrodynamics; I would only transfer the name of cardinal formula to another due to him... .

Heaviside, in his own unique manner, had even earlier paid tribute to Ampere. In 1886 he wrote a hilarious letter (which probably offended many of the older electricians) to *The Electrician* about the names of the various electrical units.[5] In one paragraph he declared:

> Ohm and volt are admirable; farad is nearly as good (but surely it was unpractical to make it a million times too big—the present microfarad should be the farad); erg and dyne please me; watt is not so good, but it is tolerable. But what about those remarkable results of the Paris Congress, the ampere and the coulomb? Speaking entirely for myself, they are very unpractical. Coulomb may be turned into coul, and is then endurable; this unit is, however, little used. But ampere shortened to am or amp is abominable. Better make it pere; then it will do. Then an additional bit of sentiment comes in to support us. Was not Ampere the father of electrodynamics?

In this letter Heaviside also wrote "Mac, tom, bob, and dick are all good names for units. Tom and mac (plural, Max) have sentimental reasons for adoption [obviously in reference to William Thomson and Clerk Maxwell], bob and dick may also at some future time." Heaviside's irreverence didn't fail to attract unfavorable notice, with a writer[5] to *The Electrician* declaring melodramatically, "It is much to be regretted that Mr. Oliver Heaviside, who has done so much for the improvement of electrical nomenclature, should talk of such things as *Macs*! *Mac* has nothing to do with the name Maxwell whatever, and even if it had it would be quite as great a barbarism as *Am*, which has come into use. Life is surely not so short that we have to contract mere great names into what are dangerously like grotesque and familiar nicknames, which one hardly cares to associate with the great men whose memory it is designed to perpetuate by association with electric and magnetic units. If it is, another reason has been added for thinking life not worth living."

Action-at-a-Distance

Ampere was, indeed, a man of genius, but even after his work there were still great, obvious puzzles to be addressed. The major difficulty was that of the *nature* of the interaction between currents (or electrically charged bodies). This interaction involved, obviously, forces on bodies, but these forces were not produced by anything obviously pushing or pulling on the bodies. It was *action-at-a-distance*. Action-at-a-distance was nothing new in physics by Ampere's time, as Newton himself had been faced with the same concern in his theory of masses interacting gravitationally and instantaneously across the empty vastness of space. Newton was not happy about action-at-a-distance, and indeed his gravitational theory was attacked by many who claimed it was a reversion to "explaining" Nature by invoking occult powers. Unable to suggest anything else in place of it he contented himself with his famous passage[6]:

> That Gravity should be innate, inherent and essential to Matter, so that one body may act upon another at a Distance thro' a *Vacuum*, without the Mediation of anything else, by and through which their Action and Force may be conveyed from one to another, is to me so great an Absurdity, that I believe no Man who has in philosophical Matters a competent Faculty of thinking, can ever fall into it. Gravity must be caused by an Agent acting constantly according to certain Laws; but whether this Agent be material or immaterial, I have left to the Consideration of my Readers.

The Luminiferous Ether

The nature of all forces known to Newton and his contemporaries by direct, earthly experience seemed always to be that of contact, i.e., a push or a pull by the intimate mechanical interaction of one thing (via a rope, or a stick, or one's hand, etc.) with another. Gravitational action-at-a-distance (whether instantaneous or not) is most mysterious if acting in a mechanical way through a vacuum which is truly empty. But suppose that even a vacuum is filled with a substance that can transmit forces, a substance something like air but ever so much thinner and penetrating, a substance that can slip through all of ponderable matter and fill every nook and cranny of the universe. Suppose the universe is embedded in an ocean of this mist called *ether* (or *aether*)[7]—then what?

Interacting bodies, even though *apparently* separated by the empty gulf of a vacuum, could then be imagined as actually still in mechanical communion via stresses and strains induced in the ether. So attractive is this idea, in fact, that the ether concept can be traced to ancient times, at least as far back as Aristotle. The price paid for this imaginative idea, however, was a high one—for every sort of apparent action-at-a-distance phenomenon it was necessary to postulate a corresponding ether until, as Maxwell complained,[8] "Aethers were invented for the planets to swim in, to constitute electric atmosphere and magnetic effluvin, and so on, to convey sensations from one part of our bodies to another, and so on, till all space had been filled three or four times over with aethers." All of these various ethers were viewed with a mixture of wonder and suspicion, but the one that survived the longest was perhaps the most amazing of all. To explain the ability of light to travel through space, a special ether, the *luminiferous ether*, was conceived, and it was required to be truly magical stuff.

This ether was thought able to transmit wave motion (from the interference experiments of Thomas Young it was generally known by 1801 that light is a wave phenomenon), much like a gas conducts sound waves. Sound waves, however, are longitudinal waves, with the medium "waving" back-and-forth along the direction of wave propagation. The initially puzzling fact that light can be polarized was, however, incompatible with longitudinal or "back-and-forth" compression waves. Then Young and Augustin Fresnel, in 1817–18, showed how polarization can be explained by *transversal* waves, with the medium "waving" in a direction *perpendicular* to the direction of the wave propagation. This, in turn, made a gaseous ether unthinkable, as it would be unable to support the shear stresses required by a transversal wave. The luminiferous ether could not, in fact, be a gas at all, but instead must be an elastic, jelly-like solid, a bit of imagery due to William Thomson's old friend, G. G. Stokes.[9] The required mechanical properties of such an ether are fantastic, to say the least.

This jelly had to be both thin enough for "the planets to swim in" without any observable retardation or deviation from Newton's laws of motion and rigid enough to propagate waves (light) at a speed of 186,000 miles per second. To imagine such a substance is not easy, yet in 1854 William Thomson wrote,[10]

> That there must be a medium forming a continuous material communication throughout space to the remotest visible body is a fundamental assumption in the undulatory Theory of Light. Whether or not this medium is (as appears to me most probable) a continuation of our own atmosphere, its existence *is a fact that cannot be questioned* [my emphasis]... .

Thomson was not alone in his absolute belief in the ether, and it wasn't until after Einstein's work that the concept finally died. Even then Heaviside never wavered in *his* faith in the ether, and neither did most, if not all, of his fellow Victorian electrodynamicists.[11] Their position was perhaps put best by Heaviside when he wrote[12] in 1893,

> As regards the ether, it is useless to sneer at it at this time of day. What substitute for it are we to have? Its principal fault is that it is mysterious. That is because we know so little about it. Then we should find out more. That cannot be done by ignoring it. The properties of air, so far as they are known, had to be found out before they became known.

This passage shows an increase in either Heaviside's optimism or desperation, as earlier in 1885 he had written,[13]

> Ether is a very wonderful thing. It may exist only in the imaginations of the wise, being invented and endowed with properties to suit their hypotheses; but we cannot do without it... . But admitting the ether to propagate gravity instantaneously, it must have wonderful properties, unlike anything we know.

and then a few months later[14]

> The actual constitution of the ether is unknown. It never can be *known*.

Faraday and Lines of Force

So the situation just before Faraday and his conception of *fields* was that the underlying basis for all of electrodynamics was action-at-a-distance, with all forces transmitted *instantaneously* (the fact that light had a finite velocity was known, but it wasn't known until Maxwell that light was electrodynamic in nature) between interacting bodies, and that all such forces are *central* forces. The nature of the intervening space (the medium) between interacting bodies was taken to be Heaviside's "mysterious" ether. Then, on August 29, 1831, Michael Faraday (1791–1867) found the long-sought second link between electricity and magnetism, electromagnetic induction, in an experiment sometimes claimed to be "one of the few really great experiments in the history of science."[15] With this discovery Faraday showed how to generate electricity by a purely *mechanical* (as opposed to chemical) process, which was quickly implemented in the construction of the first d-c generator (the very next year the French instrument maker Hippolyte Pixii built the first a-c generator).

How Faraday's field ideas influenced Maxwell is discussed in Tech Note 2 at the end of this chapter.

William Thomson

As one writer has put it,[16] "The Victorian period was the golden age of classical science, and perhaps the most original since the scientific quest was undertaken." One reason for this praise is the work of William Thomson. He was, in many ways, the "iron man" of 19th century science and technology. It seemed he could do everything, and almost always better than anybody else. Even before he died in 1907 (as Lord Kelvin) he was practically a mythical giant (but, as would be Einstein's fate decades later, younger men came to believe his body far outlived the scientific agility of his mind). He was buried in Westminster Abby, along with Huxley and Darwin (with whom he had debated the age of the Earth, as we'll see in Chapter 11), to lie forever next to Newton, an honor reserved by England only for her *supremely* great heroes. If an even greater honor could have been bestowed, there is little doubt that it would have been granted. If obituary page counts mean anything, then Thomson surely holds top place among all the Fellows of the Royal Society—his ran to a nearly book-length seventy-six pages (with three photographs) in the Society's *Proceedings*. Heaviside got two pages (and no photograph) and the passing of Sir William Henry Preece was, except for a brief note, ignored by the *Proceedings*.

Thomson's ability was realized early, and while only twenty-two (in 1846) he became a professor at Glasgow University. He remained there for more than half a century, twice declining (in favor of Maxwell and Rayleigh) the position of Cavendish Professor at Cambridge. His work was literally in every field of the physical sciences of his day: elasticity, optics, hydrodynamics, thermodynamics, electrodynamics, engineering (e.g., in addition to his *electrical* analysis of submarine cables, he also studied the *mechanical* problems of laying long, heavy cables in deep water), and the application of mathematics to such mysterious and wonderful problems as calculating the ages of the sun and the Earth. It seems he studied everything from the structure of cosmic nebulae to the nature of atoms.

Thomson's genius first flourished in electricity. When barely out of his teens (1845) he invented the ingenious method of images, today a mainstay of every undergraduate physics course in electricity. And even earlier, in the spring of 1840 while still a student at Glasgow, he had read Fourier's *Theorie analytique de la chaleur* ("In a fortnight I had mastered it—gone right through it."[17]) and had been tremendously impressed by the "mathematization" of heat via the heat equation (recall from Chapter 3 how Thomson used it fourteen years later in his submarine cable analyses). Later, after leaving Glasgow for Cambridge, he became familiar with Gauss' mathematical work in potential theory, and George Green's application of the theory to electrostatics.

Thomson was greatly excited by Green's work,[18] and used the concept of the *electrostatic scalar potential function* to calculate forces due to distributed electricity. This potential function is usually introduced as a mathematical artifice that is calculated by an integration, and *then* the force on a unit charge at any point is found by a differentiation of the potential. In mathematical form the procedure is to first calculate

$$V(x, y, z) = \int_{\text{all space}} \frac{\rho}{r}\, dv$$

where ρ is the spatial distribution of electrical charge, and r is the distance from the differential volume element dv to the point where the potential is to be found. *Then* the force[19] on a unit charge is the vector $\vec{F}$ with components

$$F_x = -\frac{\partial V}{\partial x}, \quad F_y = -\frac{\partial V}{\partial y}, \quad F_z = -\frac{\partial V}{\partial z}$$

or, in modern vector-differential notation[20]

$$\vec{F} = -\text{gradient } V = -\vec{\nabla} V.$$

The $\vec{\nabla}$ symbol (a vector-differential *operator*, in modern terminology) plays a central part in Maxwell's equations, as we will soon see.

But for all the importance of these and other achievements yet to come, perhaps with the hindsight of a century it is fair to say it was Thomson's realization that a Faraday field could in some way *store energy* that was his greatest idea. It was Thomson, in fact, who first calculated[21] in 1853 the expression $\frac{1}{2}LI^2$ (an expression so well known today it ranks with Ohm's law as a "technical cliché") for the energy stored by an inductor of value L conducting a current I. His general expression for this energy is in the form of a three-dimensional volume integral taken over all space, which might have suggested to him that the energy is *distributed* throughout space (but a careful reading of Thomson shows he never took this last step). It remained for Thomson's friend, Clerk Maxwell, to make the idea of distributed energy in a field the central concept in electrodynamics.

Maxwell

James Clerk Maxwell (1831–1879) was certainly *the* most influential physical scientist of the 19th century, and together with Newton and Einstein, one of the three most important of all time (and that may well hold true for some time to come). It is clear from Heaviside's writings that Maxwell was his hero. Maxwell's equations for the electromagnetic field are today what they were 100 years ago, being one of the rare theories that have resisted the erosion and corrosion of progress. Special relativity had no effect on the equations (as magnetic phenomena are, in fact, relativistic effects, relativity is actually "built into" the equations!), and they were hardly perturbed at all by the arrival of quantum mechanics.[22]

Maxwell's enduring contribution to electrical physics is his famous set of equations, and historians of science have rightfully devoted much effort in their attempts to reconstruct the evolution of his thought.[23] If we use $\vec{E}$ and $\vec{B}$ to denote, respectively, the electric and magnetic vector fields, then the state of electromagnetic knowledge just before Maxwell is quite easy to express (in modern notation). We shall not actually *do* much with these equations (but see the end-of-chapter Tech Notes), except to see precisely what Maxwell's contribution was. One is of course free to admire them for their obvious power to sum up so succinctly the entire subjects of electricity, magnetism, and optics (and one does not need either to know or to recall any mathematics to appreciate this point).

So, just before Maxwell, people in essence knew that if ρ and $\vec{J}$ denote, respectively, the *net* electric charge and the electric conduction current vector (both densities), and if ϵ and μ are scale factors describing the electric and magnetic properties of the ether, then:

$\vec{\nabla} \cdot \vec{E} = \rho/\epsilon \rightarrow$ Electric field lines *diverge from* positive charge, and *converge onto* negative charge—this is called Gauss' Law.

$\vec{\nabla}\cdot\vec{B}=0\rightarrow$ Magnetic field lines have zero divergence everywhere (that is, there is no such thing as "magnetic charge"), and so they always form closed loops with no beginning or end—this has no special name.[24]

$\vec{\nabla}\times\vec{B}=\mu\vec{J}\rightarrow$ An electric current creates[25] a magnetic field—this is called Ampere's Law.

$\vec{\nabla}\times\vec{E}=-\frac{\partial\vec{B}}{\partial t}\rightarrow$ A time-varying magnetic field creates an electric field—this is just Faraday's Law of Induction.

$\vec{\nabla}\cdot\vec{J}=-\frac{\partial\rho}{\partial t}\rightarrow$ A *change* in the net charge density at a point in space requires a conduction current (the minus sign means a *reduction* in charge which implies a current flowing *away* from the point)—this is called the Continuity of Charge Law (or better, *local* conservation of charge).

Essentially all this was in the ken of William Thomson (as well as Maxwell) who also eagerly sought an understanding of the inner workings of electricity and magnetism. Indeed, when Maxwell was about to begin his quest along research lines he had reason to believe Thomson was also pursuing, he felt compelled to write a letter[26] (dated September 13, 1855) to the older man to be sure he wouldn't be thought a poacher (a word he himself used):

> I have got a good deal out of you on electrical subjects, both directly & through the printer & publisher & I have also used other helps, and read Faraday's three volumes of researches. My object in doing so was of course to learn what had been done in electrical science...and to try to comprehend the same in a rational manner by the aid of any notions I could screw into my head.

and then soon after, perhaps in the hope of prompting some information on how far along Thomson might be,

> As there can be no doubt that you have the mathematical part of the theory in your desk... .

As we know today, Thomson had no such papers in his desk in 1855, and in fact would *never* write such papers. Indeed, even after Maxwell's death, and up to his own death, Thomson never quite grasped what the new theory was all about.[27] This was so despite the fact that Maxwell's approach was in the mold of Thomson's, that of using mechanical models to deduce a dynamical theory of electrical phenomena.[28] Maxwell had been greatly impressed with Thomson's analogy of electricity with heat and he became quite inventive with his own analogies. To quote one writer[29]

> Maxwell's papers teem with gushing ideal fluids, gears, idle wheels and such hardware. One French commentator was led to remark upon reading Maxwell's magisterial *Treatise on*

> *Electricity and Magnetism* that he thought he was to enter the quiet groves of electromagnetic theory only to discover that he had walked into a factory!

But, and perhaps somewhat paradoxically, Maxwell did not commit himself to these models, and they were merely aids for the evolution of his thinking. He strived, in fact, to achieve a physical theory *without making assumptions* about the underlying details of the physics (such as hypothesizing what might be the precise structure of atoms, or trying to describe the force interactions deep inside matter, an approach others *did* use and which he criticized[30]). Maxwell came early to this philosophy, and in a letter[31] to Stokes (dated October 15, 1864) he wrote,

> I have now got materials for calculating the velocity of transmission of a magnetic disturbance through air founded on experimental evidence, *without any hypothesis about the structure of the medium or any mechanical explanation of electricity or magnetism* [my emphasis].

Heaviside very clearly adopted this point of view. As he wrote[31] in 1885:

> I am not objecting to use of the imagination. That would be absurd; for most scientific progress is accomplished by the free use of the imagination (though not after the manner of professional poets and artists when they touch upon scientific questions). But when one, by the use of the imagination, has got to a definite result, and then sees a stricter way of getting it, it is perhaps as well to shift the ladder, if not to kick it down. For I find that practically, in reading scientific papers, in which fanciful arguments are much used, it gives one great trouble to eliminate the fancy and get at the real argument. Nothing is more useful than to be able to distinctly separate what one knows from what one only supposes.

and then ten years later, in a penciled note[31] in his copy of *Nature*, next to an article by Boltzmann which used Hertz's phrase of the "draped figure of nature":

> Better put it in the other way; we should not identify the naked skeleton we assume Nature to be with the only real thing, the gay-coloured vesture of Nature.

The Displacement Current

Where did this frugal philosophy eventually lead Maxwell? After years of pondering and endless tinkering with his mechanical analogies, he was led to accepting as they stood the five equations given previously, with the exception of Ampere's Law. To the right-hand side of that, to the term involving the conduction current $\vec{J}$ which represents the flow of electricity (whatever *that* is) in a conductor, he added another term involving a new, *fantastic* current. This he called the *displacement* current (because he was, for a while, thinking of it as arising from a polarization or displacement of charges in the dielectric), and it is fantastic for two reasons.

First, this new displacement current is not associated with the flow of electricity in a conductor, but it still *acts* like a current. Second, there was at the time absolutely no experimental evidence to justify such a modification to Ampere's Law. Historians of science have scholarly analyses[32] to explain this, but I like the direct, simple (and probably simplistic, I'll admit) explanation—Maxwell was a genius and maybe it is a bit too much to hope to rationalize why he did what he did. This is what genius is, after all, the ability to take the unexpected, even seemingly irrational step, but yet the correct step. With hindsight, of course, lesser mortals have been able to partially rationalize Maxwell's leap of insight (see Tech Note 1 at the end of this chapter). Heaviside said[33] it was Maxwell's invention of the displacement current (which made *all* electric circuits *closed* ones, even those that seemed to be open—such as those with series capacitors) that "boldly cut the Gordian Knot of electromagnetic theory."

The new equations, as is well known these days, predict the existence of *propagating* electric and magnetic fields (the traveling electromagnetic wave). Further, the equations specify the speed of propagation to be that of light. It was this "coincidence" which Maxwell refused to believe was really a coincidence that led him to declare light *is* electromagnetic in nature. Indeed, one historian of science has advanced the interesting idea[34] that these two events actually took place in reverse order:

> How did Maxwell hit on...his displacement current? The best I can suggest by way of an answer is this: Once Maxwell had stumbled on the possible identity of the magnetic and luminiferous ethers...his main objective became the deduction of a truly electromagnetic theory of light that was independent of the details of any mechanical model. *He then juggled with his theory until he found the form of the displacement current that would enable him to derive a wave equation* [my emphasis].

Today we have a very dramatic means at hand to demonstrate the electromagnetic nature of light. Lasers can now be made to create focused light beams so intense their electric fields (more than a billion volts per centimeter) exceed the breakdown potential of air and create sparks before our eyes. Maxwell would have loved it!

Still, it is one thing for a theory to *predict* and another for the predictions to be verified. It wasn't until 1888, nearly a decade after Maxwell's death, that Hertz finally offered experimental evidence so strong it couldn't be ignored. But before that there were other electrodynamic theories besides Maxwell's, most notably the action-at-a-distance ones proposed first by Karl Friedrich Gauss and then later, in 1846, by Wilhelm Weber, that incorporated velocity- and acceleration-dependent forces.[35] This competition, as well as Maxwell's own often difficult presentation, clunking with gears, ratchets, and other distracting machinery, made a clear-cut choice difficult.

But of course Maxwell did prevail, and soon all men of science knew that a giant intellect had been at work among them. Heaviside wrote,[36] toward the end of his life, a moving expression of his unbounded admiration for Maxwell and what he had wrought:

> A part of us lives after us, diffused through all humanity—more or less—and through *all* nature. This is the immortality of the soul. There are large souls and small souls. The immortal soul of the "Scienticulists" is a small affair, scarcely visible. Indeed its existence has been doubted. That of a Shakespeare or Newton is stupendously big. Such men live the best part of their lives after they are dead. Maxwell is one of these men. His soul will live and grow for long to come, and hundreds of years hence will shine as one of the bright stars of the past, whose light takes ages to reach us.

Postscript: Just What *Is* Electricity, Anyway?

Now that we have followed events through to the end of Maxwell's life perhaps it is reasonable to expect an answer to this fundamental question. Unfortunately, even after Maxwell, an understanding of the nature of electricity remained beyond human knowledge. The great French scientist Henri Poincaré caught the correct spirit of gloom when he said,[37]

> One of the French scientists who has probed Maxwell's work the most deeply said to me one day, "I understand everything in this book, except what is meant by a charged sphere."

Oliver Lodge was equally blunt when he said,[38]

UNFORTUNATE FOR BODGER.

This cartoon by W. S. Gilbert (later of "Gilbert and Sullivan" operatic fame), published in *Fun* (May 3, 1862), carried the subcaption: "The new steel collars and wristbands are capital things in their way; but be warned by poor Bodger's fate, and never, on any account, wear them in a thunderstorm." By the mid-19th century there was no doubt that electricity was a wonderful thing (and yet slightly terrifying, too), and one had to keep its powers in mind, even in matters of fashion.

> Now then we will ask first, What is Electricity? and the simple answer must be, We don't know.

But of course not everybody had doubts; a year earlier, in 1880, William Henry Preece had answered our question with confidence. First declaring[39] electricity must be either matter or force, he then argued his way to the conclusion

> Electricity is therefore not a form of matter. Hence, according to our reason, it must be a form of force.

A little earlier Preece had, in fact, declared both heat and light to be forces, too. Later, in 1900, Preece went so far as to assert (in a lecture at the Institution of Civil Engineers) that[40]

> Maxwell proved the *identity of electricity and light* [my emphasis] by showing that they moved through the aether with the same velocity—196,400 miles per second—and in the same undulatory fashion.

Sir William, in fact, said many astonishing things about electricity during this lecture, but perhaps the *most* astonishing was his explanation of electrolysis:

> In liquid conductors the [molecular] motion [which Preece said is electric current] probably

> becomes revolution. The result is decomposition by the activity of the centrifugal force [!] overcoming chemical affinity. The atoms fly away in fixed determined lines, and collect at opposite poles.

Preece wasn't alone in this nonsense, and in the same year the eminent electrical engineer Reginald Fessenden (to become best known for his work in radio broadcast engineering) actually claimed[41] to have shown that gravitation is merely "a secondary electric effect," and that "the nature of electricity and magnetism may now be considered to be definitely settled."

This intense *scientific* curiosity was mirrored in an almost romantic fascination by the general public, except that it was more interested in what the mysterious stuff could *do*. Electricity could light your home, but it could also burn it down, and the image of electricity had a distinct dark side to it. The following passage is from one of the more vivid Victorian essays[42]:

> Day by day we read that men are killed by the very force which flashes the news of their death across the world. One is at work on a telegraph cable: he touches a lightning wire; instantly he hangs lifeless and burning above the head of a horror-stricken, shrieking crowd. (It is said that a woman died of the sight.) A few days later a shopman moving a metal show-case was done to death; in his shoe a projecting nail pierced the stocking, and thus helped on his fate.

Electricity could kill just as surely as a bullet through the heart, but of course one could actually hold a bullet between finger and thumb and examine it at leisure. Electricity, though, is gone in a twinkling, leaving nothing behind but its victim. So even the ordinary fellow on the street was asking, "Just what *is* electricity?" William Thomson knew the rewards the answer would bring—in 1862 he declared to his friend Peter Tait, "If you will tell me what electricity is, I will tell you everything else." So, here we now are more than a century after Maxwell—just what the devil *is* electricity? Oliver Lodge's response is still valid. Nobody knows. Perhaps we will never know. But we do have Maxwell's equations, and as Heaviside once put it,[43] "Good old Maxwell!"

Tech Note 1: A Technically Nice, Often Taught, but Historically False "Explanation" of the Displacement Current

For *any* vector field (not just electric or magnetic fields) it is a *mathematical* truism that the divergence of the curl is always zero. That is, if $\vec{F}$ is any vector field at all, then

$$\vec{\nabla}\cdot(\vec{\nabla}\times\vec{F})=0.$$

Many professors of electrical engineering and physics cannot resist using this pretty identity to make the following "derivation" of the displacement current for their students (perhaps some of the professors even believe this is historically correct, but we can all hope they are few in number!). Maxwell (so goes this false tale) noticed that Ampere's Law and the Continuity of Charge are not compatible. That is, from Ampere's Law we have

$$\vec{\nabla}\cdot(\vec{\nabla}\times\vec{B})=0=\vec{\nabla}\cdot(\mu\vec{J})=\mu\vec{\nabla}\cdot\vec{J}$$

or

$$\vec{\nabla}\cdot\vec{J}=0,\ \text{always.}$$

But charge continuity requires

$$\vec{\nabla}\cdot\vec{J} = -\frac{\partial\rho}{\partial t}.$$

For steady-state currents in closed circuits (the only sort of experimental setups studied up to Maxwell's time) ρ is a constant and thus both derivations for $\vec{\nabla}\cdot\vec{J}$ agree. But for non-steady-state situations ρ *can* vary, and something is obviously not right. Genius that he was, Maxwell realized (remember now, this is all a myth) a patch was needed to Ampere's Law so that it would give the $\vec{\nabla}\cdot\vec{J}$ demanded by charge continuity. So, he added a second term (let's call it $\vec{D}$), i.e.,

$$\vec{\nabla}\times\vec{B} = \mu(\vec{J}+\vec{D}).$$

Then

$$\vec{\nabla}\cdot(\vec{\nabla}\times\vec{B}) = 0 = \vec{\nabla}\cdot\mu(\vec{J}+\vec{D}) = \mu(\vec{\nabla}\cdot\vec{J}+\vec{\nabla}\cdot\vec{D})$$

or

$$\vec{\nabla}\cdot\vec{J} = -\vec{\nabla}\cdot\vec{D}$$

or, to make this agree with charge continuity,

$$\vec{\nabla}\cdot\vec{D} = \frac{\partial\rho}{\partial t}.$$

But, from Gauss' Law we have (assuming, as electrical engineers almost always do, that the time and space differentiation operators can be interchanged),

$$\frac{\partial\rho}{\partial t} = \epsilon\frac{\partial}{\partial t}(\vec{\nabla}\cdot\vec{E}) = \vec{\nabla}\cdot\left(\epsilon\frac{\partial\vec{E}}{\partial t}\right)$$

and therefore

$$\vec{\nabla}\cdot\vec{D} = \vec{\nabla}\cdot\left[\epsilon\frac{\partial\vec{E}}{\partial t}\right]$$

which "suggests" the needed patch is just exactly what Maxwell ended up with for his displacement current,

$$\vec{D} = \epsilon\frac{\partial\vec{E}}{\partial t}.$$

This is all very pretty and Maxwell no doubt would have loved it, but it is, I repeat, *not* the way he actually arrived at the displacement current. And that, when all is said and done about historical integrity, is too bad, because it really *is* very pretty. Even a skilled historian can fall into this particular quasi-history trap.[44]

TECH NOTE 2: ACTION-AT-A-DISTANCE, FIELDS, AND FARADAY'S ELECTROTONIC STATE

Maxwell did not accept instantaneous action-at-a-distance. Even action-at-a-distance with force propagated at a *finite* speed (an idea explored by Gauss in 1845) was rejected because

Maxwell believed it would result in a violation of conservation of energy. In a letter[45] (dated March 12, 1868) to his old schoolboy chum Professor Peter Guthrie Tait (who will appear again in this book as yet another adversary for the rambunctious Heaviside), Maxwell outlined how, given finite-speed action-at-a-distance, one could make "a locomotive engine fit to carry you through space with continually increasing velocity." As described by Maxwell, this device would propel itself forward forever without any source of energy! Rejecting this as absurd, he instead developed Faraday's primitive field theory concepts into the mathematical equations discussed previously in this chapter (although he used different notation).

The key idea underlying field theory is that "something real" (but *not* force) propagates through space at finite speed, and that this "something real" can be described by partial differential equations. This "something real" is commonly taught today to be the fields (electric and magnetic). Einstein, himself, stated[46] this to be Maxwell's unique contribution: "the description of Physical Reality by fields which satisfy without singularity a set of partial differential equations."

In actuality, however, the fields were not the primary reality to Maxwell at all (that is an idea that developed after Maxwell's death and is due to Hertz and Heaviside), but rather it was Faraday's "electrotonic state" that he thought to be the *real* thing. Like Faraday, Maxwell believed that electromagnetic effects are observable results of an altered *state* of the ether. The mathematical formulation of this electrotonic state, for Maxwell, is what we today (as he did) call the *vector potential*. This is usually presented to students today as a mere mathematical aid to making field calculations, and as not having any physical reality (another idea due to Hertz and Heaviside). To Maxwell, however, the vector potential[47] had a most definite physical meaning.

To see how this works, recall the Maxwell equation that says there is no such thing as magnetic charge,

$$\vec{\nabla}\cdot\vec{B}=0$$

and, from Tech Note 1, that for *any* vector field $\vec{A}$ it is always true that

$$\vec{\nabla}\cdot(\vec{\nabla}\times\vec{A})=0.$$

This means that for any given magnetic field $\vec{B}$, there is an $\vec{A}$ such that

$$\vec{B}=\vec{\nabla}\times\vec{A}.$$

$\vec{A}$ is called the vector potential. So far this is nothing but mathematics, but Maxwell showed that $\vec{A}$ has a deep *physical* meaning, too. To see this, recall now another Maxwell equation, the one that expresses Faraday's Law of Induction,

$$\vec{\nabla}\times\vec{E}=-\frac{\partial\vec{B}}{\partial t}.$$

Substituting the vector potential gives

$$\vec{\nabla}\times\vec{E}=-\frac{\partial}{\partial t}(\vec{\nabla}\times\vec{A})=\vec{\nabla}\times\left(-\frac{\partial\vec{A}}{\partial t}\right)$$

which suggests

$$\vec{E}=-\frac{\partial\vec{A}}{\partial t}.$$

We now do something Maxwell loved—we make an analogy. Recall from Newton's mechanics that if $\vec{P}$ is the momentum vector of a particle, then the force vector $\vec{F}$ acting on the particle is given by

$$\vec{F}=\frac{d\vec{P}}{dt}.$$

In electrodynamics we have a force law that says if a traveling $\vec{E}$-wave sweeps over a particle with charge q, then that particle will experience a force

$$\vec{F}=q\vec{E}=-q\frac{\partial\vec{A}}{\partial t}.$$

By analogy, then, we are almost irresistibly led to the conclusion that $\vec{A}$ is intimately associated with a momentum. The momentum of what? Why, the momentum of the traveling wave! Since $\vec{E}$ propagates, so does $\vec{A}$, and we conclude that electromagnetic waves carry momentum (this means light *pushes* on whatever it hits; this is the mechanism behind the imaginative concept of space travel by "solar sailing," using the radiation pressure of the sun as a sailboat uses the wind[48]). Traveling waves also carry energy, but Maxwell died before he could pursue these ideas, and they were left for John Henry Poynting and Heaviside to develop.

Maxwell called[49] $\vec{A}$ the "Electrokinetic Momentum," and specifically said it "may even be called the fundamental quantity in the theory of electromagnetism." Indeed, modern thought seems to be returning to this idea, and turning away from the view of Hertz and Heaviside that the *fields* are the real thing.[50] What a man Maxwell must have been to have seen all this so clearly that even now, a century later, we are still learning from him! I am reminded of Professor Tait's description[51] of his friend's ability:

> [a] man of real power, though (to all seeming) perfectly unconscious of it—who goes straight to his mark with irresistible force, but neither fuss nor hurry—reminding one of some gigantic but noiseless "crocodile".

Notes and References

1. H. Hertz, *Electric Waves,* New York, NY: Dover, 1962, p. 21 (originally published in 1893). Heaviside used almost the same words to say the same thing in the same year, 1893, in the Preface to EMT 1 (pp. vi–vii).
2. *Physical Review,* vol. 11, pp. 60–64, 1900.
3. J. C. Maxwell, *A Treatise on Electricity and Magnetism,* vol. 2, New York, NY: Dover, 1954, p. 175.
4. EP 2, p. 501.
5. *The Electrician,* vol. 16, pp. 227–228, January 29, 1886, reprinted in EP 2, pp. 25–27. Heaviside's critic appeared in vol. 29, p. 433, August 19, 1892.
6. In a letter dated February 25, 1693, to Dr. Richard Bentley, Bishop of Worcester, and reprinted in I. B. Cohen, *Isaac Newton's Papers & Letters on Natural Philosophy,* Cambridge, MA: Harvard University Press, 1978, pp. 302–303.
7. The hoary metaphysical assertion that "Nature abhors a vacuum" was enough in itself for some to believe in an ether, i.e., with an ether the concept of a vacuum becomes vacuous (no pun intended). As Oliver Lodge described the feelings of others, "empty space could hardly exist, that it would shrink up to nothing like a pricked bladder unless it were kept distended by something material." See "The ether and its functions," *Nature,* vol. 27, pp. 304–306, January 25, 1883, and pp. 328–330, February 1, 1883, as well as the intelligent reply to Lodge by S. T. Preston, vol. 27, p. 579, April 19, 1883.
8. W. D. Niven (Ed.), *The Scientific Papers of James Clerk Maxwell,* vol. 2, Cambridge: Cambridge University Press, 1890, p. 763.
9. See D. B. Wilson, "George Gabriel Stokes on stellar aberration and the luminiferous ether," *The British Journal for the History of Science,* vol. 6, pp. 57–72, June 1972. A solid ether was necessary to explain the

THE MARCH OF SCIENCE

"Interesting Result Attained, with Aid of Röntgen Rays, By a First-Floor Lodger When Photographing His Sitting-Room Door."

While Heaviside and his fellow scientists debated the technical questions concerning x-rays, more popular concerns were mostly along the lines of this 1896 *Punch* illustration. Another similar issue was related to the need, perhaps, to develop a process to manufacture women's undergarments from lead! And as *The Electrician* observed: "So long as individuals of the human race continue to professionally inject bullets into one another, it is well to be provided with easy means for inspecting the position of the injected lead... ."

observed transversal waves of light, but that did not preclude such an ether from supporting longitudinal waves as well. The failure to observe such waves was long a puzzle, one that many at first thought was solved when Röntgen discovered the very short wavelength x-rays in 1895. Röntgen himself initially believed that he had stumbled onto the long missing longitudinal ether waves, and Silvanus Thompson went so far as to call them "ultra violet sound"! Others thought the "new radiation" might even be the key to understanding gravity, and Thomas Edison thought x-rays could be used to let the blind see again. Maxwell's equations predict only transverse waves, but Oliver Lodge speculated (*The Electrician,* vol. 36, pp. 471–473, February 7, 1896) that the longitudinal waves might be a "disturbance requiring some generalization or modification of these equations." One person who claimed to know how to make this modification was Professor Gustav Jaumann at the German University of Prague—see his mathematical paper "Longitudinal light," *The Electrician,* vol. 36, pp. 629-631, March 6, 1896, and pp. 656–657, March 13, 1896, and pp. 685–688, March 20, 1896. Lord Kelvin's imagination was particularly struck by the "new radiation" and he wrote a curious paper about it ("On the generation of longitudinal waves in the ether," in *Mathematical and Physical Papers,* vol. 6, J. Larmor, Ed., Cambridge: Cambridge University Press, 1911, pp. 54–57). See Note 43 and the next chapter for Heaviside's reaction to this debate.

10. W. Thomson, "Note on the possible density of the luminiferous medium and on the mechanical value of a cubic mile of sunlight," in *Mathematical and Physical Papers,* vol. 2, J. Larmor, Ed., Cambridge: Cambridge University Press, 1884, pp. 28–33.

11. S. Goldberg, "In defense of ether: The British response to Einstein's special theory of relativity, 1905–1911," in *Historical Studies in the Physical Sciences,* Philadelphia, PA: University of Pennsylvania Press, 1970, pp. 89–125. One of the most energetic English believers in the ether was Oliver Lodge, who tried to detect it experimentally. In his autobiography (*Past Years,* New York, NY: Charles Scribner's Sons, 1932) Lodge called these experiments "the most important" of his life (p. 195) and included a photograph of his "ether whirling machine" (opposite p. 200). This was a rather large gadget, big enough for Lodge to sit inside; the visual impact of the photo is much like watching H. G. Wells' Time Traveler about to set out on his journey to the year 802,701. His experiments were of course a failure; as J. J. Thomson said at the B.A. meeting of 1896 at Liverpool, "You are all doubtless acquainted with the heroic attempts made by Prof. Lodge to set the ether in motion, and how successfully the ether resisted them."—see *The Electrician,* vol. 37, pp. 672–675, September 18, 1896. Heaviside was, of course, sympathetic with Lodge's efforts and, on October 30, 1893, wrote to Lodge expressing how glad he was "in these days of dynamos" that there were still men interested in "abstruse questions of science"—see B. Hunt, "Experimenting on the ether: Oliver J. Lodge and the great whirling machine," *Historical Studies in the Physical and Biological Sciences,* vol. 16, part 1, pp. 111–134, 1986.
12. EMT 1, p. 321.
13. EP 1, p. 433. The passage first appeared in *The Electrician,* vol. 14, p. 150, January 3, 1885. When reprinted in *Electrical Papers,* Heaviside's very next words were cut: "Similarly in the propagation of the transverse vibrations of light, it is not easy to conceive ether acting as an elastic solid. We may know some day. But if not, what matter? Things will go on all the same whether we know how it is done or not. What vanity is this the philosophers are striving after? Is it not 'Science, falsely so called,' when there is the infinite blank of the unknowable behind? And, speaking of blanks, might there not be a great blank in space, to which we shall arrive some day, to find all theories confuted?" It is curious to note that today's astrophysicists are definitely in the spirit of Heaviside, still wondering about great blanks in space where all theories may fail—only now they call them "black holes"!
14. EP 1, p. 420.
15. L. P. Williams, "Faraday's discovery of electromagnetic induction," *Contemporary Physics,* vol. 5, pp. 28–37, October 1963. The discovery of how magnetism could create electricity (the reverse of Oersted's discovery) had been missed by even the great Ampere, who had been looking for such an effect since 1822—all during those years Ampere and Faraday were very friendly and corresponded regularly. See K. R. Gardiner and D. L. Gardiner, "Andre-Marie Ampere and his English acquaintances," *The British Journal for the History of Science,* vol. 2, pp. 235–245, June 1965.
16. This is the opening sentence of a really fascinating essay by L. P. Williams, "The historiography of Victorian science," *Victorian Studies,* vol. 9, pp. 197–204, March 1966.
17. H. T. Sharlin, *Lord Kelvin,* University Park, PA: The Pennsylvania State University Press, 1979, p. 16. Interesting personality insights can also be found in A. Kent, "Lord Kelvin, the young professor," *Philosophical Journal,* vol. 5, pp. 47–53, 1968.
18. George Green (1793–1841), born the son of a Nottingham baker, left his name scattered all about in mathematical physics in the various forms of *Green's theorem* and *Green's functions.* Despite this, however, it is safe to say that no textbook writer today tells his readers *anything* about Mr. Green himself. The only extensive biographical sketch on this unsung genius appears to be the essay by H. G. Green, "Biography of George Green," in *Studies and Essays in the History of Science and Learning in Honor of George Sarton,* New York, NY: Henry Schuman, 1944, pp. 545–594. The measure of the brilliance of Green's work is related in Sharlin (Note 17), pp. 54–57.
19. M. N. Wise, "William Thomson's mathematical route to energy conservation: A case study of the role of mathematics in concept formation," in *Historical Studies in the Physical Sciences,* vol. 10, Baltimore, MD: The Johns Hopkins University Press, 1979, pp. 49–83. Lagrange had made similar use of the scalar potential earlier in gravitational calculations. See also Wise's later paper, which includes a scholarly study of Thomson's analogy between electricity and heat, "The flow analogy to electricity and magnetism, Part I: William Thomson's reformulation of action at a distance," *Archive for History of Exact Sciences,* vol. 25, pp. 19–70, 1981.
20. The vector-differential operator $\vec{\nabla}$ is called *Nabla,* a name adopted by Maxwell (after the ancient Egyptian harp which had the same shape). Heaviside once suggested it be called the *vex* operator! As indicated in the text, if V is a scalar potential that is a function of the three dimensions, then $\vec{\nabla} V$ is a vector. A second application of $\vec{\nabla}$ to $\vec{\nabla} V$ leads to $\nabla^2 V$ which is now a new scalar function called the *scalar Laplacian* of V (it is the algebraic sum of the three *second*-order partial derivatives of V). It is an operation that occurs countless times in mathematical physics and electrical engineering. As Heaviside wrote (EP 2, p. 531), "Physical mathematics is very largely the mathematics of $\vec{\nabla}$." Maxwell, himself, called the negative of $\nabla^2 V$ the concentration of V, and a very nice historical treatment is in J. E. McDonald, "Maxwellian interpretation of the Laplacian," *American Journal of Physics,* vol. 33, pp. 706–711, September 1965.
21. "On the mechanical values of distributions of electricity, magnetism, and galvanism," in *Mathematical and Physical Papers,* vol. 1, J. Larmor, Ed., Cambridge: Cambridge University Press, 1882, pp. 521–533.
22. The original, particular physical visualizations about electricity due to Maxwell *have,* however, been largely discarded. This occurred primarily because of the rejection of the ether and the discovery of the electron, i.e., the

discovery that electric charge comes in tiny, discrete quantities. For a masterfully written account of this conceptual transition, see J. Z. Buchwald, *From Maxwell to Microphysics,* Chicago, IL: University of Chicago Press, 1985.

23. See, for example, A. F. Chalmers, "Maxwell's methodology and his application of it to electromagnetism," *Studies in History and Philosophy of Science,* vol. 4, pp. 107–164, August 1973, and P. M. Heimann, "Maxwell and the modes of consistent representation," *Archive for History of Exact Sciences,* vol. 6, pp. 171–213, 1970. Chalmers is very mathematical in his treatment, while Heimann manages to do quite well with a mere handful of equations (and even then only in the footnotes). Both of these very long articles are not the sort of thing one reads in a single sitting. For the "reader in a hurry," I highly recommend A. M. Bork, "Maxwell and the electromagnetic wave equation," *American Journal of Physics,* vol. 35, pp. 844–849, September 1967. This paper also contrasts our modern notation with the symbols Maxwell actually used.
24. When I say there is no magnetic charge, what is meant is that none has yet been observed. If magnetic charges (called monopoles) do exist, then all we would need to do is alter the right-hand side of $\vec{\nabla}\cdot\vec{B} = 0$. Indeed, when Heaviside wrote Maxwell's equations that is exactly what he did, imagining there *are* magnetic charges, which he called "magnetons" (he did this because he liked the symmetrical-looking equations—he called this the "duplex" form of the equations; see EP 1, p. viii and EMT 3, pp. 57–58), and when done with his analyses he just set the magnetic charge value to zero. This approach possibly had an important impact on Paul Dirac's development of monopole theory (Dirac—who won the 1933 Nobel prize in physics for his work in quantum mechanics—did his undergraduate work in electrical engineering and was familiar with and a great admirer of Heaviside's writings). See the paper by J. Hendry, "Monopoles before Dirac" (*Studies in History and Philosophy of Science,* vol. 14, pp. 81–87, March 1983) which was written in reply to an earlier one by H. Kragh, "The concept of the monopole. A historical and analytical case study" (*Studies in History and Philosophy of Science,* vol. 12, pp. 141–172, June 1981). For Dirac the fascination in monopoles came from his proof that the existence of *just one monopole anywhere in the entire universe* would explain why electricity comes in discrete lumps (the electronic charge)—see, for example, J. Schwinger, "A magnetic model of matter," *Science,* vol. 165, pp. 757-761, August 22, 1969.
25. Since $\vec{\nabla}$ is a vector operator (see Note 20), we can calculate both its dot (scalar) product with a field vector (which gives the two divergence expressions for $\vec{E}$ and $\vec{B}$), or its cross (vector) product (which gives the two *curl* equations, Ampere's and Faraday's Laws). The word *curl* is due to Maxwell. The divergence equations describe the space rate-of-change of the field vectors *along* the vector directions, while the curl equations describe the space rate-of-change at *right angles* to the vector directions. This explains what is meant by saying a current *creates* a magnetic field—a nonzero $\vec{J}$ means a nonzero curl of $\vec{B}$ (and hence, by extension, a $\vec{B}$!).
26. Sir Joseph Larmor (Ed.), *Origins of Clerk Maxwell's Electric Ideas,* Cambridge: Cambridge University Press, 1937, pp. 17–19.
27. Oliver Lodge, for example, recalled how at the contentious meeting of the B.A. at Bath in 1888 (in which Preece figured so prominently), with the announcement of Hertz's discovery of the electromagnetic waves as predicted by Maxwell's theory, that "Lord Kelvin who had hitherto been hostile to Maxwell's theory, began to be shaken in his opposition to it, and went about with the second volume [of Maxwell's *Treatise*] under his arm, every now and then appealing to [George Francis] FitzGerald to explain a passage. The theory hardly fitted in with his own conceptions, and I should say that he was never an enthusiastic admirer of Maxwell, who was a younger man than he, and had admittedly assimilated much knowledge from his senior." Quoted from *Advancing Science*, New York, NY: Harcourt, Brace, 1932, p. 98. FitzGerald was a good choice for this role as tutor, as three years earlier he had written a critical assessment entitled "Sir Wm. Thomson and Maxwell's electro-magnetic theory of light," *Nature,* vol. 32, pp. 4–5, May 7, 1885.
28. D. F. Moyer, "Continuum mechanics and field theory: Thomson and Maxwell," *Studies in History and Philosophy of Science,* vol. 9, pp. 35–50, March 1978.
29. L. P. Williams, *The Origins of Field Theory,* New York, NY: Random House, 1966, p. 122. Three excellent essays on this issue are J. Turner, "A note on Maxwell's interpretation of some attempts at dynamical explanation," *Annals of Science,* vol. 11, pp. 238–245, 1956; M. J. Klein, "Mechanical explanation at the end of the nineteenth century," *Centaurus,* vol. 17, pp. 58–82, 1972; and D. R. Topper, "Commitment to mechanism: J. J. Thomson, the early years," *Archive for History of Exact Sciences,* vol. 7, pp. 393–410, 1971.
30. R. Kargon, "Model and analogy in Victorian science: Maxwell's critique of the French scientists," *Journal of the History of Ideas,* vol. 30, pp. 423–426, July–September 1969.
31. J. Larmor (Ed.), *Memoir and Scientific Correspondence by the Late Sir George Gabriel Stokes,* vol. 2, Cambridge: Cambridge University Press, 1907, pp. 25–26. Heaviside's words can be found in EP 1, p. 423, and in AN, February 28, 1895, p. 413.
32. J. Bromberg, "Maxwell's displacement current and his theory of light," *Archive for History of Exact Sciences,* vol. 4, pp. 218–234, 1967; A. M. Bork, "Maxwell, displacement current, and symmetry," *American Journal of Physics,* vol. 31, pp. 854–859, November 1963; and D. M. Siegal, "Completeness as a goal in Maxwell's electromagnetic theory," *Isis,* vol. 66, pp. 361–368, September 1975.
33. EMT 1, p. 29.
34. A. F. Chalmers, "Maxwell and the displacement current," *Physics Education,* vol. 10, pp. 45–49, January 1975.

35. A. E. Woodruff, "Action at a distance in nineteenth century electrodynamics," *Isis,* vol. 53, pp. 439–459, 1962, and his later paper, "The contributions of Hermann von Helmholtz to electrodynamics," *Isis,* vol. 59, pp. 300–311, 1968.
36. E. J. Berg, "Oliver Heaviside, a sketch of his work and some reminiscences of his later years," *Journal of the Maryland Academy of Sciences,* vol. 1, pp. 105–114, 1930.
37. Quoted from J. Bromberg, "Maxwell's electrostatics," *American Journal of Physics,* vol. 36, pp. 142–151, February 1968. An analysis of Maxwell's models for electricity, and their sometimes perplexing aspects, is in P. M. Heimann, "Maxwell, Hertz, and the nature of electricity," *Isis,* vol. 62, pp. 149–157, 1971.
38. "The relation between electricity and light," *Nature,* vol. 23, pp. 302–304, January 27, 1881.
39. "The nature of electricity," *Nature,* vol. 21, pp. 334–338, February 5, 1880.
40. *The Electrician,* vol. 45, pp. 18–20, April 27, 1900. Historians of science rightfully scorn those who laugh at the ignorance of the past, but Preece's comments were seen to be absurd *by those present at the scene of the crime,* and the April 27 *Electrician* editorial page was devoted to poking fun at Preece's statements.
41. "An explanation of gravitation," *Electrical World and Engineer,* vol. 36, pp. 478–479, September 29, 1900.
42. C. W. Vincent, "The dangers of electric lighting," *The Nineteenth Century,* vol. 27, pp. 145–149, January 1890.
43. "On compressional electric or magnetic waves," *The Electrician,* vol. 40, pp. 93-96, November 12, 1897; reprinted in EMT 2, pp. 493–506. The opening line is his response to the x-ray debate (see Note 9): "There are no 'longitudinal' waves in Maxwell's theory analogous to sound waves."
44. H. G. J. Aitken, *Syntony and Spark: The Origins of Radio,* Princeton, NJ: Princeton University Press, 1985, p. 21.
45. A. Koslow (Ed.), *The Changeless Order,* New York, NY: George Braziller, 1967, pp. 245–246.
46. *James Clerk Maxwell, a Commemoration Volume 1831–1931,* New York, NY: Macmillan, 1931, pp. 67–73. A quite interesting analysis of the evolution of the concept of "electromagnetic field" from Faraday to Einstein, in great detail, is N. J. Nersessian, "Aether/or: The creation of scientific concepts," *Studies in History and Philosophy of Science,* vol. 15, pp. 175–212, September 1984.
47. A. M. Bork, "Maxwell and the vector potential," *Isis,* vol. 58, pp. 210–222, 1967, and P. F. Cranefield, "Clerk Maxwell's corrections to the page proofs of 'A dynamical theory of the electromagnetic field'," *Annals of Science,* vol. 10, pp. 359–362, 1954.
48. Light seems such a "flimsy" thing it is probably difficult to imagine it pushing on anything, or at any rate pushing very hard. And yet, the radiation force on the Earth due to the sun is on the order of an impressive 100,000 *tons*! This is equivalent, for example, to a fully loaded aircraft carrier.
49. J. C. Maxwell, *A Treatise on Electricity and Magnetism,* vol. 2, New York, NY: Dover, 1954, p. 187 and p. 232.
50. For example, Feynman wrote: "In the general theory of quantum electrodynamics, one takes the vector and scalar potentials as the fundamental quantities. $\vec{E}$ and $\vec{B}$ are slowly disappearing from the modern expression of physical laws..."—see *Lectures on Physics,* vol. 2, Reading, MA: Addison-Wesley, 1964, p. 15-14. Heaviside strongly rejected $\vec{A}$, writing in an 1888 letter to Oliver Lodge of his desire to "murder" Maxwell's "monster"—see B. J. Hunt, *The Maxwellians* (Ph.D. dissertation), Baltimore, MD: The Johns Hopkins University, 1984, pp. 116–117.
51. This is from an unsigned book review of Maxwell's *Treatise* in *Nature* (vol. 7, pp. 478–480, April 24, 1873), but a few months after Maxwell's death Tait revealed his authorship in another *Nature* essay reviewing his friend's lifework (vol. 21, pp. 317–321, February 5, 1880). Tait ended this review with a moving expression of his emotional anguish: "I cannot adequately express in words the extent of the loss which his early death has inflicted not merely on his personal friends, on the University of Cambridge, on the whole scientific world, but also, and most especially, on the cause of common sense, of true science, and of religion itself, in these days of much vain-babbling, pseudo-science, and materialism. But men of his stamp never live in vain; and in one sense at least they cannot die. The spirit of Clerk-Maxwell still lives with us in his imperishable writings, and will speak to the next generation by the lips of those who have caught inspiration from his teachings and example."

7
Heaviside's Electrodynamics

It will be understood that I preach the gospel according to my interpretation of Maxwell, ...

— Oliver Heaviside, EP 2, 1888

I see that Hertz is not a Maxwellian though he is learning to be one. By a Maxwellian I mean one who follows Maxwell as interpreted by O.H.

— Oliver Heaviside in an 1889 letter to G. F. FitzGerald

Now, there are spots on the sun, and I see no good reason why the many faults in Maxwell's treatise should be ignored.

— Oliver Heaviside, EMT 1, 1891

It is, perhaps, not too much to say that the author has a clearer grasp and insight into electromagnetic theory than any living writer.

— From a review of EMT 1 in *The Electrical Engineer* (New York), 1894

The ability to follow Mr. Oliver Heaviside in his solitary voyages "on strange seas of thought" is given to few. Most of us do get but glimpses of him when he comes into some port of common understanding for such fresh practical provisions as are necessary for the prosecution of further theoretical investigation. These obtained, he steams fast to sea again. Some of us in our puny way paddle furiously after him for a little distance, but we are rapidly left astern, and, exhausted, laboriously find our way back to the land through the fog created of our own efforts.

— *The Electrician,* 1903

Now all has been blended into one theory, the main equations of which can be written on a page of a pocket notebook. That we have got so far is due in the first place to Maxwell, and next to him to Heaviside and Hertz.

— H. A. Lorentz, in a lecture at the California Institute of Technology, 1922

In 1891, Heaviside summed up his work on Maxwell's theory in a single paper printed by the Royal Society in 1892. This was the most important and the most ambitious paper Heaviside ever wrote. It is fairly safe to say that no one yet born has been able to understand it completely.

— W. E. Sumpner, in his 1932 Kelvin Lecture, helping to spread the legend of Heaviside

The Conversion of a Skeptic

Heaviside may have felt he was "preaching the gospel," but for some time after Maxwell's death the congregation was very limited in size. Maxwell's theory most definitely did *not* strike the scientific world like a thunderbolt. The German and French scientists on the Continent had their own ideas on how to treat electrodynamics[1] theoretically. William Thomson (as mentioned in the previous chapter) never really quite believed in Maxwell's ideas, and others with little or no mathematical ability (like Preece) could not have understood Maxwell no matter how hard they tried.

But then there also were those who *could* rethink their initial position of skepticism, and who had the analytical and intellectual skills to eventually appreciate the power of Maxwell's creation. An interesting and important example of this evolution can be found in Heaviside's friend at Trinity College in Dublin, Professor George Francis FitzGerald. Writing in November 1879 concerning displacement currents, FitzGerald declared,[2]

> ... however these [displacement currents] may be produced by any system of fixed or movable conductors charged in any way, and discharging themselves amongst one another, they never will be so distributed as to originate wave disturbances propagated through space outside the system.

Two years later, almost to the day, he was still the pessimist, writing[3]

> It may be worth while remarking that no effect except light has ever yet been traced to the displacement currents assumed by Maxwell ... and until some such effect of displacement currents is observed, the whole theory of them will be open to question.

But a few months later (May 1882) he had done nearly a complete about-face (in an 1889 letter to Heaviside, FitzGerald recalled his earlier ideas and called them "more or less idiotic"!), now saying[4]

> ... equations founded on Maxwell's theory certainly lead to the conclusion ... energy is gradually transferred to the medium ... .

He even suggested an experimental approach for generating Maxwell's propagated disturbances: " ... discharging condensers through circuits of small resistance" to "obtain sufficiently rapid alternating currents."

The Maxwellian conversion of FitzGerald was complete by 1883 when, in what may be the shortest important technical paper ever printed, he predicted Leyden jar discharges might be used to generate radiation at a wavelength of ten meters, and performed calculations[5] showing the radiation from a small circular antenna increases as the fourth power of the frequency. As Heaviside put it a decade later,[6] after FitzGerald came all the others who were finally convinced by Hertz's experiments, until a veritable "electrical boom" occurred. Heaviside also claimed, however, that while such experimental evidence is important (my own reaction is that it is *crucial*!), "other evidence was convincing to a logical mind," meaning the persuasiveness of the underlying mathematical beauty of Maxwell's theory was sufficient to convince Oliver. As he wrote[7] in 1891,

> ... to one who had carefully examined the nature of Maxwell's theory, and looked into its consequences, and seen how *rationally* [Heaviside's emphasis] most of the phenomena of electromagnetism were explained by it, and how it furnished the only approximately satisfactory (paper) theory of light known; to such a one Hertz's demonstration came as a matter of course—only it came rather unexpectedly.

and at about the same time,[8]

> Maxwell's theory is no longer entirely a paper theory, bristling with unproved possibilities. The reality of electromagnetic waves has been thoroughly demonstrated by experiments ... .

Even earlier, in a letter to Hertz (July 13, 1889), he wrote,

> I look upon Maxwell's theory as so true, that we should not be seeking to verify it, but should turn the tables, and make use of it the other way! I.e. assume its truth, & conclude therefrom, & experiment, the physical constants involved.

But if Heaviside was initially ahead of FitzGerald in seeing the strengths of Maxwell's theory, FitzGerald showed more flexibility than his friend in seeing the flaws. Heaviside was a *Maxwellian* from start to finish. FitzGerald, however, after once twisting around his conceptions of how things work, was able to do it again at the turn of the century with the new concept of the electron, as introduced by Joseph Larmor (who thought of the electron as a sort of singularity in the ether, while the modern idêa of a charged *atomic particle* is due to J. J. Thomson and Hendrik Lorentz (1853–1928)). Heaviside, on the other hand, as we will see, wasn't really able to take this giant step beyond Maxwellian, Victorian ideas. In a letter to Heaviside (December 27, 1893), for example, FitzGerald quoted Oliver's words back to him that atoms are "human inventions." Heaviside would in a certain sense remain intellectually trapped in the 19th century.

The Electrician

By a double quirk of timing (the dates of Heaviside's birth and Maxwell's premature death), Heaviside was one of the first to be able to pick up the reins dropped by the Scottish genius. As we've seen, Heaviside had already by 1879 been publishing for some time, and in 1882 he began a long-term, amazingly fruitful association with *The Electrician.*

Heaviside actually first appeared on the pages of *The Electrician* in 1878 soon after the journal had resumed publishing after a false first start.[9] This initial paper was a reprint of one originally published in the *Philosophical Magazine.* Over the next twenty years he appeared many more times with original work. Many other men of great importance in developing electrical theory and practice appeared on these same pages, too, including Lodge, FitzGerald, S. P. Thompson, and once even the great Maxwell himself.[10]

The Electrician was a weekly trade journal, not a purely scientific publication like *Nature* or the *Philosophical Magazine.* Its primary readership included electrical manufacturers, working electricians, and interested lay-persons (and the occasional quack), not professional scientists and academics (although such men flipped through it, too). It carried regular columns on legal activities related to electrical matters (e.g., court actions dealing with the latest accidental electrocutions, bankruptcies, and liquidations) and financial reports on companies doing business of interest to electricians. Also included were reports on the most recent murders of electrical personnel in foreign lands (the latest reports on Mr. Tom London,

The view along Fleet Street toward Ludgate Hill (with St. Paul's dome clearly visible), much as it must have looked outside *The Electrician*'s offices. This area, which was home for 2000 daily, weekly, and monthly publications (including some of the more awful "Penny Dreadful" and "Shilling Shocker" horror magazines which delighted the Victorian middle class), was heavily bombed during the London Blitz and after. (From an 1872 woodcut by Dore.)

telegraph engineer who "met with foul play" in Mombasa, East Africa, appeared like soap opera updates in 1907–8 right up to the hanging of five murderers on the very spot where they had "hacked him to death with knives"), a reader's query section,[11] tutorial essays for "students," and a regular editorial page. These editorials usually addressed practical matters of a timely interest, and not surprisingly reflected the nationalistic and cultural biases of male-dominated Victorian England.

Whatever Oliver Heaviside thought of all this is unknown. What mattered was the journal paid for his writings. This was obviously an important consideration for the unemployed

Heaviside, but some of his friends still wondered at the wisdom of publishing his highly technical papers in such a commercially oriented journal. Oliver Lodge, for example, wrote to Oliver in 1888 that "it was hardly a suitable place for them to appear," and that his (Lodge's) own contributions to *The Electrician* were "mostly waste paper." Heaviside's reasoning[12] about this was quite simple: "I just tried to get as much of my work published as I could." Still, after Heaviside sent FitzGerald a copy of EMT 1, FitzGerald replied (January 4, 1894):

> Thank you very much indeed for the copy of your book you have so kindly sent me. I am very glad to get your papers collected in this way. It is like getting oxygen for combustion: it works much more energetically than when diluted with the nitrogen of the *Electrician* in general.

However, as we'll see, without *The Electrician* there might never have been the book!

In March 1881, John Pender, M.P. and Chairman of the Eastern Telegraph Company which owned *The Electrician,* hosted a social dinner at the popular Criterion restaurant[13] for the staff and contributors. Present at this first of many such dinners to follow were Oliver Lodge, Desmond FitzGerald (one of the original founders of the journal), John Perry, David Hughes, Latimer Clark, and William Preece; all were men who would have been greatly interested in Heaviside's work and might have aided him (even Preece, at this early date), if only they had met him. Oliver, of course, never attended *any* social gathering[14] outside those of his family, much to his own loss. Also dining those long-ago evenings in Piccadilly Circus was a man who would play an enormously important role in Oliver's life, Mr. Charles Henry Walker Biggs, the editor of *The Electrician.* By 1884 Biggs would be reduced to making a futile personal appeal to get Heaviside to show his face. In a letter dated April 22, 1884 he wrote, "Last year you sent us an Excuse—this year you must come and let one see you personally ... . I look upon your presence ... as a great personal favor." Later, however, the journal could only print its regrets at Oliver's absence from that year's dinner, too.

Biggs also played an active role in "The Dynamicables" (originally called "The Electric-Arc Angels"!), a small group of electricians whose major objective was "to meet and wyne." Started in 1883, in addition to Biggs its membership included David Hughes, Latimer Clark, Fleming Jenkin, Preece, and Ambrose Fleming. Its President in 1884 was Sir William Thomson, who remained an active member right up to his death. With membership rules that demanded an annual fee of half a guinea (a not insignificant amount of money in the 1880s) and the pleasure of paying for one's own dinner at the regular meetings, Biggs probably knew it would be hopeless to try to get Heaviside to join. At least there is no record of such an invitation, and Heaviside would surely have ignored it.

The Importance of Mr. Biggs

Usually the role played by a technical editor in the publication of the work of a scientist is a behind-the-scenes one, having meaning only insofar as the act of publication itself is concerned. The editor's contribution, while important, is in general not of historical interest a century after the fact. C. H. W. Biggs, however, was much more than just Heaviside's editor, being also his champion within a generally indifferent, even perhaps hostile, editorial staff. Powerful establishment forces would have preferred Heaviside to remain silent (which will be discussed in detail in the next chapter), and without the intervention of Biggs they might have gotten their wish. In the end, in fact, this cost Biggs his job. Just *why* Biggs became Heaviside's supporter in the face of significant opposition is unclear. Certainly he was, on the surface, an unlikely candidate for being able to appreciate the genius underlying the manuscripts arriving at his Fleet

C. H. W. Biggs (?–1923)

Heaviside's editor at *The Electrician* until 1887, he published some of Oliver's most advanced, theoretical work. This is all the more extraordinary as Biggs himself was a self-proclaimed "practical man." As he wrote in an 1889 editorial in another journal, "electrical engineers do not care a dump (use any other expression if this is not suitable) about theories: they want something that will assist them to gain bread and cheese."

Street office from an obscure, unseen (there is strong reason to believe Heaviside never met anybody connected with *The Electrician*) correspondent living in a seedy part of London.

Prior to joining *The Electrician* in 1878 Biggs had worked as a tutor and had authored a children's grammar book, but had no involvement with electrical matters. He probably was hired mostly for writing skills and an administrative ability to help get the journal produced each week on time. Years later he would write and edit several texts on electricity (none of which met with much acclaim), probably based on what he learned of the subject during his tenure as editor. We can get an idea of what was needed to work for *The Electrician* from a reminiscence[15] written by W. G. Bond, who was editor from 1895 to 1897, and who had joined the journal the year after Biggs was fired. Bond was hired by the editor who had replaced Biggs, W. H. Snell (and who had an important negative role in Heaviside's life, as we'll see in the next chapter):

> I drifted on to the editorial staff of *The Electrician,* so far as I can remember, in 1888, and left the paper in the autumn of 1897. The mode of my introduction was in itself a tribute to the value of the paper as an advertising medium, inasmuch as I found my way to the editorial heart by reason of an advertisement of my manifold qualifications which I had inserted in the space sacred to out-of-works. The bait that caught the Editor was the unusual combination of

> a modest knowledge of science, a decided knowledge of French and German, and the ability to write phonetic shorthand at the rate of 90 words a minute—for one minute.

Bond also gives us a glimpse of working conditions at the journal when, at his promotion to editor, he "was induced into the editorial chair in that noisy, evil-smelling back room." Even a casual reading of *The Electrician* shows it was a well-prepared journal, one that received a lot of care in its production. Heaviside, himself, appreciated this. There is a note in the unpublished memoirs (in the Archives of the Institution of Electrical Engineers) of A. P. Trotter, who was editor in the early 1890s, of a letter from Heaviside declaring, "Compositors are very intelligent, read mathematics like winking and carry out all instructions by author."

The precise manner in which Heaviside first made contact with *The Electrician* is not known, but one story[16] circulated at the time of Heaviside's death is almost certainly not true:

> ... many years ago [he] submitted by mail to Sir John Pender, the editor of the London *Electrician,* a communication which Pender at once recognized as of very unusual value. In writing to the London paper Heaviside directed that any remuneration should be sent to a certain grocery which he designated. Apparently at that time he was living in extreme poverty. He continued to send in articles to the *Electrician* and to receive remuneration in the same manner, but persistently refused to answer letters or to reveal his abode.

This passage makes a good tale, but it strikes many false chords. Heaviside never made a secret of his address, signing St. Augustine's Road to several early letters in *The Electrician*; Pender was a highly successful financier at home with budgets and investments, and to imagine him sitting in an editor's chair reading a Heaviside manuscript while exclaiming at its brilliance is ludicrous—in any case, Biggs had been editor since the first issue; and for Heaviside not to have answered a letter would have been like a bird refusing to fly.

What *is* known is that on September 1, 1882 Biggs wrote to Heaviside

> I should be greatly obliged if you would let me have a paper or two for *The Electrician.* Pray choose your own subject. The sooner I can have them the greater will be my obligation to you.

It is interesting to speculate on how Biggs became aware of Heaviside—perhaps it was through conversation with Preece during the first Criterion dinner. If this is so, then Preece would eventually come to rue the day he put Heaviside's name in the editor's ear! In any case the request worked, and on December 5, 1882 Biggs wrote again to say

> Many thanks for your recent communications which are just what I want ... . So long as you remain good to write, so long shall I be pleased to receive and insert your MS ... . I may say that I hope this will be a period of infinitely long duration.

It wouldn't last that long, but over the next five years Biggs found room enough for Heaviside's writings, sufficient to fill over 500 pages when reprinted ten years later as part of the two-volume *Electrical Papers.*[17] Biggs wrote his letter just three weeks after the appearance of Heaviside's tutorial essay "The Earth as a return conductor." The opening words of this article[18] had given Biggs fair warning as to Heaviside's style:

> The daily newspapers, as is well known, usually contain in the autumn time paragraphs and leaders upon marvelous subjects which at other times make way for more pressing matter. The sea-serpent is one of these subjects.

Of course, it may well have been precisely Heaviside's sense of humor that brought him to Biggs' attention; it is easy to believe he found this an attractive aspect of his otherwise decidedly odd contributor (certainly no one else writing for *The Electrician* managed to work sea-serpents into their essays!). Biggs, the former grammar book author turned editor, must have particularly enjoyed Heaviside's linguistic contributions to electrical science (*inductance* and *reactance* are Heaviside creations, for example) a few years later (1885, 1886, and 1887) in "Notes on nomenclature," which appeared in five parts. There he discussed[19] the pros and cons of various terms such as resistivity, conductivity, and permeability, and then concluded with "All these things will get right in time, perhaps. Ideas are of primary importance, scientifically. Next, suitable language." Perhaps sensing he was becoming too serious, Heaviside's next words on the subject were really written tongue in cheek:

> In the beginning was the word. The importance of nomenclature was recognized in the earliest times. One of the first duties that devolved upon Adam on his installation as gardener and keeper of the zoological collection was the naming of the beasts.
>
> The history of the race is repeated in that of the individual. This grand modern generalisation explains in the most scientific manner that fondness for calling names displayed by little children.
>
> Passing over the patriarchal period, the fall of the Tower of Babel and its important effects on nomenclature, the Egyptian sojourn, the wanderings in the desert, the times of the Kings of the Babylonian captivity, of the minor prophets, of early Christianity, of those dreadful middle ages of monkish learning and ignorance, when evolution worked backwards, and of the Elizabethan revival, and coming at once to the middle of the 19th century, we find that Mrs. Gamp [see Note 20] was much impressed by the importance of nomenclature. "Give it a name, I beg, Sairey, give it a name!" cried that esteemed lady.

Perhaps Biggs gave Heaviside such a free hand in expressing himself on this issue because he believed, as did Oliver, that the British were the proper choice to perform this important task. Oliver knew his efforts weren't always appreciated (in a letter to Hertz dated April 1, 1889 he admitted he had been called "the interminable terminologist"), but even so felt he had a mission to perform. As Heaviside put it " ... Consider what frightful names might have been given to the electrical units by the Germans."[21]

Getting Off to a Bad Start

During his years with *The Electrician* Heaviside accomplished much of his greatest work. As he himself somewhat overstated[22] things, "I wish to say that practically nearly all my original work was done before 1887, and is contained in my *Electrical Papers.*" He published in other places, too, such as the *Philosophical Magazine* and (after his election as a Fellow of the Royal Society in 1891) the *Proceedings* and the *Philosophical Transactions* of the Royal Society, but it is in *The Electrician* that we find the reformulation of Maxwell's theory into modern form, as well as the development of energy flow in the electromagnetic field; the skin effect for alternating currents in conductors; analytic speculation on charged particles traveling faster than light; and the discovery of the theory of distortionless signal transmission, along with the articulation of the concept of inductively loaded circuits.

All the while these outstanding technical accomplishments flowed from his pen, each *individually* being enough to mark his as a superior intellect, he lived with his parents in their London home on St. Augustine's Road. He was not a hermit in the sense of remaining locked in his room to be seen only rarely through a window by curious neighbors. Oliver, in fact, was a physically active man who at one time was a gymnast who took an interest in the state of his body.[23] He would regularly emerge from his house to walk the streets of London to observe the people around him (and occasionally to insert his reaction to what he saw in letters to *The Electrician*); and later, at the turn of the century, he even became caught up with the multitudes (as was Sir William Preece) in the bicycling craze that swept over England.

But despite all that he did effectively cut himself off, in a social sense, from all technical men. They knew him only through his published articles and a willingness to correspond by mail. His brother, Arthur, in the 1880s already a well-known and respected electrician in his own right, was Oliver's only direct human contact with the daily activities of the real world of electrical engineering. This isolation led to a certain lack of sophistication on his part, and may help explain some of his more outrageous literary acts. Consider, for example, the following episode, in which he succeeded in offending the religious sensibilities of many, including those of Sir William Thomson. While writing a highly technical, multipart article on "The energy of the electric current," he concluded one installment with the following words (which were later deleted from the first volume of *Electrical Papers*), words[24] that poke fun at those who put their faith in a Supreme Being:

> Before leaving Ohm's law... I cannot help unburdening myself of a conviction that has long been forcing itself on my mind with ever-increasing persistency. Ohm's law, although, singularly enough, its recognition was slow in the first place (for it is said that Ohm, who was a shy, retiring man, received the cold shoulder, and was shunted off into the obscurity of his garret with frozen grate), long ago became of immense practical importance, owing to the commercial value of the applications of electricity. But in the last few years its use has been multiplied enormously, through electric lights and one thing and another. In fact, for once it is used by the telegraph engineer it is used twenty times or more by an electric light man. Conceive, then, if you can, the electric light young man having to consult at every step volumes of tables and intricate formulae to ascertain what current corresponded to a given E.M.F. in a given wire under given circumstances if its resistance were not, as it is, practically constant; if it were to vary, for instance, greatly with how long the current had been on, or with the E.M.F., or in innumerable other ways; for who shall say what might not have been? The electrician would be in a perpetual state of harassment, which would certainly tend to shorten his days though it lengthened his day's work. Now, is not this fitness of things as regards Ohm's law a powerful argument in favor of design? Can it be doubted that the constancy of R was providentially arranged to meet the requirements of man? I think this can only be contested by those wicked writers of dreary articles in monthly magazines whose pernicious stuff appears to be deliberately intended to undermine all our inherited faiths, and who seem determined to leave no place in the world for its own maker. At any rate, this argument in favor of a providence might be added (to keep it up to the times) as an appendix to the next edition of *Paley's Evidences* [see Note 25], a work whose perusal I can cordially recommend to all who, having had their peace of mind destroyed by the scoffers, have a difficulty in getting to sleep at night. As for myself, I may add that such was its soothing and calming effect on the nerves that, like Ingoldsby's [see Note 26] shirt that never was new, it was often begun but never got through.

If this weren't bad enough, the very next week Professor George Minchin of the Royal Indian Engineering College at Cooper's Hill (a personal friend of FitzGerald's) wrote in response and thereby gave Heaviside an opportunity to continue to offend. Minchin's letter[27] first addressed one of Heaviside's technical points on current energy, and then turned to the theological issue. Minchin was a great admirer of Heaviside, and obviously understood he meant nothing very serious with his spoofing. Minchin's words are in the same irreverent spirit:

> Mr. Heaviside then goes into matters which, though not deducible from the equations of the subject, possess a strong interest, at any rate for me. The design argument can be vastly extended and illustrated in even pure mathematics just as Mr. Heaviside (to whom, therefore, the theologians are deeply indebted) has shown that it can be in the consideration of Ohm's law, and the fearful consequences which would result from its invalidity. Mr. Heaviside ... says, "Conceive, then, if you can the electric light young man." I can conceive

> him, but I prefer at present to refer to an archbishop, who two or three years ago delivered to his clergy a charge which was meant to be a supreme criticism of science based on the most authoritative knowledge, to assure his clergy and the public in general that the heads of the church kept pace with the foremost scientific ideas. There was, however, a poverty of illustration, at least in the electrical department of the charge; for the archbishop, mounting to the summit of eloquence and scientific knowledge, exclaimed, "I may admire your electroscope as a scientific wonder, but beyond it there is something." If bishops and archbishops take Mr. Heaviside's hint they can compose magnificent charges with very little trouble by taking up a few of the less known laws and facts of science—say conical refraction and law of Lenz—in the way pointed out by Mr. Heaviside. Electroscopes are, no doubt, mighty instruments in the hands of science, but they are not now sufficiently impressive; for it is possible that even some school girls may be present who will recognise the name when it is trotted out for effect.

With this as an opening, Heaviside needed no more encouragement to go finally too far. After getting the technical issues out of the way, he then proceeded with[28]

> As to Professor Minchin's remarks on the archbishop, I must say I do not think them quite respectful, I really do not. Archbishops are privileged. They have reached the highest summit of respectability in an ultra-respectable country, to say nothing of their powers of intercession to turn a drought into damp. I should hardly dare to criticise an archbishop, not if he fired off a whole volley of electroscopes and Leyden jars in defense of the Faith. I had the great pleasure of *seeing* a live archbishop this summer, the new one, entering Westminster Hall, no doubt on his way to his parliamentary duties, perhaps to keep men from marrying their deceased wives' sisters, and a nicer looking man for an archbishop I could not conceive. He wore the most *delightful* pair of trousers, buttoned tight all the way up the calves, and my respect for the cloth, always great, was much enhanced thereby. His appearance at the door and dignified entrance will not be soon forgotten.

These sophomoric remarks resulted in no replies to *The Electrician*, or at least none that were printed. It is curious that Biggs printed Heaviside's strange comments; perhaps Biggs failed to serve Heaviside as an *editor* in this instance. Later, as we'll see, Biggs was more cautious about printing anything and everything Oliver sent him. Heaviside did pay a price for his lapse into juvenility. It called his judgment into question, and many never forgot it. There is an entry in one of Heaviside's research notebooks,[29] made years after the fact, that shows how dear was that price (and how little he had learned since):

> Heard from O.J.L. [Oliver Joseph Lodge] in 1889 that Sir W.T. [William Thomson] was very much disgusted at my remarks about Archbishop. Really, however, there was nothing to be disgusted about. It was simply a bit of fun.

At least it is gratifying to know that Heaviside understood and appreciated what Biggs had done for him. As he wrote in an April 1892 letter to Oliver Lodge,[30] "This gentleman was entirely free from that degrading superstition which is prevalent, that an editor should not admit into his journal anything that he cannot understand himself." And of course most of what Heaviside published was not about archbishops but of electrical science, and in sum Biggs surely did have much to be proud of when he looked back over his years as editor. As Heaviside implied, it is unlikely Biggs really knew quite what Oliver was doing, but to his supreme credit, publish it he did.

Reformulating Maxwell's Equations

When Maxwell died in 1879 he left his theory not as we discussed at the end of the last chapter (with the two concise curl and two divergence equations), but rather as *twenty equations in twenty variables*! Professor A. M. Bork (now at the University of California, Irvine) performed a real service for students of the history of science when he prepared[31] three charts showing how these equations evolved in Maxwell's writings, and how they correspond with the modern formulation given in the last chapter. These original equations bear little resemblance to those of Chapter 6, which use modern vector-operator notation. Maxwell wrote his equations out in cartesian component form, and in his *Treatise* he also mixed in quaternionic concepts (about which Heaviside had much to say, none good, and which we'll take up in Chapter 9).

Maxwell's original equations display no obvious symmetry in their form, an idea that has come to play an important role in all modern physical theories. To Maxwell, the magnetic vector potential, not the fields, played the central role in electrodynamics (an idea enjoying a resurgence these days). Both of these ideas were rejected by Heaviside (and Hertz). Heaviside could be particularly stirred up by the potentials, writing,[32]

> ... a function called the vector potential of the current, and another potential, the electrostatic, [work] together not altogether in the most harmoniously intelligible manner—in plain English, muddling one another. It is, I believe, a fact which has been recognized that not even Maxwell himself quite understood how they operated ... .

Heaviside considered the *fields*, $\vec{E}$ and $\vec{B}$, as having primacy as they represent the propagated *physical state* of an electromagnetic wave, while the potentials were not (in his opinion) physical but, rather, *metaphysical*.[33] The electrostatic potential was called[34] a "*physical* inanity."

From a mathematical point of view, Heaviside strongly advocated the use of vector methods[35] in all physical investigations (this will be treated at length in Chapter 9); and since a vector field is completely specified by knowledge of its curl and its divergence (if it vanishes sufficiently fast—at least as fast as $1/r^2$—at infinity, a condition that is always satisfied in "real-life"), to Heaviside it was clear that one needed at most just *four* equations (not twenty) to describe the two fields. For him the potentials added nothing. As he wrote,[36]

> ... the two potentials [vector and electrostatic], if given everywhere, are *not sufficient* to specify the state of the electromagnetic field. Try it; and fail.

He summarized[37] his views of Maxwell's potentials in terms that allow no misinterpretation, saying it "best to murder the whole lot," a comment one reviewer of *Electrical Papers* thought[38] "boisterous"!

The reduction of Maxwell's twenty equations to the four of Chapter 6 (with the banishment of the potentials) was done by Heaviside and, independently, by Hertz in Germany. We remember Hertz today mostly for his experimental discovery of Maxwell's predicted waves, but he was a gifted theoretician, too. The two men arrived at the same end point by two entirely different paths. For Heaviside, Maxwell's fundamental ideas of *fields* (as opposed to action-at-a-distance) were just fine, but their expression in a symmetrical form (minus the potentials), and in vector notation, was of primary importance for Oliver. How he did this is sketched in Tech Note 1 at the end of this chapter; the end result was the compact equation-pair

$$\vec{J} = \vec{\nabla} \times \vec{H}$$

$$\vec{M} = -\vec{\nabla} \times \vec{E}$$

which relate two vector fields (the electric $\vec{E}$ and the magnetic $\vec{H}$, proportional to $\vec{B}$) to two vector current densities, $\vec{J}$ and $\vec{M}$, where $\vec{J}$ and $\vec{M}$ are, in turn, also interconnected. The $\vec{M}$

contains a special term due to Heaviside, $\vec{m}_c$, that represents the flow of free *magnetic* particles (then, as now a hundred years later, still not observed). These two equations are Heaviside's two "circuital laws,"[39] which he called the *duplex* form of Maxwell's theory. They are an enormous compression of Maxwell's twenty equations! Heaviside included the $\vec{m}_c$ term to make the equations symmetrical, and carried it along in all his calculations. Only at the end of each analysis, in deference to the *experimental* fact that there are no (as yet) observed free magnetic particles, would he set $\vec{m}_c = 0$.

Heaviside considered this formulation one of his fundamental contributions, and emphatically claimed credit for it. Writing in the Preface of EMT 1 he outlined the book's presentation of "the fundamentals of electromagnetic theory from the Faraday–Maxwell point of view," and then stated,

> It is also done in the duplex form I introduced [Note 40] in 1885, whereby the electric and magnetic sides of electromagnetism are symmetrically exhibited and connected, whilst the "forces" and "fluxes" are the objects of immediate attention, instead of the potential functions, which are such powerful aids to obscuring and complicating the subject, and hiding from view useful and sometimes important relations.

The reactions to Heaviside's reformulation were all across-the-board. The *Glasgow Herald*[35] said of it "Here the author introduces a new method of treating the subject, which he terms the duplex method, but which might more appropriately be termed the 'heavy method'." Men like FitzGerald and Minchin, however, appreciated what Heaviside had done and said so in reviews of his books. Writing of *Electrical Papers,* FitzGerald said,[41]

> Maxwell, like every other pioneer who does not live to explore the country he opened out, had not had time to investigate the most direct means of access to country nor the most systematic way of exploring it. This has been reserved for Oliver Heaviside to do. Maxwell's treatise is cumbered with the *debris* of his brilliant lines of assault, of his entrenched camps, of his battles. Oliver Heaviside has cleared these away, has opened up a direct route, has made a broad road, and has explored a considerable trace of country. The maze of symbols, electric and magnetic potential, vector potential, electric force, current, displacement, magnetic force and induction, have been practically reduced to two, electric and magnetic force. Other quantities it may be convenient, for sake of calculation, to introduce, but they tell us but little of the mechanism of electromagnetism. The duality of electricity and magnetism was an old and familiar fact. The inverse square law applied to both, every problem in one hand its counterpart in the other. Oliver Heaviside has extended this to the whole of electromagnetism. *By the assumption of the possibility of magnetic conduction he has made all the equations symmetrical. Every mathematician can appreciate the value and beauty of this* [my emphasis].

Minchin, writing[42] on the first volume of *Electromagnetic Theory,* simply said "Mr. Heaviside is the Walt Whitman of English Physics," and James Swinburne elaborated[43] with, "The style is that of Whitman, except that Mr. Heaviside is not affected, and has something to say."

A Friend in Germany

At essentially the same time as Heaviside's ideas were developing, Heinrich Hertz in Germany was blazing a similar path. True to the Continental tradition of action-at-a-distance, Hertz in 1884 (three years before his experiments of 1887–88 confirming Maxwell's prediction of propagating waves) arrived at a reformation of Maxwell's equations by applying an infinite

Heinrich Rudolf Hertz (1857–1894)

His discovery of electromagnetic energy propagating through space showed the truth of Maxwell's theory. In a letter to Hertz (July 13, 1889) Heaviside wrote of the pre-Maxwell theories: "I recognized that those theories were nowhere, in the presence of Maxwell's, and that he was a heaven-born genius. But so long as a strict experimental proof was wanting [of Maxwell's theory] so long would these absurd speculations continue to flourish. You have given them a death blow."

sequence of corrective terms to a short-range (or local) action-at-a-distance theory. The details[44] of how Hertz did this are not pertinent to Heaviside's story (or even to the mainstream development of electromagnetic theory). What *is* important, however, is that his final results were those of Heaviside, including the dismissal of the potentials.

Heaviside and Hertz were effectively in a dead-heat race but Hertz, a gracious man, wrote,[45]

> ... I have been led to endeavor for some time past to sift Maxwell's formulae and to separate their essential significance from the particular form in which they first happened to appear. The results at which I have arrived are set forth in the present paper. Mr. Oliver Heaviside has been working in the same direction ever since 1885. From Maxwell's equations he removes the same symbols [the potentials] as myself; and the simplest form which these equations thereby attain is essentially the same as that which I arrive. In this respect, then, Mr. Heaviside has the priority ... .

For some years the reformulated equations were called the "Hertz–Heaviside equations." Later the young Einstein referred to them as the "Maxwell–Hertz equations." Today, of course, they are just ''Maxwell's equations,'' but to quote two mathematicians who carefully studied this historical evolution:

> The great service which Heaviside now rendered to science was ... to base the theory on what he called the ''duplex'' equations ... which modern writers generally call ''Maxwell's equations''—though they are not found in Maxwell's *Treatise,* and the modern writers have in fact copied them from Heaviside. [Note 46]

and

> Heaviside is recognized by physicists as one of the first to use Maxwell's electromagnetic theory successfully on numerous individual problems; among others, the differential equations (named for Maxwell) of the electromagnetic field, which he himself had only described in words, are found explicitly written down for the first time in Heaviside's writings. [Note 47]

Heaviside and Hertz became good friends through an extensive correspondence, but never met face-to-face. That they didn't was Heaviside's choice. When Hertz was awarded the Royal Society's Rumford Medal in 1890 and traveled to London for the ceremonies (he died,[48] in great agony, just three years later in January 1894, a few weeks short of his 37th birthday, from jaw cancer complicated by blood poisoning from surgery) he met such men as David Hughes, William Thomson, Stokes, and Rayleigh. One evening he had a private dinner with Lodge and FitzGerald at the Langham Hotel. One would think Heaviside would have walked miles to have joined these three men who so shared his interests and appreciated his talents (and there can be no doubt that both Lodge and FitzGerald would have been almost speechless with astonishment and pleasure to have had him present), but absent he was.

But this odd shyness of Oliver, this lack of willingness to meet socially with men he knew to be sympathetic in their views (and under circumstances that were literally guaranteed to be free of any possible unpleasantness), was nothing new for him. Just the year before Hertz's visit, for example, in January 1889, he had received a letter from the eminent electrical engineer Silvanus P. Thompson, inviting him to dinner. It was a warm and gracious invitation, one that today (at least to my perhaps overly romantic imagination) conjures up an image of the exciting Victorian conversation about a mystery of Nature among Wells' Time Traveler and *his* dinner guests:

> ... I know that you live a very quiet life. Do you mind for once a little scientific excitement? On March 7th Oliver Lodge will be staying with me—we live very quietly, and my household is a quiet one—would you join us that evening at dinner, and we would have a three-corner chat afterwards on electric matters.

How could Heaviside turn this down? At first he didn't, replying that it was too far in the future for him to accept, that he'd think it over and let Thompson know. As of March 5, however, Thompson still didn't know if Heaviside was coming to dinner. He wrote again, this time adding the enticement that Professor John Perry would also be there, and again promised ''We shall have plenty of time during the evening for a lively chat on electrical matters.''

But skip it he did, missing what surely would have been a most enjoyable and important interaction with men who could have done so much (they did a lot, anyway) to aid his cause. A 1907 note from Thompson to Heaviside, asking for permission to visit and ''to make your personal acquaintance'' makes it clear that Thompson, Lodge, and Perry dined alone that evening in 1889. Meanwhile, in another part of London in the house on St. Augustine's Road, the strange Heaviside perhaps sat in his pipe-smoke-filled room, bent over his lonely work through the night while keeping his thoughts and soul to himself.

More Germans: Föppl, Boltzmann, and Planck

The blurring of Heaviside's and Hertz's common theoretical work took place fairly quickly, and on June 30, 1897 August Föppl, a professor of engineering mechanics at the Technical University in Munich, wrote to Heaviside in an attempt to clarify the issue of priority with respect to the duplex equations. Föppl, in 1894, had authored a well-received book on Maxwell's theory (there is strong evidence[49] that it was from Föppl's book that Einstein learned his electrodynamics), and along with his letter he enclosed a copy of the book. Föppl thought well of Heaviside's work ("As you will find, I have there frequently made reference to your work, of which I am a true admirer."), and in the preface wrote,

> The works of [Heaviside] have in general influenced my presentation more than those of any other physicist with the obvious exception of Maxwell himself. I consider Heaviside to be the most eminent successor to Maxwell in regard to theoretical developments ... .

Perhaps this praise played a part in an episode that got Oliver entangled in a debate with the tragic German genius, Ludwig Boltzmann (he committed suicide in 1906). Heaviside's reply has been lost, but we can get an idea of what he said from a second letter Föppl wrote (July 22, 1897):

> I regret very much that I can in no way agree with your views of the relative merits of Hertz and earlier German writers. Helmholtz—it is true—has had but little success in this branch of science [electrodynamics]; but I think Weber and Kirchhoff did very valuable work. It was in my opinion a necessary step for the progress of science, to show the impossibility of reducing all electromagnetic phenomena to a action-in-distance law and this impossibility could only be recognized after the full development of this theory. As relating to Hertz [his work] comes very near to your duplex system ... . There can be no doubt that you stated this much better; he failed to insist upon the extreme importance of these equations then and he has fairly acknowledged your priority in this respect ... . It was also, as I believe, *not merely a lucky hit* [my emphasis] that he found the waves experimentally ... .

A few months after Föppl's letters Heaviside wrote[50] a book review on Charles Emerson Curry's book, *Theory of Electricity and Magnetism,* which included a preface by Boltzmann.[51] Heaviside was not impressed with Curry's book, which he felt was not faithful to Maxwell but rather was an attempt to retain the old action-at-a-distance theories ("If Maxwell is really taught this way generally on the Continent, I think it is a great pity."). Of Hertz, Heaviside wrote he "became quite Maxwellian after his *great hit* [the substitution of *great* for *lucky* perhaps means Föppl's letter had an effect on Heaviside] ... ." Heaviside then inserted a nice plug for Föppl's book, obviously written after reading the copy Föppl himself had sent him:

> Then there is Dr. A. Föppl, whose excellent *Einführung in die Maxwell'sche Theorie der Electricität* deserves to be read by all who can understand it. It is the least academical of the German works I have seen, and also the clearest and most advanced. Dr. Föppl thoroughly appreciates that the best way [i.e., Heaviside's way!] of exhibiting electric and magnetic relations is not by potentials, but by the electric and magnetic fluxes ... .

In contrast, Heaviside interpreted Curry's book to be advocating Helmholtz's rival electrodynamic theory[52] which predicted longitudinal waves *in addition* to the transverse waves predicted by Maxwell. Many thought these longitudinal waves might be the recently discovered x-rays. Indeed, Boltzmann was among those who so thought (and there is some evidence that Hertz, who was Helmholtz's student, believed this, too). Heaviside took violent exception to the modifications introduced into Maxwell's theory to give these extra waves ("No one has the right to trifle with Maxwell's equations this way.''), called them ''*hocus pocus,* and nothing more,'' and said they ''misrepresented and perverted'' Maxwell. He ended his review with the promise '' ... as the matter cannot be properly treated in this review, I will write a separate article about it.''

Before it could appear Boltzmann replied.[53] It was a cordial reply (''The excellent scientific work by the worthy critic of Dr. Curry's book has always been accorded my highest appreciation, and especially since he has *further developed* Maxwell's theory exactly in the spirit of the originator.''). To show, in fact, how much he appreciated Maxwell, Boltzmann quoted a passage from Goethe's *Faust*: ''He wrapped the power of Nature around me and filled my heart with joy.'' But there was no retreat on technical issues, and Boltzmann wrote, ''Whether the longitudinal oscillations and the other generalizations, which Helmholtz had added to Maxwell's theory, are of great importance or not, is a question that the present stage of science is unable to decide.''

Heaviside disagreed, and his promised ''separate article'' was his reply,[54] beginning with the flat rejoinder, ''There are no 'longitudinal' waves in Maxwell's theory ... ,'' and later adding that such waves ''spoil his [Maxwell's] work.''

This ended Heaviside's exchange with Boltzmann,[55] but he quickly found another soon to be famous German to shoot at—Max Planck. Time has shown that Heaviside's electrodynamic ideas have prevailed over those of Boltzmann, but when he stepped outside his area of special expertise things could turn against him. This happened when he became peripherally involved in a nasty squabble (on the concept of entropy) between James Swinburne and John Perry. At the time (1902) Swinburne was President of the Institution of Electrical Engineers and Perry had just served a term in the same office. These two eminent men, both of whom greatly admired Heaviside, made only token efforts to be civil to one another, with Swinburne (in my opinion) having the better of their exchanges in *The Electrician*, often reducing Perry to words that practically sputter in print.

In his Presidential address Swinburne severely took to task the usual textbook definition of entropy, calling it ''fundamentally wrong,'' claiming authors were pushing the ideas of *reversible* thermodynamic processes far beyond their domain of applicability. This prompted a reply from Perry, who wrote,[56] '' ... When Mr. Swinburne gravely informs an audience that two and three do not make five, but something else, everybody takes him to be joking, but when he tries to upset some other well-established scientific principles ... he may do much harm ... ,'' and later stated that Swinburne ''does not know what he is talking about.'' Things got even worse after this, and another writer who was equally upset by Swinburne's comments on entropy went so far as to suggest[57] ''$\int_0^\infty JS\,dt = 0$ conveys a profound truth''!

In an effort to settle the dispute, the editors of *The Electrician* asked Oliver Lodge to write[58] a tutorial on entropy, but his essay was so clearly an attempt to offend no one that it satisfied no one. Finally, Swinburne himself appealed directly to outside authorities, and *The Electrician* published the comments of Henri Poincaré and Max Planck (who must have embarrassed Lodge when he wrote[59] of his ''astonishment [at seeing] a man so well known and so eminent in science as Sir Oliver Lodge putting forward ideas on thermodynamics which I have combated ever since the commencement of my studies in that science'').

Planck called Swinburne's writing on entropy ''excellent'' and stated, '' ... He has written one of the best and clearest expositions of the subject that has ever been written, especially when he points out that Nature never undertakes any change unless her interests are served by an increase of entropy ... .'' This metaphysical-sounding statement caught the eye of Heaviside, and prompted a letter[60] which tells us, explicitly, something about his thermodynamic ideas, and of energy (which will lead us nicely into the next section):

> I should like to ask Professor Max Planck whether the view he expresses that "Nature never undertakes any change unless her interests are served by an increase of entropy" is to be taken with or without any particular reservation or with any special interpretation of "her interests". My thermodynamical ideas are somewhat old-fashioned—viz., that there is invariably a dissipation of energy or loss of availability of energy due to imperfect or total want of reversibility in natural processes. This entirely agrees in effect with the way of expressing things in terms of increase of "entropy", although that subtle quantity is certainly "ghostly", and is somewhat too evasive to be regarded as a physical state even though it be a function of the physical state referred to a standard state. But the question is how the interests of Nature are served by imperfect reversibility? Professor Planck's words suggest a choice on Nature's part, as if Nature had any choice. Goethe said God Himself could not alter the course of Nature. That was truly scientific. Then, again, what are to be considered the interests of Nature? Are we to take things exactly as we find them, and define the interests in that way? If so, it carries us no further. Or is there a theorem of greatest entropy, showing how any variation from the proper course of Nature would tend to reduce the rate of increases of the entropy?

Planck replied[61] promptly with a sharp rejection of Heaviside. After first dismissing the "ghostly" business ("Whether entropy has any 'ghostly' attributes, is a question I will not open, but I am for the present quite content to know that it is a quantity which can be measured without ambiguity"), he continued with "I do emphatically deny, and always have combated the proposition adduced by Mr. Heaviside, of the universal dissipation of energy." Planck then explained his position, and there is no record of a Heaviside reply.

This Heaviside interest in energy and *changes* in energy was actually not just idle chatter at a philosophical level, of course, but was deeply rooted in his electrodynamics. But for a matter of months, in fact, it would be Heaviside's name, instead of another's, attached to the first important result to come from Maxwell's theory since the prediction of traveling electromagnetic waves—the energy flow theorem of Poynting.

Energy and Its Flux

By 1884 the principle of the conservation of energy was well established, but it hadn't been many years before that the idea of just energy, alone, was a new and strange one. The concept of *force* was the prominent one as late as the 1850s, for example, and it seemed to be intuitively the "thing" that should be the hinge pin of dynamics, whether the system under consideration be mechanical *or* electromagnetic. The development of thermodynamics in the early and mid parts of the 19th century, however, began the process of elevating *energy* and *changes* in energy to the level of importance we attach to them today. Writing[62] in 1887 Heaviside expressed this as, "There are only two things going, Matter and Energy. Nothing else is a thing at all; all the rest are Moonshine, considered as Things."

The ability to *store* what seemed to be astonishing amounts of energy in the newly perfected (1881) lead-acid battery (by Camille Faure) led to a special flurry of interest in the matter, among even the general public. There was something about storing and transporting electrical *energy* (although a battery is really a box of *chemicals*) that was special, and particularly appealing to the Victorian mind. Coal was just a dirty rock out of the ground, while electricity was modern!

So, with all this interest in electrical energy, it is not surprising that people were also paying attention to its more abstract properties such as its conservation and even how it moves about. There is more to the conservation of energy, however, than may be apparent at first glance. As Heaviside put it[63] in 1891,

> The principle of the continuity of energy is a special form of that of its conservation. In the ordinary understanding of the conservation principle it is the integral [total] amount of energy that is conserved, and nothing is said about its distribution or its motion. This involves continuity of existence in time, but not necessarily in space also. *But if we can localize energy definitely in space* [my emphasis—this is a most important idea, one we'll pursue with interest], then we are bound to ask how energy gets from place to place. If it possessed continuity in time only, it might go out of existence at one place and come into existence simultaneously [see Note 64] at another. This is sufficient for its conservation. This view, however, does not recommend itself. The alternative is to assert continuity of existence in space also, and to enunciate the principle thus: When energy goes from place to place, it traverses the intermediate space.

And then a little later in the same passage, writing of the mathematical result that precisely specifies just how electromagnetic energy "traverses the intermediate space," he said,

> This remarkable formula was first discovered and interpreted by Prof. Poynting, and independently by myself a little later. It was this discovery that brought the principle of continuity of energy into prominence.

Heaviside was referring, of course, to John Henry Poynting (1852–1914). Professor of physics at the University of Birmingham, Poynting combined his considerable ability in physics with that of a skilled mathematician (he was 3rd Wrangler in the 1876 *Mathematical Tripos*[65] competition at Cambridge), and this double edge to his powers led to the writing of many papers which Oliver Lodge called[66] "sledge-hammer communications." This certainly was the right way to describe the impact of Poynting's powerful paper "On the transfer of energy in the electromagnetic field," published[67] by the Royal Society in its *Philosophical Transactions* in 1884. Starting with the Maxwellian idea of localized field energy (see Tech Note 2) Poynting was able to derive the elegantly simple expression $\vec{E} \times \vec{H}$, now called the *Poynting vector,* for the *flow* of electromagnetic energy through space (see Tech Note 3).

Poynting's paper, as well as some of the odd implications of the result, attracted a good deal of attention. Oliver Lodge, in particular, was tremendously impressed by it and wrote a curious paper (with a very long title[68]!) in response. Lodge was particularly fascinated by the idea of being able to track an individual "bit of energy," writing " ... The route of the [bit of] energy may be discussed with the same certainty that its existence [is] continuous as would be felt in discussing the route of some lost luggage which has turned up at a distant station in however battered and transformed a condition." This semi-metaphysical paper seems not to have had much impact, but its opening words, describing Poynting's paper, were prophetic, calling it "a paper which cannot but exert a distinct influence on all future writings treating of electric currents."

One of its most profound influences was the complete overthrowing of how people think of energy flowing in a wire carrying an electric current. In fact, according to $\vec{E} \times \vec{H}$, the electromagnetic energy doesn't flow through the wire, but *into* it, *sideways,* from the fields surrounding the wire! This seemingly "crazy" conclusion was not greeted with Lodge's excitement by many of the "old-time" electricians. In 1891, for example, Silvanus Thompson and John Sprague (see Chapter 5) became embroiled in a dispute over the nature of energy flow in electric circuits. Sprague held to the old view of energy transfer through a wire, while Thompson argued for the revolutionary new viewpoint. The debate appeared over an extended period of time in the *Correspondence* section of *The Electrician,* and finally the journal felt it necessary to terminate the issue with an editorial[69]:

> ... although we undoubtedly side with Prof. Thompson's views, there is no doubt much

John Henry Poynting (1852–1914)

A versatile man, physicist Poynting was also a close student of statistical economics. In the same year that his famous paper on the flow of energy in the electromagnetic field appeared, he also published the somewhat lesser known "A comparison of the fluctuations in the price of wheat and in the cotton and silk imports into Great Britain." Neither of these papers was what most Victorians were reading in the summer of 1884, however. Of much more interest were the newspaper stories about a nasty little war in the Egyptian Sudan, and the growing peril faced by General Charles Gordon at Khartoum (where, six months later, he literally lost his head to Dervish warriors).

which appears, at first sight, highly artificial in the elaborate structure of lines of electric and magnetic force and induction, complicated still further, as it is, by the more recently discovered lines of energy-flow ... the idea that energy is located at all, and that, when it changes its position, it must move along a definite path, is quite a new one. The law of the conservation of energy implies that energy cannot disappear from one place without appearing in equal quantity somewhere else; but, although this fact has long been accepted, it is only within the last few years that the idea of transference of energy has been developed, or that anyone has attempted to trace out the actual path along which energy flows when it moves from place to place. The idea of an energy current is of more recent date than the electro-magnetic theory, and is not to be found explicitly stated anywhere in Maxwell's work. We believe that the first time it was applied to electrical theory was in the pages of *The Electrician,* by Mr. Oliver Heaviside, to whom so much of the extension of Maxwell's theory is due. The idea was also independently developed and brought to the notice of the Royal Society in a Paper by Prof. Poynting.

In fact, *The Electrician* was perfectly correct in this proud claim for the priority of Heaviside. Poynting's paper certainly did not appear in print until sometime *after*[67] June 19, 1884 and yet, in the June 21, 1884 issue of *The Electrician* Heaviside wrote[70] (in a passage entitled "Transmission of Energy into a Conducting Core"):

> The direction of maximum transference [of energy] is therefore perpendicular to the plane containing the magnetic force and the current directions, and its amount per second proportional to the product of their strengths and to the sine of the angle between their directions.

These words are not remembered today, and it wasn't until January 10, 1885 that Heaviside published[71] the same result as is found in Poynting's paper (which is why historians today always write of Heaviside's discovery as dating from "the year after" Poynting's). Heaviside took a somewhat different view of history, however, and while he never disputed Poynting's credit, he also took care to remind his readers of the June 21 date, as when he wrote[72] (in March 1885):

> The transfer of energy in a conductor (isotropic) takes place not with the wire, but perpendicular thereto, as I showed in *The Electrician* for June 21, 1884, thus being delivered into a wire from the dielectric outside.

It is not clear when Heaviside first learned of Poynting's paper, but there is an interesting note on one of his copies of *Nature*[73] (dated March 26, 1885) which was prompted by a report on a mechanical model (made of wheels and rubber bands) invented by FitzGerald "illustrating some properties of the ether." In particular, this model showed how "the energy of the medium was conveyed into" a wire and "not along its length in accordance with what Prof. Poynting has recently shown to be the case in *all electric currents* [my emphasis]." Heaviside's note shows he was by then most familiar with Poynting's work, and thought his own more comprehensive:

> But it is only true for conduction currents, not for all currents. Not true in the dielectric [where the displacement current cannot be ignored]. The general formula for energy current ... [was] proved by me for conductors in the summer of 1884, and in January 1885 extended to all media non-homogeneous as regards capacity, conductivity and permeability.

While Poynting may have beaten Heaviside into *recognized* print, and while Poynting's mathematics is impeccable, it is curious to note that his *physics* has a flaw which seems to have gone unnoticed, or at least unremarked upon, for the last one hundred years, except for Heaviside's own comments about it (see Tech Note 4). Even *with* impeccable mathematics, however, many found Poynting's (and Heaviside's) ideas on energy transfer hard to believe, and not all of the skeptics were "old timers" like John Sprague. As Professor J. J. Thomson wrote[74] two years after Poynting's paper,

> This interpretation [the Poynting vector] of the expression for the variation in the energy seems open to question. In the first place it would seem impossible *a priori* to determine the way in which the energy flows from one part of the field to another by merely differentiating [see Tech Note 3] a general expression for the energy in any region with respect to time, without having any knowledge of the mechanism which produces the phenomena which occur in the electromagnetic field ... .

These words show Professor Thomson, the bright young academic star of English physics at the time, was still committed to the Maxwellian goal of mechanical model-building. But eventually even Thomson came around and in 1893 he called[75] Poynting's result "a very important theorem'' and ''of great value.'' There was no mention of Heaviside's contributions to the energy flux theorem by Thomson, and I find this particularly ironic because this slighting by Thomson was to be *his* fate, too, with another equally important result. And to make it doubly ironic, Heaviside was *also* involved in this (Heaviside's role is again forgotten, along with Thomson's).

Moving Charges

We are, today, so used to thinking of electricity as coming in tiny little lumps called *electrons*[76] that it is difficult to realize (if we do at all) that not much more than one hundred years ago there *was no such concept.* To us electrons *are* electricity, and electrons in copper wires are the same as electrons jumping across space as a spark. The internal nature of electric current was a mystery, and nobody thought of it as charged subatomic particles in motion. In the early 1880s, when Heaviside began his electromagnetic studies, nobody had this picture of the electric current in his head.

The idea of electricity as the flow of an incompressible fluid of some sort was still common, and there was no obvious reason to believe such a flow would behave the same inside a conductor as it does outside (e.g., a spark), just as the flow of water *inside* a pipe is very different from what it is after being discharged into (for example) an open sewer. It was the supposed incompressibility of the ''electric fluid'' model, in fact, that lay behind Edwin H. Hall's famous experiment of 1879 which showed that a current-carrying conductor (Hall used a gold leaf) develops a potential *difference across itself, perpendicular* to the current direction, if immersed in a magnetic field. Ironically, it is the Hall Effect that essentially struck the death blow at the fluid model.

Hall (1855–1938) was a student of Henry A. Rowland (1848–1901) at The Johns Hopkins University, and his famous experiment was the culmination of Rowland's longtime thinking about the nature of the electric current.[77] Rowland was trained as a civil engineer, and today is mostly remembered for his construction of astonishingly precise diffraction gratings (up to nearly 15,000 parallel lines *per inch*!), but he also performed a delicate and beautiful experiment that played an important role in subsequent ideas about electric currents.

Maxwell himself had written in the *Treatise* (Article 770) of his ''supposition that a moving electrified body is equivalent to an electric current.'' Indeed, the idea is easily traced all the way back to Faraday, who wrote,[78]

> ... if a ball be electrified positively in the middle of a room and be then moved in any direction, effects will be produced, as if a current in the same direction had existed: or, if the ball be negatively electrified, and then moved, effects as if a current in a direction contrary to that of the motion had been formed, will be produced.

After first visiting with Maxwell in the summer of 1875, Rowland traveled to Berlin where he showed experimentally that moving charge (on a disk spinning at 3660 rpm) does indeed create a magnetic field just like an ''ordinary'' current in a wire. This was followed in 1881 by J. J. Thomson at Cambridge University, who made the first *theoretical* study[79] of the magnetic effect of a moving electric charge. Thomson had followed with great interest recent experiments involving electric discharges in a vacuum, and started his paper with

> ... particles of matter highly charged with electricity and moving with great velocities form a prominent feature in the phenomena ... . It seems therefore to be of some interest ... to take some theory of electrical action [Thomson ''took'' Maxwell's theory] and find what, according to it, is the force existing between two moving electrified bodies, what is the magnetic force produced by such a moving body ... .

Yours very truly

J. J. Thomson.

Joseph John Thomson (1856–1940)

Third Cavendish Professor of Experimental Physics and "discoverer" of the first known subatomic particle (the electron). As with Lodge and Heaviside, Thomson's belief in the ether did not waver with the new century. As late as the 1909 meeting of the B.A. in Winnipeg he declared "the study of this all-pervading substance is perhaps the most fascinating and important duty of the physicists." (From a steel engraving published by *The Electrician* in 1896.)

In this paper Thomson deduced the form of the force experienced by a charge q as it moves with velocity $\vec{v}$ through a magnetic field $\vec{B}$, i.e., $(1/2)q\vec{v} \times \vec{B}$, what we today call the "Lorentz-force."[80] The scale-factor of 1/2 is wrong (this was corrected by Heaviside in 1889, when he showed[81] the factor is unity), but the really exciting outcome of Thomson's analysis was its initial promise of *explaining mass*!

Thomson first argued (in prose, before he did the very extensive math) that a moving sphere with a uniform surface charge would, according to Maxwell, produce a magnetic field which then, as we have seen, would imply *energy* stored in space. This energy must come from somewhere—namely, the mechanical motion of the sphere. This, in turn, means the moving sphere must experience a resistance as it moves through the dielectric medium, as it is through this resistance that the mechanical energy of motion is converted to electromagnetic energy. If, however, the sphere is moving through a nonconducting dielectric medium (e.g., the vacuum of space) then this resistance cannot be a mere frictional drag. Instead, as Thomson wrote, it

> ... must correspond to the resistance theoretically experienced by a solid moving through a perfect fluid. In other words, it must be equivalent to an increase in the mass of the charged moving sphere.

Thomson's analysis for the mass increase is limited to the case of a *slowly* moving charge, and his result is independent of the velocity. He found it to be

$$\frac{4}{15}\mu\frac{q^2}{a}$$

for a sphere of radius a carrying a charge q through a dielectric with magnetic permeability μ.

In 1885 Heaviside replaced[82] the scale-factor 4/15 with 2/3, but again he agreed with the functional form of Thomson's result. Thomson then performed an interesting calculation, in which he took the sphere to be the Earth itself, charged to the maximum possible, i.e., to the point where the electric field at the Earth's surface would be just on the verge of ionizing the atmosphere. His result was an increase of "about 650 tons, a mass which is quite insignificant when compared with the mass of the earth"; perhaps because of this relatively trivial value Thomson did not pursue this matter in his paper.[83] But Heaviside did.

In 1889 he wrote a paper for the *Philosophical Magazine*,[84] "On the electromagnetic effects due to the motion of electrification through a dielectric," in which he corrected calculation errors in Thomson's 1881 paper.[85] Heaviside extended Thomson's analysis of a *slowly* moving charge, in which all results were velocity independent, to that of a *point* charge moving at *any* speed, u, right up to that of the speed of light (which he denoted by v) *and beyond*! His results include velocity-dependent terms, for sub-light speeds of the form $1 - (u/v)^2$, which look remarkably like Einstein's results. But of course Einstein's work was of a vastly more general nature than Heaviside's (Heaviside began with moving charged matter and then applied complicated mathematics and Maxwell's electrodynamics while Einstein used nothing but algebra and the two relativity principles—Einstein saw the whole forest while Heaviside and Thomson were looking through a magnifying glass at the bark on a single tree).

When Heaviside tried to extend his analysis from a point charge to Thomson's spherical surface charge, he stumbled when he assumed the electric field lines would be perpendicular to the sphere's surface. This was a fortuitous error, in one sense, however, as it prompted a letter (dated August 19, 1892) from the young Demonstrator of Experimental Physics in the Cavendish Laboratory at Cambridge, G. F. C. Searle (1864–1954). In this letter Searle (who had been working with Thomson at Cambridge since 1888) wrote, "I quite agree with your solution for the motion of an electrified point, but I do not understand how it will apply when we deal with a *sphere*." Searle went on to show Heaviside that as long as "u^2/v^2 is utterly negligible" Heaviside's assumption is permissible, but if "u^2/v^2 is not neglected" that it is not, and then Searle used some of Heaviside's *own previously published work* to show Heaviside how to do things right![86]

A print from one of several negatives (dated 1893) found in an old cardboard box with a note in Heaviside's hand: "The one with hands in pockets is perhaps the best, though his mother would have preferred a smile."

A Friend at Cambridge

This letter was the start[87] of a personal friendship that would continue until Heaviside's death thirty-three years later. Searle, was, perhaps, the only *true* friend he had other than FitzGerald, one who faithfully wrote, visited (after Searle married in 1904 his wife would accompany him and she and Oliver became quite fond of each other), loaned him money, and most importantly, *told him when he was wrong or behaving in a stupid manner.* They certainly made a curious pair, not ones on the surface likely to strike up a friendship. With one the devout son of a clergyman, born in a vicarage, and the other a skeptic who loved to tease archbishops, they were definitely an "odd couple." But when it came to mathematical ability, Searle was very much in Heaviside's league. While his rank in the 1887 *Mathematical Tripos* was "only" that of 28th Wrangler (his father,[88] the Reverend W. G. Searle, had been a Wrangler, too), he was no mathematical second-rater. Later in life Searle would make profound impressions on his students, with one of them recalling[89] "Searle looked exactly like the professor [the "perfectly impossible" Professor Challenger] in Conan Doyle's *The Lost World.*"

While stern with his students, Searle was not a humorless man, and the nature of his dry wit is, I think, displayed in the following lines from one of his many letters to Heaviside: "I have got a ... student to take up the mathematics for a revolving ring of electrons. He ought to succeed for he once calculated a thing with 20,000 terms in it. He wrote it on a roll of paper." Searle and Heaviside became intimate enough over the years, in fact, to feel comfortable writing of the current states of their health. In another letter, for example, there is a long, funny passage by Searle on his recently having been bothered by "trouble of the large bowel," and that as a palliative he "had 1/2 pint of oil injected and in the afternoon 2-1/2 pints of soap and water." He later closed the letter with "I am afraid this is a rather rambling letter, but the oil rumbles a good deal and works about inside me in a way which is not conducive to well connected literary effort."

Perhaps one reason, at the subconscious level, for Heaviside's particularly friendly and close feelings toward Searle was that the younger man had once actually met Maxwell, Oliver's hero. Even Searle's mentor at the Cavendish, J. J. Thomson, who was at Cambridge before Maxwell died, had never actually met the great man. As a birthday treat (there is some uncertainty as to which one it was, 12th, 13th, or 14th) Searle's father had taken him to the Cavendish Laboratory where he got an hour's tour, with Maxwell as his guide. This made such an impression on the young boy that after his college graduation in 1887 he "instinctively" returned to the Cavendish where, in October 1888, Professor J. J. Thomson gave him a junior post. It was then that Searle became interested in the problem of moving charges, an interest that sparked the fateful 1892 letter to Heaviside. Heaviside learned much from Searle (and Searle would become famous for having been Oliver's friend).

One thing Heaviside learned from Searle was the correct way to go from point charges to

Searle in his laboratory.

extended charges. In Searle's words, those he used when he wrote it all up for publication[90]:

> ... the surface which gives rise to a field the same as that due to a point-charge is an ellipsoid of revolution, whose minor axis, which is also the axis of the figure, lies along the direction of motion, and whose axes are in the ratio $1{:}1{:}(1 - u^2/v^2)^{1/2}$ where u is the velocity of the point and v is the velocity of light through the dielectric ... . The charge is distributed in the same way as if the ellipsoid were statically charged ... . This surface I call the ''Heaviside ellipsoid''.

This result looks very ''Einsteinian'' and, of course, what Searle and Heaviside were doing, without realizing it, was probing at the edges of relativistic electrodynamics. This work was to take them into very exciting places; some of it would eventually lead to Nobel prizes (for others).

Faster-than-Light

Heaviside's velocity-dependent solutions for the field effects of a moving charge took Thomson's electromagnetic mass increase result of 1881 one step further, predicting a mass increasing with velocity. The possibility that at least part of the mass of a charged body could be attributed to the electrodynamic effects of its motion held out the exciting possibility that *perhaps all* of it could be so explained. If so, and if all matter is ''merely'' a vast number of atoms that are themselves ''merely'' complex constellations of charged bodies (i.e., electrons), then the ultimate interpretation of Nature would be *electromagnetic,* rather than the clunking mechanical one of Kelvin, and even of Maxwell. What a heady thought for the true believers in Maxwellian philosophy! This hope eventually proved to be flawed,[91] however, and in one of his notebooks[92] Heaviside claimed never to have been seduced by it:

> I will not go so far as to say that the view which is popular now, that ''mass'' is due to electromagnetic inertia, is a mere Will o' the Wisp. I will however say that the light it gives is somewhat feeble and uncertain, and that it eludes or evades distinct localisation. The mere *idea,* that electromagnetic inertia *might* account for ''mass'', occurred to me in my earliest work on moving charges, but it seemed so vague and unsupported by evidence, that I set it on one side. It explains too much, and it does not explain enough.

One curious feature of the predicted mass variation with speed is the infinity that results at the speed of light. This is a result we interpret today to mean that nothing with mass can travel as fast as light (in a vacuum). Light *can* travel (obviously) as fast as light because photons are massless. Searle studied the electromagnetic effects of moving charges for decades and accepted this conclusion, but Heaviside did not.[93] Scattered all through his books, in fact, are analyses on charges moving *faster*-than-light (with the earliest dating from 1888[94]). This caused Searle much frustration as, for example, when he wrote a note in his 1896 paper declaring

> Mr. Heaviside has *stated* [see Note 95] the result where u is greater than v [the speed of light], but has not up to the present divulged the manner in which he has obtained the solution in this case. I confine this paper to the case in which u is not greater than v.

The following year Searle published a continuation[96] of his 1896 paper in which he worked out the energy associated with moving ellipsoids. His closing words were,

> In all these cases it will be found that when $u = v$ the energy becomes infinite, so it would seem to be impossible to make a charged body move at a greater speed than that of light.

Heaviside would have none of Searle's caution. In 1898 he wrote a reply[97] in which he attempted to refute Searle's position. I don't really find Heaviside's prose arguments (there is not a single line of mathematics) very convincing—here is his opening "proof" that Searle was wrong:

> The argument ... seems to be that since the calculated energy of a charged body is infinite ... at the speed of light, and since this energy must be derived from an external source, an infinite amount of work must be done, that is, an infinite resistance will be experienced. There is a fallacy here. One easy way of disproving the argument ... is to use not one, but two bodies, one positively and the other negatively charged to the same degree. Then the infinity disappears, and there you are, with finite energy when moving at the speed of light.

What Heaviside was imagining (I *think*!) was the replacement of the q^2 factor in his energy-increase expressions with $[q + (-q)]^2$ to get zero. He ignored the alternative view of $q^2 + (-q)^2 = 2q^2$. This was all pre-Einstein, of course, and Heaviside didn't know the infinity comes from much deeper considerations than just whether the body carries a net charge or not.

The appearance of infinites in his work, an event that makes most analysts stop to ponder at what they might *mean,* hardly ever caused Heaviside to do anything but dip his pen into the well for more ink to write why there was no need for concern. We'll see a *lot* of this in Chapter 10. For example (writing on a problem different from moving charges), he once declared,[98] "When mathematicians come to an infinity they are nonplused, and hedge around it ... . We must not be afraid of infinity." And the concluding words of his reply to Searle about moving at the speed of light sent the same message: "The moral is—don't be afraid of infinity!"

Heaviside took his own advice and spent years thinking about charges moving at light and hyperlight speeds. In 1903 he wrote[99] of the visual image he had of charged bodies moving at these speeds:

> The photographs taken some years ago by Prof. Boys of flying bullets showed the existence of a mass of air pushed along in front of the bullet. Is there anything analogous to this in the electromagnetics of an electron? Suppose, for example, that an electron is jerked away from an atom so strongly that its speed exceeds that of light ... . So long as its speed is greater than that of light, it is accompanied in its motion by a conical wave.

The conical acoustic shock waves revealed in Boys' photographs[100] were *not* the original inspiration for Heaviside's imagination, however, as they appeared in print in 1893 while Heaviside first wrote[94] of this effect of faster-than-light charged particles in 1888. The shock wave is created, in both cases, by a body moving through a medium at a speed faster than it can propagate a disturbance. For bullets in air, the crucial speed is that of sound, and an acoustic shock wave is the well-known "sonic boom" produced by jet aircraft "breaking the sound barrier." High-speed boats moving faster than the wave-speed in water produce conical bow wakes. For charged atomic particles, the crucial speed is the speed of light in the medium.

The astonishing thing is that, in a certain physical sense, Heaviside was absolutely right about his claims for hyperlight motion *if* the medium is something other than a vacuum, such as water. Then the speed of light is *less* than it is in a vacuum, charged particles *can* exceed the "lower" speed of light, and, in fact, Heaviside's conical, electromagnetic shock wave *is observed*! Today we call it Cherenkov radiation,[101] after the Russian physicist Pavel A. Cherenkov who exhaustively studied it experimentally in the 1930s, although Madame Curie was apparently the first to notice, in 1910, this radiation effect in radium solutions (but she did not appreciate its true origin). Along with his theoretical colleagues I. M. Frank and I. E. Tamm, Cherenkov received the 1958 Nobel physics prize. In their Nobel acceptance talks[102] none of the three Russians so much as mentioned Heaviside, even though he had anticipated them by decades and his books are full of mathematical treatments of conical, electromagnetic shock waves caused by faster-than-light charged particles.[103]

Heaviside discussed the question of faster-than-light motion with others besides Searle (including Hertz), but they were no more encouraging. FitzGerald, for example, wrote in an undated letter (most likely from early 1889),

> You ask "what if the velocity be greater than that of light?" I have often asked myself that but got no satisfactory answer. The most obvious thing to ask in reply is "Is it possible?"

And years later in a postcard (July 4, 1897) he asked

> Do you agree with Searle that the energy of an electrified sphere [note the importance in FitzGerald's mind of the sphere being charged] moving with [the] velocity of light is infinite? I had not appreciated it.

According to modern thought, of course, Heaviside was wrong in believing faster-than-light motion *in a vacuum* to be possible, although the idea has enjoyed a resurgence in recent years with many articles appearing on tachyons. One of the modern objections to such superspeed particles is that they seem to imply the possibility of time-travel to the past, an exotic concept full of apparently mind-numbing paradoxes.

Just what Heaviside would have had to say about time-travel would unquestionably be interesting to read; even though he seems not to have written anything along those lines, maybe we can get a glimmer of his likely reaction from a few words he scribbled in a copy of *Nature*. Time-travel speculations are often presented in terms of the "fourth dimension" and Heaviside didn't like that idea at all. In a note[104] written next to a book review of Dr. J. W. Withers' *Euclid's Parallel Postulate* which quotes Withers as writing "We may some time gain experience of a new kind, presenting itself as spatial and requiring us to assume more than three dimensions in space.", Heaviside snapped in reply "Can't see it, and never could." Perhaps, if Heaviside had had the opportunity to think about hyperlight speeds in a vacuum in terms of traveling backward in time, he might have reconsidered his position.

Dr. Heaviside, F.R.S.

Searle ended his 1896 paper[90] with the following words of acknowledgment:

> I have much pleasure in expressing my best thanks to my friend, Mr. Oliver Heaviside, F.R.S. Besides giving me some personal instruction in Electromagnetic Theory on several occasions, he has constantly encouraged me during the process of this investigation.

Heaviside had, indeed, transformed himself from the nobody of 1881 to a Fellow of the Royal Society of 1891, an honor of tremendous importance in those days. Even today, with an enormous increase in important scientific bodies and honors, being able to write "F.R.S." after your name is nothing to sniff at. Heaviside's voluminous articles on electrodynamics in *The Electrician* and elsewhere had not gone unnoticed (at least not by a select few first-rate scientists), and the official "Certificate of a Candidate for Election" form of the Royal Society for Oliver included the signatures of such luminaries of British science as Oliver Lodge, William Thomson, John Perry, Charles Boys, George Francis FitzGerald, and John Poynting. These important men signed their names beneath the following testimonial to Heaviside's contributions: "Learned in the science of Electromagnetism, having applied high mathematics with singular power and success to the development of Maxwell's theory, of electromagnetic wave propagation, and having extended our knowledge of facts and principles in several directions and into great detail."

An odd character like Heaviside, however, couldn't just take his election in June 1891 in graceful stride, but rather felt compelled to make a show of being above it all. Upon receiving a copy of the Society's statutes he immediately wrote to Oliver Lodge, "The Secretary of the Royal Society has sent me a sort of habeas corpus." And in one of his notebooks is the following funny rhyme declaring his amusement at the rites of initiation for the Royal Society:

> Yet one thing More
> Before
> Thou perfect Be
> Pay us three Poun'
> Come up to Town
> And then admitted Be
> But if you *Won't*
> Be Fellow, then *Don't*.

Of course he didn't go, and of course they still elected him a Fellow (as Oliver certainly knew they would). Many years later, however, this humor gave way to not a little bitterness at what he felt to be a lack of respect from his family and his neighbors. As he wrote in a 1918 letter[105] to a correspondent in America, " ... F.R.S. means nothing at all, being less than F.R.H.S. [Fellow of the Royal Horticulture Society], to which our respected gardening townsman belongs by paying a guinea."

Other honors came his way. In 1905 he received one that particularly pleased him (at the time) when the University of Göttingen[106] presented him with an honorary Doctor's degree, again in recognition of his work in electrodynamics. The diploma reads (in translation from the original Latin) as follows:

> That Eminent Man
> Oliver Heaviside
> An Englishman by nation, dwelling at Newton Abbot,
> Learned in the artifices of analysis,
> Investigator of the corpuscles which are called electrons,
> Persevering, fertile, happy though given to a solitary life,
> Nevertheless among the propagators of the Maxwellian science
> easily the first.

But this, too, he eventually brushed aside. In 1924 he wrote to an American correspondent[107] that he had once requested a friend in France to "stick my ... diploma from the University of Göttingen up on the door of the Academy of Science! ... I should like it blacked out, because I have given up all idea of accepting any more titles, and would be glad to get rid of the few old ones." This was all to come later, however—initially these honors all brought much pleasure to Oliver.

And so at the age of fifty-five he was no longer the unemployed[108] former telegraph operator who had taught himself at home. Now, at last, he had credentials and was Dr. Oliver Heaviside, F.R.S. But the ten years from 1881 to 1891 were not as easy as this technical discussion of his work has perhaps made it appear. That decade contained periods that were among the worst of his life.

The battle with William Henry Preece was at the bottom of it all, both the hard times and the glorious vindication that, despite the glory, Heaviside never forgot took a long time coming.

TECH NOTE 1: THE DUPLEX EQUATIONS

Recall from the last chapter the equations connecting the electric and magnetic fields:

$$\vec{\nabla}\times\vec{B}=\mu\left(\vec{J}_e+\epsilon\frac{\partial\vec{E}}{\partial t}\right),\qquad \vec{\nabla}\cdot\vec{B}=0$$

$$\vec{\nabla}\times\vec{E}=-\frac{\partial\vec{B}}{\partial t},\qquad \vec{\nabla}\cdot\vec{E}=\rho_e/\epsilon$$

where μ and ϵ are the magnetic and electrical parameters of the medium (the square root of the reciprocal of their product is the speed of light in the medium), $\vec{J}_e$ is the electric conduction current density (due to the motion of electric charge), ρ_e is the net electric charge density (or *electrification,* as it was called in Heaviside's day), and $\vec{E}$ and $\vec{B}$ are the fields. Now just as $\vec{J}_e = \sigma_e\vec{E}$ where σ_e is the electric conductivity of the medium, suppose there are also *free magnetic charges* with density ρ_m and medium conductivity σ_m. Then, just as the *total* electric current density is

$$\vec{J}=\vec{J}_e+\epsilon\frac{\partial\vec{E}}{\partial t}$$

we could also write the *total* magnetic current density as

$$\vec{M}=\frac{\partial\vec{B}}{\partial t}+\vec{m}_c$$

where $\vec{m}_c$ is the magnetic conduction current due to the motion of our hypothetical magnetic charges. It is mathematically convenient to define a new auxiliary field, $\vec{H}$, proportional to $\vec{B}$, i.e., $\vec{B} = \mu\vec{H}$, and to write $\vec{m}_c = \sigma_m\vec{H}$. This is nice because then

$$\vec{\nabla}\times\vec{B}=\vec{\nabla}\times\mu\vec{H}=\mu\vec{J}$$

and the μ-factor cancels out(!), i.e.,

$$\vec{J}=\vec{\nabla}\times\vec{H}.$$

This goes well with the second curl equation which is just

$$\vec{M}=-\vec{\nabla}\times\vec{E}.$$

Of these two duplex equations Heaviside wrote in the Preface of EP 1, "The duplex method is eminently suited for displaying Maxwell's theory, and brings to light many useful relations which were formerly hidden from view by the intervention of the vector-potential and its parasites." And in a letter to FitzGerald (January 30, 1889), "I never made any progress till I threw all the potentials overboard, and made $\vec{E}$ and $\vec{H}$ the objects of attention … ." FitzGerald needed no convincing of Oliver's "progress," although he did have *some* reservations. Just a week later (February 8) he wrote to Hertz to say of Heaviside: "His investigations are very interesting but he is such a master of symbols that he is difficult to follow and is a little carried away by them to study non-existing phenomena. However he has done a great deal to advance our knowledge of the subject and is one of the leaders in the campaign against the old classical theories … ."

TECH NOTE 2: THE LOCALIZATION OF ELECTROMAGNETIC FIELD ENERGY

The idea that energy is stored in what seems to be "empty" space just by virtue of the presence of electric and magnetic fields, is one of the great continuing (and, to my mind, *mysterious*) issues of electromagnetics. The Maxwellians, including Heaviside, could think of this energy as somehow being mechanically stored in a strained and stressed state of the ether, much like energy stored in a bent twig or a stretched rubber band. But with today's dismissal of the ether to the same wastebasket that we have swept phlogiston into, we don't have this invisible, convenient storage vault to make the field energy easier to "visualize." The field energy is, in this sense, a greater mystery for us today than it was for the Victorians. I am, of course, referring to the questions of *where* and *how* the energy is stored—the calculation of the value of the stored energy (wherever it might be) is without controversy.

The analytic formulation of the idea of energy *in a field* is due to Maxwell, himself, who wrote (in Article 631 of the *Treatise*),

> Hence, the electrostatic energy of the whole field [due to electrical charge] will be the same if we suppose that it resides in every part of the field where electrical force [our $\vec{E}$] and electrical displacement [our $\epsilon\vec{E}$] occur, instead of being confined to places where free electricity is found. The energy in unit of volume is [Maxwell now uses quaternion notation, equivalent to $(1/2)\epsilon\vec{E}\cdot\vec{E}$].

and later (Article 636) for magnetic energy,

> According to our hypothesis, we assume the kinetic energy [Maxwell thought of magnetic energy as the kinetic energy of rotational vortices in the ether] to exist wherever there is magnetic force, that is, in general, in every part of the field. The amount of this energy per unit of volume is [again Maxwell uses quaternion notation, equivalent to $(1/2)\mu\vec{H}\cdot\vec{H}$], and this energy exists in the form of some kind of motion of the matter in every portion of space.

In a passage published by *The Electrician* in March 1883 (EP 1, p. 249) Heaviside demonstrated this result (magnetic energy density in space is proportional to B^2) and wrote that the "conclusion is irresistible that we have got an expression for the energy which may correctly locate it in amount at different places." Heaviside did admit to the possible ambiguity in localizing energy (EP 2, p. 570), but he qualified that caution by saying of $(1/2)\vec{H}\cdot\vec{B}$ that it is "perhaps the least certain part of Maxwell's scheme ... yet only in this way are thoroughly consistent results apparently obtainable."

According to Maxwell, then, the total field energy density is

$$W=\frac{1}{2}\mu\vec{H}\cdot\vec{H}+\frac{1}{2}\epsilon\vec{E}\cdot\vec{E}$$

and this density *localizes* the energy in space.

TECH NOTE 3: HEAVISIDE'S DERIVATION OF THE ELECTROMAGNETIC ENERGY FLOW VECTOR IN SPACE

The total field energy per unit volume in a region of space with $\vec{E}$ and $\vec{H}$ fields present is given by, as shown in Tech Note 2,

$$W=\frac{1}{2}\mu\vec{H}\cdot\vec{H}+\frac{1}{2}\epsilon\vec{E}\cdot\vec{E}.$$

The rate of energy flow out of this region is just the negative of the time rate of change of W, which in turn is the divergence of the power vector, $\vec{P}$ (the energy decrease per unit time). Thus,

$$\vec{\nabla} \cdot \vec{P} = -\frac{\partial W}{\partial t} = -\mu \vec{H} \cdot \frac{\partial \vec{H}}{\partial t} - \epsilon \vec{E} \cdot \frac{\partial \vec{E}}{\partial t}.$$

Now, from Tech Note 1 we have the Heaviside duplex equations

$$\vec{J} = \vec{\nabla} \times \vec{H}$$

$$\vec{M} = -\vec{\nabla} \times \vec{E}.$$

For the case of charge-free (electric and magnetic) space we have the electric and magnetic currents equal to zero, and the duplex equations reduce to

$$\epsilon \frac{\partial \vec{E}}{\partial t} = \vec{\nabla} \times \vec{H}$$

$$\frac{\partial \vec{B}}{\partial t} = -\vec{\nabla} \times \vec{E} = \mu \frac{\partial \vec{H}}{\partial t}.$$

Substituting these into the $\vec{\nabla} \cdot \vec{P}$ expression gives

$$\vec{\nabla} \cdot \vec{P} = \vec{H} \cdot (\vec{\nabla} \times \vec{E}) - \vec{E} \cdot (\vec{\nabla} \times \vec{H}).$$

Heaviside next recalled a basic theorem in vector mathematics which says that for *any* two vectors $\vec{A}$ and $\vec{C}$

$$\vec{\nabla} \cdot (\vec{A} \times \vec{C}) = \vec{C} \cdot (\vec{\nabla} \times \vec{A}) - \vec{A} \cdot (\vec{\nabla} \times \vec{C}).$$

Associating $\vec{A}$ with $\vec{E}$ and $\vec{C}$ with $\vec{H}$, we have

$$\vec{\nabla} \cdot (\vec{E} \times \vec{H}) = \vec{H} \cdot (\vec{\nabla} \times \vec{E}) - \vec{E} \cdot (\vec{\nabla} \times \vec{H})$$

which is precisely equal to $\vec{\nabla} \cdot \vec{P}$! That is,

$$\vec{\nabla} \cdot \vec{P} = \vec{\nabla} \cdot (\vec{E} \times \vec{H}).$$

Thus, $\vec{P} = \vec{E} \times \vec{H}$, *plus* any arbitrary vector that has zero divergence, i.e., a vector field of *closed loops*; call it $\vec{G}$. Then, at last,

$$\vec{P} = \vec{E} \times \vec{H} + \vec{G}.$$

This simple vector derivation is as it was done by Heaviside (but I have simplified it, as his is much more general, including regions where there are currents) and, indeed, is the way it is presented in modern texts. It is called *Poynting's* theorem, however, because Poynting published before Heaviside, but his derivation is much less nice (it is full of *triple* integrals). Heaviside's derivation appeared in *The Electrician* on February 21, 1885 (EP 1, pp. 449–450). I have, however, always thought it most appropriate that both *power* and *Poynting* start with the same letter, and of course what could be a better name for the vector *pointing* in the direction of energy flow than the *Poynting* vector? Even Heaviside, with his often professed belief in the divine inspiration behind such "coincidences," would surely have agreed!

One interesting difference of substance between Poynting and Heaviside is the issue of the divergence-free energy flux term, $\vec{G}$. Poynting said nothing at all about the possibility of $\vec{G}$, and in this sense his paper is less general than Heaviside's. Heaviside said a great deal about $\vec{G}$, but only to dismiss it (EMT 1, pp. 247–248) as

> ... of no moment whatever in the practical use of the idea of a flux of energy for purposes of reasoning. We should introduce an auxiliary circuital flux [only] when some useful purpose is served thereby.

Oliver Lodge, on the other hand, disagreed, and in a rebuttal wrote,[109]

> Mr. Heaviside ... seems to consider circuital fluxes of energy as strange and useless phenomena. But I see no reason in this at all. The circulation of matter—for instance in the inner circle of the Metropolitan railway—is, I suppose, considered useful. The circulation of commodities is the essence of commerce. So does the circulation of energy constitute the activity of the material universe. It is the act of transfer that is beneficial (or the reverse); what becomes of a conservative quantity is a minor matter. It must go somewhere, and may very well, after a series of transfers, ultimately return to its starting point.

Lodge concluded his letter with an invitation for Heaviside to counter in the public record, but there is no evidence he ever did so.

The presence of the seemingly arbitrary $\vec{G}$ has, for the last one hundred years, puzzled and troubled (at least for a while) essentially every physicist and electrical engineer who has studied the issue. Such a divergence-free term, which represents energy flowing endlessly around in closed loops, seems to introduce ambiguity into the actual energy flow. But surely there can be nothing ambiguous about a flow of energy! In fact the accepted practice from the very first has been to take $\vec{G} = 0$ and to just write $\vec{P} = \vec{E} \times \vec{H}$. This simplest of all possible choices for $\vec{G}$ leads, in fact, to entirely consistent physical results. Ironically, however, there are situations where this $\vec{P}$ can *still* give seemingly strange energy flow paths that are counter to most people's intuition, including divergence-free, closed-loop paths!

Consider, for example, the case of the Earth with electric field lines directed radially downward toward the surface (the potential *rises* with increasing altitude at a rate of about 100 volts per meter through a clear atmosphere). Passing through the atmosphere are also magnetic field lines, directed from the magnetic North Pole to the South. These two fields combine in the $\vec{E} \times \vec{H}$ expression to give a perpetual energy flow around the magnetic polar axis of the Earth from east to west, in closed loops! So here, even with $G = 0$, we still have a divergence-free energy flow. Is such a flow *real,* i.e., does it have any physical significance? Or is the whole idea just, as one author[110] once called it, "absurd"? In a typically brilliant piece of exposition[111] Professor Richard Feynman has presented a thought experiment that shows just such an energy flow is actually *required* to maintain the conservation of angular momentum in the world. There is, in fact, nothing imaginary, arbitrary, or absurd about divergence-free energy flows at all, and the fact that our intuitions are sometimes shocked by the Poynting vector just means our intuitions are sometimes not very good. In a very interesting paper[112] the *quantitative* details of the field angular momentum for a charged magnetic sphere (e.g., the Earth) have been evaluated and the authors have shown how Feynman's thought experiment really does work out precisely right *in the numbers.*

As Oliver put it (EMT 1, p. 381), "However mysterious energy (and its flux) may be in some of its theoretical aspects, there must be something in it, because it is convertible into dollars, the ultimate official measure of value." And as FitzGerald put it in a letter (September 26, 1892) to Heaviside concerning belief in the energy flow theorem—"it seems to me to distinguish the Physicist from the Old Tory who sticks to the old ideas."

Tech Note 4: Poynting's Physics (and Oliver's Objection)

When writing of the sideways entrance of energy into a current-carrying wire it appears that Poynting believed the electric field is *always* parallel to the wire. *Inside* the wire this is correct, as the current and electric field vectors are parallel (Ohm's law tells us $\vec{J} = \sigma\vec{E}$, where the wire's conductivity is some positive quantity and certainly the current vector is parallel to the wire!). But *outside* the wire the electric field is, in fact, nearly *perpendicular* to the wire. This surprises most people, often even electrical engineers, perhaps because it isn't a topic discussed much in texts. An exception is the classic book by Skilling[110] which has a very nice explanation of the nature of the external electric field.

Assuming Poynting's erroneous conception of the external $\vec{E}$, and combining it with the circular $\vec{H}$ around the wire, then according to Poynting nearly the entire $\vec{E} \times \vec{H}$ energy vector is pointed inward toward the wire. The whole idea of the Poynting vector, of course, is that there is indeed sideways energy flow into the wire, but as Heaviside pointed out only a *small* component of the external $\vec{E} \times \vec{H}$ is inward directed. The rest of $\vec{E} \times \vec{H}$ is parallel to the wire. But let Heaviside speak for himself. In his words (EP 2, p. 94), written in 1887:

> It [the energy transfer] takes place, in the vicinity of the wire, very nearly parallel to it, with a slight slope towards the wire ... . Prof. Poynting, on the other hand [here Heaviside references Poynting's second Royal Society paper[113] on energy transfer, which appeared in 1885] holds a different view, representing the transfer as nearly perpendicular to a wire, i.e., with a slight departure from the vertical. This difference of a quadrant can, I think, only arise from what seems to be a misconception on his part as to the nature of the electric field in the vicinity of a wire supporting electric current. *The lines of electric force are nearly perpendicular to the wire* [my emphasis]. The departure from perpendicularity is usually so small that I have sometimes spoken of them as being perpendicular to it, as they practically are, before I recognized the great physical importance of the slight departure. It causes the convergence of energy into the wire.

Inside the wire $\vec{E} \times \vec{H}$ *is* entirely inward pointing, and represents the energy dissipation ascribed to Joule heating, i.e., the "I^2R loss." Thus, all energy propagation *along* the direction of the wire takes place *outside* of the wire. Crazy sounding perhaps, but true (or, at least, practically everybody today *believes* it's true!).

Notes and References

1. Henri Poincaré, the greatest French mathematician of the day, in his 1901 *Electricité et Optique* wrote "The first time a French reader opens Maxwell's book, a feeling of unease, and often even of defiance, mixes with his admiration."
2. In a paper entitled "On the possibility of originating wave disturbances in the ether by means of electric forces," reprinted in *The Scientific Writings of the Late George Francis FitzGerald,* London: Longmans, Green & Co., 1902, p. 92.
3. Ibid, "Note on Mr. J. J. Thomson's investigation of the electromagnetic action of a moving electrified sphere," pp. 102–107. See Note 79 for Thomson's investigation (which is discussed later in this chapter).
4. Ibid, pp. 99–101. FitzGeraid ended with an astonishingly frank and honest evaluation of his earlier position: "In conclusion, I must apologize for having ventured to investigate these matters when I was so ignorant of what had already been done as to make mistakes requiring such serious corrections as are contained in this paper."
5. Ibid, "On the energy lost by radiation from alternating electric currents," pp. 128–129.
6. EP 1, Preface.
7. EMT 1, p. 6.
8. EP 2, p. 524.
9. *The Electrician* first appeared in November 1861, but was unable to find a secure footing and ceased publication in May 1864. It started up again in May 1878, and survived for many years. It was absorbed in 1952 by the *Electrical Review,* which still publishes today. A useful review article is by P. Strange, "Two electrical

periodicals: *The Electrician* and *The Electrical Review* 1880–1890," *Proceedings of the IEE,* vol. 132 (part A), pp. 574–581, December 1985.

10. In a long Letter-to-the-Editor (vol. 2, pp. 271–272, April 26, 1879) written in response to a heated debate on the journal's pages concerning the concept of electric potential. Lodge was one of the participants in this debate, and his letters are particularly interesting as they are full of the passion of youth (in his later years Lodge, unlike Heaviside, would learn well the rewards for using temperate language).
11. This feature allowed readers to direct specific questions to the editors, who would then print answers in the journal. It was a sort of technical "Dear Abby" column.
12. Heaviside's and Lodge's comments are quoted from D. W. Jordan, "The adoption of self induction by telephony, 1886–1889," *Annals of Science,* vol. 39, pp. 433–461, September 1982.
13. It was at the Criterion Bar, during this same year, that Dr. Watson bumped into "young Stamford" who was soon to introduce him to Sherlock Holmes.
14. There is an intriguing hint that there might have been exceptions to Heaviside's shyness in R. Appleyard's *Pioneers of Electrical Communication,* London: Macmillan, 1930, p. 254. Appleyard recorded, "There is a marginal note in pencil on a stray reprint, that when he [Heaviside] first suggested loading *at a meeting* [my emphasis] of the Physical Society ... ." However, it was common for Heaviside to send *written* comments to meetings.
15. *The Electrician,* vol. 87, pp. 603–604, November 11, 1921. Being an artist was also of value, judging from two letters by Oliver Lodge in which he complained of the drafting skills of the immediate successors to Biggs (W. H. Snell and A. P. Trotter)—see O. Lodge, *Talks about Radio,* New York, NY: G. H. Doran, 1925, p. 51 and J. A. Hill, *Letters from Sir Oliver Lodge,* London: Cassell, 1932, p. 47.
16. *Electrical World,* vol. 85, p. 364, February 14, 1925.
17. The total page count of *Electrical Papers* is actually in excess of 1100, but additional papers were included besides those that originally appeared in *The Electrician.* These two volumes were brought out in 1892 (published by Macmillan and Co.) because it was felt that Heaviside's early work should be collected to provide a basis for his soon to appear *Electromagnetic Theory* (published by The Electrician Publishing Co.).
18. EP 1, pp. 190–195.
19. EP 2, pp. 23–28.
20. Mrs. Sairey Gamp is a character in Dickens' novel *Martin Chuzzlewit.*
21. Heaviside may well have been serious on this point. For example, an editorial in *Scientific American,* entitled "What the telephone might have been called" (vol. 38, p. 68, February 2, 1878), reported that the Germans settled on the word *Fernsprecher* only after considering fifty-four possible names. The editorial concluded with "one inventor collecting all his energies for one grand effort, triumphantly produced *doppelstahlblechzungen-sprecher.* The jaw can be replaced by pressing on the lower molars with the fingers, and guiding the muscles with the thumbs."
22. In a letter dated March 5, 1922, to the President of the IEE (J. S. Highfield).
23. In 1878 Oliver carefully measured his body, recording (for example) that he had a 28-inch waist, a 13-inch neck, and stood five feet four and one-half inches high. See Appleyard (Note 14), p. 217.
24. The passage appeared in *The Electrician,* vol. 11, p. 199, July 14, 1883. It was cut from the essay eventually printed in EP 1, pp. 286–291.
25. Heaviside was referring to William Paley, D. D. (1743–1805), archdeacon of Carlisle, who wrote voluminously on "evidences for Christianity."
26. This literary reference is interesting as it gives us an idea of the sort of "recreational reading" Heaviside enjoyed. Here he was referring to *Thomas Ingoldsby,* one of the pseudonyms (another was *Tim Twaddle*!) of the clergyman Richard Harris Barham (1788–1845). Barham wrote a great deal of very funny verse, collected into three volumes in 1857 as *The Ingoldsby Legends,* a Victorian best-seller. Heaviside's particular reference is to "Aunt Fanny, a Legend of a Shirt."
27. *The Electrician,* vol. 11, p. 230, July 21, 1883.
28. *The Electrician,* vol. 11, p. 253, July 28, 1883.
29. NB 3A:50. The date, 1889, is interesting, as that was the year of Thomson's inauguration as President of the Institution of Electrical Engineers, during which he publicly praised Heaviside. This was a pivotal event in Oliver's life, as will be discussed in Chapter 8. Possibly Lodge heard of Thomson's long-remembered "disgust" in private conversation with him that evening.
30. From Jordan (Note 12).
31. A. M. Bork, "Maxwell and the electromagnetic wave equation," *American Journal of Physics,* vol. 35, pp. 844–849, September 1967.
32. EMT 1, p. 69.
33. EP 2, pp. 483–485.
34. EP 2, p. 511.
35. This pioneering mathematical effort, so "obvious" to modern students who think Newton used vectors, did not meet with universal acclaim in the beginning. In an undated review of *Electrical Papers* that Heaviside pasted into his personal copy, the *Glasgow Herald* called the new notation "monstrous" and clucked that "it is a thousand pities that the notation, and even the nomenclature, used is so foreign to ordinary readers ... as to

make the papers entirely valueless." The *Manchester Guardian* (in another pasted, undated review), however, thought vectors "extremely ingenious."

36. EP 2, p. 173.
37. EP 2, p. 482. In an 1893 letter to Oliver Lodge, Heaviside said of his own work that it represented "the real and true 'Maxwell' as Maxwell would have done it had he not been humbugged by his vector and scalar potentials"—see B. J. Hunt, *The Maxwellians* (Ph.D. dissertation), Baltimore, MD: The Johns Hopkins University, 1984, p. 317.
38. *Nature,* vol. 47, pp. 505–506, March 30, 1893.
39. Nobody attaches Heaviside's name to these equations today, but I have found it so done most recently in a charming little book that every physics and electrical engineering student (and their professors!) ought to read—see L. V. Bewley, *Flux Linkages and Electromagnetic Induction,* New York, NY: Macmillan, 1952.
40. EP 1, pp. 447–448.
41. *The Electrician,* vol. 31, pp. 389–390, August 11, 1893.
42. *Philosophical Magazine,* Series 5, vol. 38, pp. 145–156, July 1894.
43. *Nature,* vol. 51, pp. 171–173, December 20, 1894.
44. H. Zatzkis, "Hertz's derivation of Maxwell's equations," *American Journal of Physics,* vol. 33, pp. 898–904, November 1965. Zatzkis discusses an error in Hertz's mathematics. See also the reply by Peter Havas the next year in vol. 34, pp. 667–669, August 1966, which argues that Hertz actually had his math correct, but his *physics* wrong! Hertz's daybook (excerpts in English are printed in the translation by D. E. Jones and G. A. Schott, *Miscellaneous Papers,* London: Macmillan, 1896) shows (p. xxv) that he arrived at his solution on May 19, 1884: "Hit upon the solution of the electromagnetic problem this morning." The correspondence Hertz had with Heaviside can be found in J. G. O'Hara and W. Pricha, *Hertz and the Maxwellians,* London: Peter Peregrinus Ltd., 1987.
45. *Electric Waves,* New York, NY: Dover, 1962, pp. 196–197. This book first appeared in English in 1893 and was reviewed by *The Electrician* (vol. 32, p. 657, April 13, 1894). This review is interesting for what it shows of how Maxwell's original mathematics struck most of his readers: "The discovery of the Rosetta stone did not more completely solve the meaning of the hieroglyphic formulae graven in the stones of Egypt, till then mere symbols devoid of physical significance, than have the researches of the late Dr. Hertz into the phenomena of the propagation of electromagnetic waves solved the physical meaning of the Maxwellian formulae. Many had gazed at those algebraic signs in wonder at their true meaning ... ."
46. E. T. Whittaker, "Oliver Heaviside," Note 1, Chapter 2.
47. F. Klein (best remembered today as the inventor of the Klein bottle, the topologically weird bottle twisted through the fourth dimension in such a way that it has no inside), *Vorlesungen über die Entwicklung der Mathematik im 19. Jahrhundert,* Part 2, New York, NY: Chelsea, 1956, p. 47.
48. Lodge wrote Hertz's obituary for *The Electrician* (vol. 32, p. 273, January 12, 1894) in which he took space to inform readers that Hertz was a corresponding member of the Society for Psychical Research. This was one of Lodge's great obsessions (the ether being another), and he was quite willing to use Hertz's death to get a plug in for it.
49. G. Holton, *Thematic Origins of Scientific Thought,* Cambridge, MA: Harvard University Press, 1973, pp. 205–212. Oliver wrote (June 19, 1897) to FitzGerald to say of Föppl's query about priority, "Nasty to have to do that. I am surprised there should be any question of comparison between my work up to 1888 and Hertz's work on the same lines ... ."
50. *The Electrician,* vol. 39, pp. 643–644, September 10, 1897; reprinted in EMT 3, pp. 503–507.
51. Boltzmann is today remembered not for his electrodynamic work, but for his thermodynamics. Engraved on his tombstone in a Vienna cemetery, in fact, is the fundamental entropy equation $S = k \log W$, where k is universally called *Boltzmann's Constant.* In a letter to Heaviside (December 22, 1893) FitzGerald called Boltzmann's electrodynamics the "weaving [of an] impossible cobweb of a beautiful mathematical structure."
52. For details on Helmholtz's generalization of Maxwell's theory, and how longitudinal solutions appear in it, see A. E. Woodruff, "The contributions of Herman von Helmholtz to electrodynamics," *Isis,* vol. 59, pp. 300–311, Fall 1968. Woodruff quoted Boltzmann to show how strong the idea of action-at-a-distance was: "We have all more or less imbibed with our mothers' milk the ideas of magnetic and electric fluids acting directly at a distance."
53. *The Electrician,* vol. 40, p. 55, November 5, 1897.
54. *The Electrician,* vol. 40, p. 93, November 12, 1897; reprinted in EMT 2, pp. 493–504.
55. Several years later, however, when Heaviside wrote a book review on vector analysis (EMT 3, pp. 135–143) he ended it with: "If Heinrich Hertz had used it [Heaviside's vector notation for a linear operator], I think it would have simplified his arguments considerably. But I hope that Prof. Boltzmann will not see this, for fear he may be led to tell another story." In a footnote Heaviside explained this curious remark by relating how Boltzmann had objected to his treatment of Hertz in the Curry review: "I spoke of Hertz's experimental establishment of Maxwell's electromagnetic waves as a 'great hit'. Prof. Boltzmann turned this into 'lucky hit', and then fell upon me!" Heaviside seems to have forgotten that in writing to Föppl he *had* used the word *lucky,* and perhaps it was from Föppl that Boltzmann heard of Heaviside's deprecative remark. Heaviside, of course, certainly *did* appreciate Hertz's work. Years before, in 1889, he had described it as "that series of brilliant experiments"—see EP 2, p. 503.

56. *The Electrician,* vol. 50, p. 398, December 26, 1902.
57. *The Electrician,* vol. 50, p. 478, January 9, 1903.
58. *The Electrician,* vol. 50, pp. 560–563, January 23, 1903.
59. *The Electrician,* vol. 50, p. 694, February 13, 1903.
60. *The Electrician,* vol. 50, p. 735, February 20, 1903. Two months after Heaviside's description of entropy as "ghostly," Perry copied him in a curious book review in *Nature* (vol. 67, pp. 602–603, April 30, 1903). This "review" was not really a review at all, but rather a statement of Perry's experiences in teaching thermodynamics to engineers. It is very funny (e.g., it is full of lines like "We know men who pet and fondle their slide rules ..."), and seems to have been written as an emotional release to the debate with Swinburne.
61. *The Electrician,* vol. 50, p. 821, March 6, 1903.
62. EP 2, pp. 91–92.
63. EMT 1, pp. 73–74.
64. Einstein's demonstration of the relativity of simultaneity destroys even this argument for a mere *global* conservation of energy law.
65. The *Mathematical Tripos* is a grueling examination that comes at the end of the honors program in mathematics at Cambridge University. The competitors are called *wranglers,* and to be third wrangler is to be third best. The top finisher is the *Senior Wrangler.* To finish last is to be the *Wooden Spoon.* Maxwell, for example, was second wrangler in the 1854 *Tripos.* See P. M. Harman, *Wranglers and Physicists,* Manchester: Manchester University Press, 1985.
66. *The Electrician,* vol. 21, p. 829, November 2, 1888.
67. This paper was reprinted in Poynting's *Collected Scientific Papers,* Cambridge: Cambridge University Press, 1920, pp. 175–193. The paper was received by the Royal Society on December 17, 1883 and *read* on January 10, 1884, but it carries a footnote added by Poynting dated June 19, 1884, showing it did not appear *in print* until after that date.
68. "On the identity of energy: In connection with Mr. Poynting's paper on the transfer of energy in an electromagnetic field; and on the two fundamental forms of energy," *Philosophical Magazine,* Series 5, vol. 19, pp. 482–487, June 1885.
69. "The transfer of energy," *The Electrician,* vol. 27, pp. 270–272, July 10, 1891.
70. EP 1, p. 378.
71. EP 1, p. 438. This material appeared in *The Electrician,* vol. 14, p. 178, January 10, 1885, and opened with "... let us consider the transmission of energy through a wire." Then came the following cautious words, cut for some reason from EP 1: "How is it done? Knock and it shall be opened unto you? No, not exactly; but we may get a peep through the door 'on the chain'." Heaviside then continued with a delightful *prose* explanation of energy transfer, using not even a single mathematical symbol.
72. EP 1, p. 420.
73. AN, March 26, 1885, pp. 498–499.
74. *The Electrician,* vol. 17, p. 263, August 6, 1886.
75. *Notes on Recent Research in Electricity and Magnetism,* Oxford: Oxford University Press, 1893, p. 308 and p. 313.
76. George Johnstone Stoney (1826–1911), who was FitzGerald's uncle, was the first to advocate the discrete nature of electricity, in 1874 when he made the first estimate of the value of the electronic charge, and in 1891 he coined the name *electron.*
77. J. D. Miller, "Rowland and the nature of electric currents," *Isis,* vol. 63, pp. 5–27, March 1972.
78. *Experimental Researches,* vol. 1, London: Taylor and Francis, 1839, p. 524 (Article 1644).
79. "On the electric and magnetic effects produced by the motion of electrified bodies," *Philosophical Magazine,* Series 5, vol. 11, pp. 229–249, April 1881.
80. It is this force, for example, that helps render *charged* particle beams (electrons, say) useless as space weapons in orbit about the Earth. In addition to beam spreading due to mutual particle repulsion, such a beam is *bent* by the Lorentz-force as it moves through the Earth's magnetic field, making aiming at a distant, fast-moving target virtually impossible. It is also the Lorentz-force, of course, that is behind the explanation of the Hall Effect.
81. EP 2, p. 506. Since Thomson first derived the $q\vec{v} \times \vec{B}$ form, and Heaviside was the first to get the scale-factor right, why (you may ask) is it the "Lorentz-force," instead of the "Thomson-Heaviside force"? Lorentz didn't write on this issue until 1892, but when he did he wrote the *total* force on a charge q moving through *both* an electric field $\vec{E}$ and a magnetic field $\vec{B}$ as $\vec{F} = q(\vec{E} + \vec{v} \times \vec{B})$, and it is this *total* force that many writers call the Lorentz-force.
82. EP 1, p. 446.
83. In 1925 Thomson was the fourth recipient of the Faraday Medal, awarded annually by the Institution of Electrical Engineers (Heaviside was the first winner, in 1922). During the presentation ceremonies, Dr. W. H. Eccles (who played a role in Heaviside's life to be discussed a little later in this book) said that "if he were asked to name two or three of Thomson's great discoveries, he should say that perhaps the principal one was his conception of electromagnetic mass, the electromagnetic inertia of a moving charge. This was the starting point of the revolution in electrical thought that culminated in Einstein's work." Quoted from *The Electrician,* vol. 94, p. 370, March 27, 1925.
84. EP 2, pp. 504–518.

85. In his 1881 paper Thomson had made two important *assumptions* without proof. First, that the surface charge distribution on a conducting sphere is unaltered by the sphere's motion. Second, that the electric field due to this charge is undistorted by the motion. The first assumption was shown to be *true* by William Blair Morton (1868–1949), who cited Heaviside, in his paper "Notes on the electro-magnetic theory of moving charges," *Philosophical Magazine,* Series 5, vol. 41, pp. 488–494, June 1896. The second assumption was shown to be *false* by Heaviside (see EMT 1, pp. 271–272), who showed the electric field is *weakened* along the direction of motion and *strengthened* in the direction *perpendicular* to the motion. The field is always *radial,* however, a fact that Heaviside rightfully called "rather remarkable"—EP 2, p. 495.
86. Heaviside received this letter in time to write a corrective footnote in *Electrical Papers* when the paper was reprinted, and in it he graciously mentioned Searle (EP 2, p. 514) and showed no irritation at having been hoisted on his own petard. He also wrote of this episode in EMT 1, pp. 269–274. Oliver truly was, I believe, interested only in the truth. As he wrote in the Preface to EP 1, "However absurd it may seem, I do in all seriousness hereby declare that I am animated mainly by philanthropic motives. I desire to do good to my fellow-creatures, even to the *Cui bonos.*"
87. This was actually the second letter Searle wrote to Heaviside. The first one, on September 14, 1891, was later called "trivial" by Searle, but in fact it was an interesting description of how Searle planned "to search for magnetic conductivity," and he wrote to Heaviside "your work is the only place in which I have seen this quantity mentioned." Whatever Oliver may have written in response, Searle did not follow up on his initial ideas.
88. Biographical information on Searle can be found in the obituary notices published by *Nature* (vol. 175, pp. 282–283, February 12, 1955) and the Royal Society (*Biographical Memoirs of Fellows of the Royal Society,* vol. 1, pp. 247–252, 1955, which includes a complete bibliography of his papers and books), and the very interesting essay by A. J. Woodall and A. C. Hawkins, "Laboratory physics and its debt to G. F. C. Searle," *Physics Education,* vol. 4, pp. 283–285, 1969.
89. J. Hendry (Ed.), *Cambridge Physics in the Thirties,* Bristol: Adam Hilger, 1984, p. 82. For more insight into why Searle must have appeared as an exacting, perhaps even a bit of a terrifying man to his students, see his "Hints on practical work in physics," in *Experimental Elasticity,* Cambridge: Cambridge University Press, 1908, pp. 178–183. These notes could well be handed out today to modern students, but they no doubt would fail to appreciate why their professors would find them not only informative but also funny.
90. "Problems in electric convection," *Philosophical Transactions of the Royal Society,* vol. 187, pp. 675–713, 1896.
91. For a history of this see R. McCormmach, "H. A. Lorentz and the electromagnetic view of nature," *Isis,* vol. 61, pp. 459–497, 1970; T. Hirosige, "Electrodynamics before the Theory of Relativity, 1890–1905," *Japanese Studies in the History of Science,* vol. 5, pp. 1–49, 1966; and (for a discussion of turn-of-the-century electron theories) J. T. Cushing, "Electromagnetic mass, relativity, and the Kaufmann experiments," *American Journal of Physics,* vol. 49, pp. 1133–1149, December 1981. Oliver Lodge tried to use the "increasing mass with speed" idea to "explain" radioactivity. The idea was that orbital electrons, by virtue of their curved paths, are accelerating. Thus they radiate energy, move in toward the center of the atom (speeding up as they do so, hence increasing their mass), and, so hypothesized Lodge, this "would disturb the balance of forces holding the parts of the system together." Hence, the atom flies apart and we have (radioactive) decay. This sequence of events opens up more questions than it answers, but all in all it has the flair of Lodge's writing and *is* interesting reading: "A note on the probable occasional instability of all matter," *Nature,* vol. 68, pp. 128–129, June 11, 1903. Lodge may well have been influenced by Heaviside as this appeared just months after Oliver had written, also in *Nature,* on the radiation from an electron orbiting the atomic nucleus (EMT 3, pp. 158–164). He followed this with two more papers in 1904 (EMT 3, pp. 169–175). Heaviside opened the second of these three papers with "The complete formula for the radiation may be useful to some of those who are now indulging in atomic speculations."
92. NB 18:326.
93. Searle published sporadically over the years on this subject, with the last such paper appearing in 1942 ("The force required to give a small acceleration to a slowly-moving sphere carrying a surface charge of electricity," *Philosophical Magazine,* Series 7, vol. 33, pp. 889–899, December 1942). In that last paper on charges Searle thought back over his long relationship with Heaviside, and of their occasional disputes on faster-than-light motion, and wrote of his then long-dead friend, "Oliver and I had many friendly combats in those days of long ago."
94. EP 2, p. 494.
95. Searle was referring to Heaviside's cryptic comments, in EP 2, p. 516, where he quickly states a result for $u > v$ and then writes "I regret that there is no space for the mathematical investigation, which cannot be given in a few words ... ." Just like Fermat's Last Theorem scribbled in a book's margin!
96. "On the steady motion of an electrified ellipsoid," *Philosophical Magazine,* Series 5, vol. 44, pp. 329–341, October 1897.
97. This essay is dated January 14, 1898, but wasn't published until the following year as an appendix in EMT 2, pp. 533–535.
98. EMT 2, pp. 112–113.
99. EMT 3, pp. 164–169.

100. Charles Vernon Boys (1855–1944), the inventor of the high-speed *Boys Camera* which used a steel mirror revolving at 60,000 rpm to take pictures with the light of electric sparks microseconds in duration. See Boys' article that appeared in *Nature,* vol. 47, pp. 415–421, March 2, 1893, and pp. 440–446, March 9, 1893, which has several photographs showing the conical acoustic shock waves produced by bullets moving at up to 1400 mph.
101. For a good technical treatment, as well as a pretty color picture of the characteristic pale-blue glow of the radiation from the faster-than-light electrons (in water) produced by a submerged nuclear reactor, see J. V. Jelley, *Cherenkov Radiation,* Elmsford, NY: Pergamon Press, 1958.
102. *Nobel Lectures (Physics) 1942-1962,* Amsterdam: Elsevier, 1964, pp. 421–483.
103. Of course Heaviside did not do everything by himself, but then neither did Cherenkov, Frank, and Tamm leave the subject complete, even on issues at the fundamental level. It wasn't until 1963, in fact, that some very peculiar properties of the electric field inside and *on* the conical wave of a hyperlight charge were finally explained. Until this work, for example, it appeared that Gauss' law failed at hyperlight speed, i.e., the integral of the electric field (which is found to point *toward* a positive hyperlight charge, rather than away as it does for sub-light speeds) over a closed surface appears not to equal the enclosed charge, but instead to diverge! This "puzzle" of the direction of the field, in particular, disturbed Heaviside who called it "an impossible electrical problem" in EP 2, p. 499. For the resolution of these questions see the very readable paper by G. M. Volkoff, "Electric field of a charge moving in a medium," *American Journal of Physics,* vol. 31, pp. 601–605, August 1963.
104. AN, July 4, 1907, pp. 220–221.
105. To Bernard Arthur Behrend, a consulting engineer in Boston and a great admirer of Heaviside who acquired some of Oliver's personal library after his death. Behrend's wife donated all of his Heaviside memorabilia of books, letters, and issues of *Nature* to the Institution of Electrical Engineers in London after her husband's suicide (under somewhat bizarre circumstances) in 1932.
106. The head of the University's world-class mathematics department, perhaps not without connection to the award, was Heaviside's great admirer, Felix Klein (Note 47).
107. Professor Ernst Berg at Union College, Schenectady, New York (the letter was dated June 27, 1924). I thank Schaffer Library at Union College for permission to quote from this and other Berg–Heaviside correspondence in its possession.
108. Heaviside couldn't get away from this part of his past. It was his present (and his future), too. On the nomination form for election to Fellow of the Royal Society, the line for *Profession or Trade* is bluntly filled in with "None."
109. *Nature,* vol. 47, p. 293, January 26, 1893.
110. H. H. Skilling, *Fundamentals of Electric Waves,* New York, NY: John Wiley, 1948, p. 132.
111. *Lectures on Physics,* vol. 2, Reading, MA: Addison-Wesley, 1964, pp. 17-5 to 17-6 and p. 27-11.
112. E. M. Pugh and G. E. Pugh, "Physical significance of the Poynting vector in static fields," *American Journal of Physics,* vol. 35, pp. 153–156, February 1967.
113. "On the connection between electric current and the electric and magnetic inductions in the surrounding field," *Collected Scientific Papers* (Note 67), pp. 194–223.

8
The Battle with Preece

Communication engineering began with Gauss, Wheatstone, and the first telegraphers. It received its first reasonably scientific treatment at the hands of Lord Kelvin, after the failure of the first transatlantic cable in the middle of the last century; and from the eighties on, it was perhaps Heaviside who did the most to bring it into a modern shape.

— Norbert Wiener, *Cybernetics,* 1948

I am unable to understand the advantage of any gain in speaking on a wire which is detrimental to telegraphic communication. Speed is of more importance than speech, and we can telegraph much faster than we can speak. In England speed is everything... .

— William Henry Preece, expressing his position on the value of the telephone in 1882

We hear sometimes of the "sweet simplicity of Nature", but to the man of science the sweet simplicity seems to be chiefly exemplified in the brains of those who employ the phrase.

— Editorial comment in *The Electrician* (October 12, 1888) directed toward W. H. Preece's ideas on how signals are transmitted by telegraph and telephone wires

...really dreadful stuff, rank quackery.

— Oliver Heaviside, expressing his opinion on Preece's ideas on the inductance of copper wires, in an 1888 letter to Oliver Lodge

If it is love that makes the world go round, it is self-induction that makes electromagnetic waves go round the world.

— Oliver Heaviside, 1904

...A WH was very afraid of P, who talked him over and convinced him that my theory was all theoretical bosh, nothing like practical conditions and A WH went over to P's side. The greatest mistake of his life. Spoiled all the rest of it.

— Oliver Heaviside, 1922

HIGH-TECH HARDWARE, LOW-TECH THEORY

As the 1870s ended and the 1880s started, the state of the electrical communication industry in England was one of impressive capability. But oddly, while the *technology* of communication was highly developed, it is a curious fact that a *theoretical* understanding of that technology was seriously deficient. The great strides forward in practice and hardware had come from clever men blessed with skilled hands and ingenious insights, guided by inspired intuitions. Theoretical considerations, in fact, had had very little to do with shaping the telecommunications business of the 1870s and 1880s, with the only mathematical theory available being William Thomson's nearly thirty-year-old analysis of very long, capacity-dominated cables (as discussed in Chapter 3). In *practice,* however, things were high technology, indeed.

By 1895 the worldwide telegraphic network so efficiently webbed the Earth that, according to William Preece, one could literally send a message around the planet.[1] At all times the message would travel by electricity only, with no messenger service required to fill in any gaps. Along its way it would pass through the wires of the English, American, Danish, Russian, Japanese, Swedish, and Australian telegraph systems, as well as others. It would, calculated Preece, take fifty-six hours and cost about eighteen dollars per word, rather expensive both in time and money, but still a very impressive *technical capability*.

From the very start, with the laying of the first successful undersea cable in 1851 (across the English Channel, between Dover and Calais), the British were preeminent in all forms of telegraphy. Even those cables not owned by them (e.g., the 335-nautical-mile Newbiggin/Sondervig cable owned by the Great Northern Telegraph Company, Heaviside's first and last employer) were almost certainly manufactured[2] and laid by Englishmen, and even often maintained by them (Oliver's job during his stay in Denmark). The underwater communication net grew so quickly that by 1896 there were 160,000 nautical miles in it (laid at an average cost of $1200 per mile).[3] Many of these Victorian cables, laid by England to unite her with far-flung political, financial, and military interests, often thousands of miles distant, continued in use until well after the Second World War. They spread out over the Earth like a gigantic spider web, with London at the center.

The situation in overland telegraphy was equally impressive. The use of relatively short (at most 400 miles inside England and Scotland) pole-strung iron wires avoided the major problem of long undersea cables—their very large capacitance. Whereas the cables might operate at a few tens of words per minute, the overland telegraphs didn't reach their limits until the *hundreds* of words per minute.[4] Consequently many messages could be sent, quickly. This, combined with the ease of maintaining such a readily accessible system, resulted in a low-cost, highly popular[5] means of private and official communication.

Then came the invention of the telephone in 1876 (William Preece's "scientific toy"), an instrument far beyond the power of the then existing theory in common use by engineers. While inductive effects could often still be safely ignored on the overland circuits, the much higher frequencies present in human speech (as compared to on–off telegraph keying) pushed Thomson's low-frequency theory past its breaking point. The severe frequency-dependent amplitude decay and propagation speed characteristics of long submarine cables rendered them virtually useless for telephone service to any significant distance, and appeared to make the thought of *transoceanic* telephone conversations unthinkable.[6]

Even overland telegraphy wasn't totally free of the problems of induction. Its very popularity was the root cause, with the ever increasing message traffic making it a common sight to see literally *hundreds* of wires crowded together on the maze of cross-arms attached to each pole (this ugly scene continued to exist in America long after the English had gone, by the late 1880s, to both underground cables and vertically stacked lines on roof-mounted poles with no cross-arms). The density of closely packed wires could continue to increase only so long before reaching the point of severe inductive cross-coupling of messages among the wires, i.e., "cross-talk." The mixing of telephone circuits with telegraph wires soon became intolerable because of cross-talk. Ten years after the fact Preece recalled,[7] "In 1884 electromagnetic disturbances were detected between telegraph circuits and telephone circuits 80 ft. apart..." and by 1887 things were sufficiently bad that he said,[8]

> I defy anybody...to speak from the top of the General Post Office down to the bottom if the wire passes through the chasings carrying the instrument leads to the Central Telegraph Station. It is impossible to speak through 100 yards of a heavy street line; ...and the reason is very simple - that the mutual induction currents from these powerful currents used for Wheatstone transmitters are probably 100,000 times greater than the currents that are used to work a telephone transmitter.

The 1884 observations were sufficiently disturbing that Preece had Oliver's brother, Arthur, conduct induction tests on the Town Moor in Newcastle to determine what the range of induction telegraphy might be. In the paper[9] reporting these experiments to the 1886 B.A. meeting at Birmingham, Preece mentioned his interest in wireless telegraphy actually dated back several years, and he reminded his audience of experiments,[10] using large metal plates immersed in the sea, for signaling across water between Southampton and the Isle of Wight. The mechanism here, of course, was neither radiation nor induction, but rather conduction currents in the water. The report at Birmingham would form the basis, as we'll see, for Preece's claim to have been the real "father of wireless," not Marconi. Preece failed to appreciate the difference between induction (near field) and radiation (far field) effects, even after the turn of the century when the difference was well known and he was told there was a difference.

Even without cross-talk problems, good-quality land-line telephony could reach out to a maximum uninterrupted distance of perhaps 100 miles. An historically interesting and unique exception to this is the work of Francois van Rysselberghe (1846–1893). Using special telephones of his own design he was able, in the winter of 1885–86, to achieve good telephony over a 1000-mile telegraph line between New York and Chicago. His system, using innovative capacitive coupling and isolation interface circuits to *existing* long-distance telegraph wires, to carry speech *as well* as telegraph signals, was an astonishing technical demonstration. In the words of a modern historian,[11]

> Not surprisingly, he was excited about it and optimistically predicted that transcontinental telephony was now feasible. However, it would be thirty years before this was achieved. Heaviside's theory of telephone transmission had not yet been published, and van Rysselberghe had no means of correct [theoretical] extrapolation from his [experimental] results.

This carly and important work was ignored by the British Postal authorities who, with their stranglehold monopoly, really had no economic incentive to innovate and were quite content with the *status quo* of their high-speed telegraphs. Indeed, Preece gave that as his reason for why the van Rysselberghe system would fail in England—van Rysselberghe's isolation circuits might work on *low*-speed telegraph lines, but would surely fail on the English high-speed circuits (and in this he was probably correct).

Induction, representing the flow and ebb of energy into and out of magnetic fields, was at the root of such communication difficulties, and to Preece all was clear—induction was a "bugaboo"[12] with "evil" effects. The solution, therefore, according to Preece, was to *eliminate*[13] induction. And if he couldn't do that, he would just ignore it! This appealingly simple approach was to mislead Preece all his life. Preece, himself, would have laughed at the ignorant people who thought telephones would result in the spread of disease by transmitting germs over the wires (a not uncommon belief at the time), but his own misconceptions were far more serious and of no laughing matter. They would have a major "blocking effect" (to use Oliver's phrase) on technical progress all through the last two decades of the 19th century.

Early Mathematical Analysis

That odd and curious things happen when currents in conductors are alternating rather than constant has been known (at least theoretically) since the mid-1850s. Long before Maxwell, Kirchhoff had discovered[14] on paper that there should be a nonuniform distribution of the current over the conductor's cross section. Maxwell, too, made specific mention of this fact in his *Treatise* (Articles 689 and 690). As the frequency of the alternation increases, the current tends to flow ever closer to the surface of the conductor, defining what has come to be called the "skin effect"—see Tech Note 1 at the end of this chapter. For sufficiently high frequencies, in fact, the conductor might as well have a ***hollow core,*** as the central region of the conductor carries essentially none of the current.

The first post-Maxwell, in-depth mathematical treatment of "what happens" in the case of alternating currents seems to have been carried out by Horace Lamb (1849–1934). Lamb, who studied under Maxwell and Stokes at Cambridge (where he was 2nd Wrangler in 1872), is remembered today mostly for his pioneering work in hydrodynamics but his powers were very broad and he was, in fact, a world-class applied mathematician of the absolute first rank. Maxwell's theory was the perfect test for his skills. As he wrote in the opening paragraph of his brilliant paper[15] published in 1883,

> This paper treats of the motions of electricity produced in a spherical conductor by any electrical or magnetic operations outside it. The investigation was undertaken *some time ago* [my emphasis—a footnote indicated Lamb was in possession of his results prior to November 1881] in illustration of Maxwell's theory of electricity. This theory is so remarkable...that it seems worth while to attack some problem in which all the details of the electrical process could be submitted to calculation... .

Lamb then proceeded to calculate a wide variety of things.[16] In particular, for a metallic sphere immersed in an oscillating field, he found that

> ...the disturbance inside the sphere consists of a series of waves propagated inwards from the surface with rapidly decreasing amplitude. Thus at a depth equal to the wavelength (ν, say), the amplitude is only 1/535 of what it is at the surface. The currents are therefore almost entirely confined to a superficial stratum of thickness comparable with ν.

Soon after, in January 1885, Heaviside generalized Lamb's result from spheres to conductors of any shape.[17] This was done in the same issue of *The Electrician* in which he developed a prose description of the flow of energy in the field *outside* of a current-carrying wire. To set the stage for what eventually came to be called the "skin effect" he first wrote (as we've seen in the previous chapter),

> Most of the energy is transmitted parallel to the wire nearly, with a *slight slant* [my emphasis—see Tech Note 4, Chapter 7] towards the wire in the direction of propagation... .

and

> Now go into the wire. A tube of energy current arriving at the surface of the wire by a long slant [i.e., almost parallel to the wire] at once turns round and goes straight to the axis. In passing from the battery to the wire through the dielectric [i.e., space] the energy current is continuous...but directly it reaches the conducting matter of the wire dissipation commences and the current begins to fall in strength, and on reaching the axis has fallen to nothing.

And then, at last, the "skin effect":

> Since on starting a current the energy reaches the wire from the medium without, it may be expected that the electric current in the wire is first set up in the outer part, and takes time to penetrate to the middle. This I have verified by investigating special cases. Increase the conductivity of a wire enormously, still keeping it finite, however. Let it, for instance, take minutes to set up current at the axis. Thus, ordinary rapid signaling "through the wire" would be accomplished by a *surface current only, penetrating to but a small depth* [my emphasis].

With the publication of the papers of Lamb and Heaviside, the *mathematical theory* of induction and the skin effect phenomenon, as they follow from Maxwell's electrodynamics, were firmly established. But it took the work of two *experimenters,* separated in time by two years, to bring the theory off the journal pages and into active discussion, and eventually into practical applications in telegraphy and telephony. Sandwiched in between these two experimental efforts, both of which concluded with noisy public debates, was what must have been the worst year of Heaviside's life. The year 1887 marked the low point of his career, when much of his most important work looked as if it would be suppressed and when, for a time, he was actually blackballed from the technical literature.

The Peculiar Experiments of David Hughes

In January 1886 David Hughes (friend of Preece, and together with him veteran of verbal warfare with Edison over the invention of the microphone) assumed the President's office of the Society of Telegraph Engineers and Electricians (which soon after became today's Institution of Electrical Engineers). As might be imagined for one elected to such a high position, Hughes was a man who enjoyed great prestige; his words were listened to with care and they carried weight. He had been awarded the Royal Medal of the Royal Society the year before (he had been elected a Fellow in 1880), and as an experimenter he was rightfully thought of as blessed with a golden touch. But for a fluke of history, for example, he would have predated Hertz *by nearly a decade* in the experimental discovery[18] of electromagnetic radiation (in 1879, the year of Maxwell's death).

For his Presidential Address Hughes presented the results of some experiments, using an induction balance, that he had performed on the behavior of conductors (of varying shapes) when carrying rapidly varying nonsinusoidal currents. The details[19] of these experiments are not directly important here except for the fact that they were greatly confusing to everybody who listened to Hughes, and most of all to Hughes, himself! A nonmathematical man, he literally had no idea of the true nature of what he had observed, and never appreciated (even after being told, more than once) that it was the skin effect he had experimentally stumbled upon. As Heaviside wrote[20] to Lodge a few years later, in 1888:

> Prof. Hughes (though he is a truly great man in his way) had no more notion of how a current begins [i.e., on the surface of a wire] than the pump.

Heaviside had not always been so charitable toward Hughes. In the late 1870s Hughes had

David E. Hughes (1831–1900)

Star-crossed indeed was Hughes, whose talent for creative experimental work was cursed with a nasty streak of bad timing. Not only did he miss, like Lodge, the discovery of electromagnetic waves in space, he also came within an eyelash of discovering (positive) feedback in 1883, with the construction of the "howling" telephone—with the mouth and ear pieces coupled. But the mathematical explanation of this waited until Harvey Fletcher at Bell Labs finally did what math-primitive Hughes couldn't (in 1926, the year before his colleague, Harold Black, discovered *negative* feedback).

become nearly obsessed with both of the amazing new electrical inventions, the Bell telephone and his own microphone. With the microphone, in particular, he succeeded in impressive fashion to make it sensitive to an unimagined degree, causing Preece to declare that the footfall of a fly could be made to sound like the thundering of a horse on a wooden bridge. But *how* the microphone worked, in detail, was open to much nonproductive speculation, with Hughes saying[21] he could understand the failures of others because "...it is as much as I can do to understand it myself." He could have learned much from a reading of Heaviside's 1878 paper[22] "On electromagnets, etc.," in which is treated (among many other topics) how "to obtain the greatest magnetic force from [sine-wave] reversals [of current]" in electromagnets, which Heaviside clearly stated "has application to the Bell telephone." This is a highly mathematical paper but it is a *circuits* paper, *not* a *field theory* one, and is heavily illustrated with practical engineering examples.

This paper was followed a few years later, in 1883, with an *experimental*(!) paper,[23] "Theory of microphone and resistance of carbon contacts." Hughes couldn't have understood much of the 1878 paper, but this one doesn't have a single equation in it. Hughes, in fact, did read a first draft of this paper when Heaviside originally submitted it to the *Journal of the Society of Telegraph Engineers.* Hughes was the reviewer, and on September 22, 1881 he returned the paper to the Society's Secretary F. H. Webb (who was Preece's brother-in-law) with the comments "it is far too long for the amount of original material it contains," "very correct, but too well known to be of any interest," "well written," and suitable only to have its *abstract* read "if you are in absolute need of a paper to fill an evening." The rejected paper eventually appeared in *The Electrician.*

To Heaviside it was all so unfair—Hughes was famous in his public ignorance and Oliver was obscure in his private brilliance. This may well be why, in July 1883, he made a bitter entry[24] in one of his research notebooks:

> We have the Sonometer [Hughes' name for his audiometer, invented in 1879 for hearing tests] of the philosopher of Gt. P. St. [Great Portland Street, where Hughes lived] with its "absolute zero." Nothing absolute about the zero at all... . The Philosopher of G.P. St. is one of the men who get more credit than they deserve *at the time.*

With Hughes once more making news in 1886, in an area Heaviside felt was "his," Heaviside was not inclined to let pass without comment that once again "it was all Hughes could do to understand" what he'd stumbled upon. Heaviside wanted it *publicly* known who knew what.

The dispute that erupted over Hughes' attempted (but wrong) explanation was important, because it dramatically made it quite clear, even to the "practical men," that the theoreticians who could analyze what was going on with rapid currents in conductors might have some value after all. As William Thomson kindly put it in *his* Presidential Address[25] just three years after Hughes' (the S.T.E. & E. had just become the IEE and Thomson was the first President under the new name):

> In the memorable Presidential Address of Professor Hughes...electro-magnetic induction was very admirably illustrated by experiments which are now more or less familiar to us all, but which have been of an immensely suggestive and stimulating character, both to mathematicians and to experimental workers. The very criticisms by mathematicians [e.g., Heaviside!] upon some of the experiments and modes of statement by Professor Hughes have...given a very large body of electric knowledge and electro-magnetic knowledge which, without such stimulus and such mathematical and experimental scrutiny as it has led to, might have been wanting for many a year.

In one way this gracious praise from Thomson might be considered Hughes' due—but he wasn't to get it until after surviving some rather blistering sarcasm from "mathematician" Heaviside. Writing in *The Electrician* of April 23, 1886, Heaviside opened with[26]

> No man has his own grammar to a greater extent than Prof. Hughes. Hence, he is liable, in a most unusual degree, to be misunderstood [see Note 27]... . Owing to the mention of discoveries, apparently of the most revolutionary kind, I took great pains in translating Prof. Hughes' language into my own...the discoveries I looked for vanished for the most part into thin air. They became well known facts when put into common language... . I have failed to find any departure from the known laws of electromagnetism.

The next week he continued his attack on Hughes' work by showing[28] how, in an elaborate mathematical analysis,

> Prof. Hughes' balance is sometimes fairly approximate, sometimes quite false... . Such a balance is damned [this is the word that originally appeared in *The Electrician*—when reprinted in EP 2, it was replaced by *condemned*] for scientific purposes.

Hughes did try to defend himself. In a letter[29] to *The Electrician* he wrote

> ...if we assume with Mr. Heaviside that all my results are contained in some mathematical formula of which I am unaware, then the mathematicians [i.e., Heaviside!] ought to feel extremely grateful for the experimental proof which I have furnished them... .

Heaviside would have none of this, and quickly replied[30] by first brushing aside Hughes' above statement, then reminding readers of *his* priority in the matter of skin effect, and finally closing by offering his appreciation for what he knew *would be* better work from Hughes, now that he (Heaviside) had straightened him out!:

> ...I said not a word about mathematicians...in my opinion, the really important part of Prof. Hughes' researches [is the] thick wire effect [see Note 31]... . Having been, so far as I know, the first to correctly describe the way the current rises in a wire, viz, by diffusion from its boundary, and the consequent approximation, under certain circumstances, to mere surface conduction; and believing Prof. Hughes' researches to furnish experimental verification *of my views* [my emphasis], it will be readily understood that I am specially interested in this effect; and I can (in anticipation) return thanks to Prof. Hughes for accurate measures of the same, expressed in an intelligible form, to render a comparison with theory... .

All in all, even a gentle man like Hughes must have felt his blood pressure rise after reading that!

A few months later Heaviside continued the assault on Hughes with very blunt language in the *Philosophical Magazine*. Beginning by reminding his readers of his previous writings on energy flow in space, and of his analyses of how a current in a conductor begins first on the surface and afterward penetrates inward, Oliver then grabbed Hughes by his coat and gave him a severe shaking[32]:

> Attention has recently been forcibly directed towards the phenomenon...of the inward transmission of current into wires by Professor Hughes' Inaugural Address to the Society of Telegraph Engineers and Electricians, January, 1886. This paper was, for many reasons, very remarkable. It was remarkable for the ignoration of well-known facts, thoroughly worked out already [by Heaviside!]; also for the mixing up of the effects due to induction and to resistance, and the author's apparent inability to separate them, or to see the real meaning of his results; one might indeed imagine that an entirely new science of induction was in its earliest stages. It was remarkable that the great experimental skill of the author should have led him to employ a method which was in itself objectionable, being capable of giving, in general, neither a true resistance nor a true induction-balance... .

And then comes page after page of analytical treatment, all in the form of Maxwell's theory under the title "New (duplex) method of treating the electromagnetic equations." Hughes couldn't have understood a word of it. But others could, and it was becoming evident that what *had* seemed strange and peculiar to Hughes could be explained, even *predicted,* by those who (like Heaviside) understood that electricity in general (and induction, in particular) required more than just Ohm's law.

Preece's "*KR*-Law" and Heaviside's Attack

At the time of Hughes' address Heaviside was involved in writing a lengthy series of articles for *The Electrician* on electromagnetic theory ("Electromagnetic induction and its propagation"). The announcement of Hughes' experiments motivated Oliver to break off this series[33] temporarily and to switch to publishing a new series on the practical implications of self-inductance for both telegraphy and telephone transmission. This work would eventually lead him to what is perhaps his most important discovery—the criterion for the distortionless transmission of signals over distributed parameter lines, and the seemingly paradoxical conclusion that induction *aids* the transmission of signals. By December 1887 he had developed the theory of uniform[34] transmission lines, including for the first time the effects of all *four* parameters: resistance, leakage, capacitance, *and* inductance. When reprinted[35] in book form it would represent nearly 300 pages of analysis! This achievement was not, however, initially greeted with applause. Certainly not by William Henry Preece.

In 1886 Preece asserted,[36]

> ...self-induction in various forms has proved a *bête noir* which required all our knowledge and all our skill, not only to master, but to comprehend; for *the effects of self-induction are invariably ill-effects* [my emphasis]... .

Preece held this view, in direct opposition to Heaviside, because of experiments he had done that seemed to support it, e.g., in a paper[37] read at the 1885 B.A. meeting at Aberdeen he observed that copper is better than iron for telegraph use because one can transmit faster over copper, and he hypothesized that induction was at the bottom of it all:

> Possibly the magnetic susceptibility of the iron is the cause of this. The magnetisation of the iron acts as a kind of drag on the currents. It is well known that telephones always work better on copper than on iron wires, doubtless for the same reason.

Since copper transmits signals better than does iron, Preece concluded that copper wires must have *less* evil inductance than do iron wires. This is true, but Preece went too far and claimed copper has essentially *no* inductance. Indeed, in 1887 he claimed[38] the inductance of copper wires to be a hundred times or more smaller than given by theory, and by 1889 he had backed himself into a corner and had to assert[39] the theory must be in error!

> Followers of Maxwell have placed too great reliance on a formula of his which gives a co-efficient of self-inductance greatly in excess of what is met with in practice, and it is questionable whether Maxwell's formula is not based on a false assumption.

Preece was partly correct, because the self-inductance of iron *is* on the order of *thousands* of times that of copper. The problem with iron as a carrier of high-frequency signals is, however, the *skin effect*, i.e., the greater magnetic permeability of iron (as compared to copper) leads to thinner conducting skins and hence to greater impedances (as compared to copper).

In February 1887 Preece finally formulated[40] in analytic terms his ideas on signal transmission. This became known as the "*KR*-law"; it completely ignores induction (as well as leakage) effects, declaring only the total resistance and the total capacitance of a circuit to be of importance. This "law" is based on nothing but an erroneous application of William Thomson's "Law of Squares" (see Chapter 3). Preece, however, asserted it applied under *all* conditions. He wrote this "law" as (see Tech Note 2):

$$x^2 = \frac{A}{kr}$$

where x is the maximum distance speech can be transmitted over a circuit having a resistance and a capacitance per unit length of r and k, respectively. Preece selected various values for A, depending on the nature of the line, and in this way attempted to adjust for the inductive and leakage effects he had ignored in deriving the "*KR*-law."

Not everybody accepted Preece's simple view of matters. Silvanus P. Thompson and W. E.

Ayrton,[41] both highly respected electrical engineers, expressed objections. Thompson, who for a while believed (incorrectly) that improvement in telephone transmission was to be found in better sending and receiving instruments and *not* in altering the nature of the transmission circuit, was attacked by Preece on this point. Thompson was nobody's pussycat, however, and he gave[42] it right back to Preece:

> I should recommend Mr. Preece for the future to avoid mathematical arguments. My only reason for replying to his extraordinary statements about mathematical theory is that I do not let them pass unchallenged for fear that other members of the society [Society of Telegraph Engineers and Electricians] would otherwise think that Mr. Preece's ideas on this topic were either authoritative or intelligible.

Such open ridicule from a man who could not be dismissed by a snap of the fingers and a sneer (as could Oliver) stung Preece to the core. By the spring of 1887 he was on the defensive ("in a rather irritated state," as Heaviside put it in an undated letter fragment) with his position on induction, and it wasn't a particularly good time for Oliver to challenge him. But that's exactly what he did.

At precisely the time Preece was penning his "*KR*-law," Oliver and his brother Arthur prepared a long paper entitled "The bridge system of telephony" based on Arthur's experimental work at Newcastle.[43] This paper required a prepublication release from the GPO (i.e., from Preece) because of Arthur's official position. In a letter (dated September 24, 1888) to Oliver Lodge, Heaviside recounted how his brother initially obtained "reluctant" approval to write the paper, with Oliver to contribute supporting theoretical analyses. The paper was long (60 pages), and included three appendices by Oliver, in which were stated for the first time the conditions for distortionless transmission. When Preece saw the finished paper, so completely and utterly at variance with his own views, he refused final permission to publish. He also did not return the paper, which in those days of no photocopy machines meant the Heaviside brothers had no backup manuscript to show others. This suppression did not come as a total surprise to Oliver, who in anticipation of it had *not* sent the appendices to Preece (apparently the main text just quoted results from Oliver's theoretical analyses, but that was enough to set Preece off).

Even though he had half expected it, the refusal to publish enraged Oliver. When somewhat altered versions of his appendices were finally published[44] in *Electrical Papers,* he added a footnote damning Preece as the "official censor":

> I was given to understand that the official censor ordered it all to be left out [Oliver's contribution], because he considered that the Society [S.T.E.&E.] was saturated with self-induction, and should be given credit for knowing all about it.

Oliver was not to be put off, however, and on June 3 he published[45] the distortionless circuit condition in *The Electrician*:

> Make $R/L = K/S$ and the distortion is annihilated... .

R, L, S, and K are the distributed per unit length resistance, inductance, capacitance, and leakage conductance of the transmission circuit; a fiendishly clever derivation[46] of this result (due to Oliver, using his operational mathematics) is given in Tech Note 1 of Chapter 10. In this paper Heaviside also gave some practical engineering suggestions on how to achieve this condition by increasing the inductance (L), as in general the "normal" values of R, L, K, and S are such that R/L is greater than K/S (see Tech Note 4). He referred to experimental evidence of good long-distance telephony using copper-covered steel wires:

> Here the copper covering practically decides the greatest resistance of the wire [R]; what current penetrates into the steel [below the skin depth] lowers the resistance and increases the inductance... .

He then went on to suggest,

> ...I think we may expect great advances in the future... . One way is with my non-conducting iron..., an insulator impregnated with plenty of iron-dust. Use this to cover the conductor. It will raise the inductance greatly... .

In addition to getting his technical work into print, he also attempted to make Preece's act of censorship public. Here he had less success. One can only wonder what he sent Biggs at *The Electrician,* but on May 30 Biggs' reply to his angry contributor was cautious, even frightened:

> I would use your letter if I could, but it is dangerous in the present state of the law, for however true it may be and however necessary, it is libelous... . I may tell you that at present six of us have two libel suits each against us—or a round dozen altogether, and I venture to predict that the cost even if we successfully defend ourselves will be considerable.

Biggs then informed Heaviside that some of the proprietors of *The Electrician* were so afraid of these suits that they had

> ...begged me over and over again to take over their shares in the concern in order that they might be free from the trouble and worries attending such suits. At present I have not been able to see my way to do this—I have therefore to be doubly cautious not to insert what even has the appearance of a libel.

Biggs' next words were blunt:

> There is no appearance however in your letter, it is straight to the point. Its intention is to show up P. and no matter how justifiable, is a libel. Candidly, then, I am afraid to use it, not personally, but for Sir James Anderson [Proprietor of *The Electrician* who had been knighted for his role as Captain of *The Great Eastern* in the laying of the 1866 Atlantic cable] whom I would not willingly give a moment's anxiety...as I esteem his friendship.

Biggs closed with the thought he was sure Heaviside would now agree with his position, but apparently Oliver didn't, as Biggs wrote again on June 1:

> I am afraid you don't quite understand the peculiar state of the law of libel, nor the...worries which the possibilities of an action give to all those concerned... . It is not the author alone then, but a number of people who suffer, no matter how true the facts of the case or how necessary the exposure.

It was Heaviside's personal attacks on Preece that Biggs feared, however, not technical ones, and he went on to counsel Oliver:

> Do you not think it rather infra. dig. in a scientist to be so moved at the doings of the scienticulist? (This is a grand word but we must retain it somehow—it conveys a lot of meaning). I really think that we shall do better to stick to the severe criticism without the personal expressions; you will gain more by critically examining Mr. Preece's statement without attacking him personally... . I have in your article marked out a few lines which are personal and phrases which I do not think should appear in an article of this character. Upon calmer consideration I think you will agree that what I have done is for the best.

What finally appeared,[47] while certainly much milder than the original, still retained a good deal of sarcasm:

> Sir W. Thomson's theory of the submarine cable is a splendid thing. His paper [circa 1855] on the subject marks a distinct step in the development of electrical theory. Mr. Preece is much to be congratulated upon having assisted at the experiments [with Faraday] upon which (so he tells us) Sir W. Thomson based his theory; he should therefore have an unusually complete knowledge of it. But the theory of the eminent scientist does not resemble very closely that of the eminent practician.

At the same time that Biggs was trying to reason with Heaviside, Oliver found that other editors were just as concerned. For example, he tried to "expose" Preece in the *Electrical Review,* too, only to receive the cautious reply:

> The question raised in your letter [of June 2] is a very important one but before deciding anything in the matter we think that an opportunity should be given us of seeing the paper to which you refer.

Oliver received this on the same day it was sent, June 3, and on that very same day (even as the distortionless circuit was revealed to the world in *The Electrician*), he wrote back

> Yours of today. "Question raised." I raised no question. But at the same time I can guess that you refer to the evil practice of burking [see Note 48] papers... . The original paper is in Mr. Preece's possession, so I cannot possibly send it to you. Nor have I a copy. You should understand that I want to keep to my *own* concerns, and not drag my brother into it. *He* will not complain publicly, as he understands his subordinate position [see Note 49]. Also, understand that I do not formally complain...though I *suggest* that I have been treated very discourteously under the circumstances. Besides the paper, with Mr. P.'s marginal notes [see Note 50], there was a lengthy document from Mr. P. to my brother savagely written. Without going into details, I may...say I think my brother was badly treated in it (my grievance is little in comparison); but this I have no right to enter upon at all and do not. I desire to expose Mr. P. in his scientific capacity... . P.S. My Appendices never went to Mr. P. ... . It [was these appendices] which Mr. P. fell foul of. Of course he fell foul of some of my brother's work, but that is not my concern.

On June 8 the *Electrical Review* declined to print Oliver's "exposure."

Heaviside hadn't been waiting to hear that decision, however, as the day before he had written of his troubles to Silvanus Thompson, a man sure to be sympathetic with such a grievance against Preece. On June 9 Thompson replied with a rather melodramatic letter:

> Do you mean to say, is it true, can it be true that a scientific paper submitted to the S.T.E. [and E.] which happened to contain some formulae that differ from Mr. Preece's opinions is rejected on the recommendation of Mr. Preece? If so, if that is true,—and I beg you to tell me briefly exactly what the truth is—then I, as a humble member of the rank and file of the S.T.E., cannot let the matter rest where it is.

Four days later, now apparently convinced of "the truth," Thompson wrote again:

> It is monstrous that an *official* of a Society should misuse his position in a Society to stifle the publication of scientific truths against which he has happened ignorantly, arrogantly, and openly to pronounce an adverse opinion, lest the publication should expose the ignorance and shallowness of his views.

Heaviside couldn't have said it better, but even with such encouragement he must have been staggered by the next blow, a few weeks later, in July. After printing over 162 pages of his series "On the self-induction of wires," the *Philosophical Magazine* decided to terminate the series. This sudden apparent stonewalling (actually, the journal had originally expected only four parts to the series and when Oliver sent in Part 8, that was too much—it was rejected)

Brothers, contractors for the electric lighting of a part of Turin, did not give the results that were expected in the public service.

"For the illumination of private parties, the contractors, instead of Gaulard transformers, installed the dynamos and transformers of Messrs. Zipernowsky and Déri.

"We have been informed that a large number of Gaulard transformers were found burned out and unfit for use, and Messrs. Bellani have decided to discontinue their employment for the public lighting. At the present moment the public service is secured by the Zipernowski-Déri system.

"The public illumination includes 80 incandescent lamps of 50 candle-power, and 3 arc lamps, all of which are run by the Zipernowsky-Déri system."

In addition, I have in my possession the results of tests made upon the Siemens cables before, during, and after the trials, which show that the failure must be ascribed to some other cause, considering, too, the fact that both systems used alternate currents, not differing greatly in electromotive force.

As the note in question contained no reference as to cause, but confined itself to facts, which Mr. Pickering has not shown to be untrue, I think he has no grounds for complaint.—Yours, &c., J. W. LIEB.

THE FIRE AT THE OPERA COMIQUE.

TO THE EDITOR OF THE ELECTRICIAN.

SIR: I was an eye witness of the catastrophe that occurred last night at the "Theâtre National de l'Opera Comique." It was about the end of the first act in "Mignon" when one of the gas jets in the cross light sputtered owing to a shock which that particular cross light received, due to some stage manipulation. The moment the light reached a sort of net decoration, in close proximity to the cross light, the net caught fire like ignited gun cotton. The decorations being so near the gas lights are always very warm and dry, so that they burn like tinder when lighted. The following is a little sketch in cross section, which shows how the fire originated :—

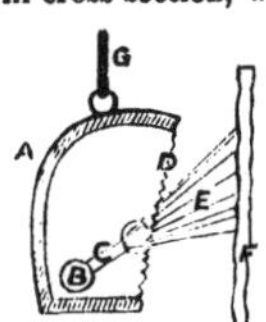

A is the iron frame of the cross light suspended by ropes G. B is the main gas tube, which has at regular intervals jets C, D is a sort of wire gauze, and F shows the decoration in front of the cross lights, E represents the sputter which ignited the decoration F, and caused the loss of so many lives and damage to property.

In a few moments the fire spread over the whole stage, and to-day the Opera Comique of Paris comes within the long list of such calamities which have occurred within the last few years, and reminds one of the Ring Theatre, Stadt Theatre in Vienna, the Arad Theatre and Szegedin Theatre in Hungary, the Nice Theatre, Rouen Theatre in France, and Alhambra in London. These are only a few of the principal ravages that our contemporary the gas light has been the cause of.

Had not the Cie. Continentale Edison been hindered from commencing their work last September, the disaster of yesterday would not have happened; but instead of receiving their contract last year they only received it a few weeks ago. They have been actively pushing their installations, and expected to commence electric lighting in July.—Yours, &c.,

FRANCIS JEHL.

25, Rue Jacob, Paris, May 26th, 1887.

SECONDARY BATTERY QUESTIONS IN BELGIUM.

TO THE EDITOR OF THE ELECTRICIAN.

SIR: The statement which I have made as to the use of minium in the construction of accumulators rests upon the authority of the late Comte Du Moncel. When Du Moncel wrote the following paragraph there can be no doubt that his information came from a reliable source. He says :—

"Planté avait bien pensé, *dès ses premières recherches*, à abréger le travail de la formation, en déposant sur les lames de plomb positives du minium et en réduisant cette couche par le courant polarisateur. . . . . mais l'adhérence de celle-ci était mauvaise et la couche déposée s'écaillait et finis sait par disparaitre.

"Quand on 1881, on pensa à employer les accumulateurs pour l'eclairage électrique et les moteurs, M. Faure *reprit l'idée* de l'application de la couch de minium sur les lames de plomb, et pour maintenir l'adhérence, il enveloppa les lames recouvertes de minium dans des sacs de feutre." "Eclairage Electrique," Du Moncel, 1883, Vol. I., pages 56 and 57.

The problem which remained to be solved after M. Planté was merely to discover a satisfactory method of attaching the minium to the lead plate, and this problem, it is very certain, every one has a right to solve in his own fashion.—Yours, &c.,

YOUR CORRESPONDENT.

ELECTROMAGNETIC INDUCTION AND ITS PROPAGATION.—XL.

BY OLIVER HEAVISIDE.

(Continued from page 51.)

Preliminary to Investigations concerning Long-distance Telephony and Connected Matters.

Although there is more to be said on the subject of induction balances, I put the matter on the shelf now, on account of the pressure of a load of matter that has come back to me under rather curious circumstances. In the present article I shall take a brief survey of the question of long-distance telephony and its prospects, and of signalling in general. In a sense, it is an account of some of the investigations to follow.

Sir W. Thomson's theory of the submarine cable is a splendid thing. His paper on the subject marks a distinct step in the development of electrical theory. Mr. Preece is much to be congratulated upon having assisted at the experiments upon which (so he tells us) Sir W. Thomson based his theory; he should therefore have an unusually complete knowledge of it. But the theory of the eminent scientist does not resemble that of the eminent scienticulist, save remotely.

But all telegraph circuits are not submarine cables, for one thing; and, even if they were, they would behave very differently according to the way they were worked, and especially as regards the rapidity with which electrical waves were sent into them. It is, I believe, a generally admitted fact that the laws of Nature are immutable, and everywhere the same. A consequence of this fact, if it be granted, is that all circuits whatsoever always behave in exactly the same manner. This conclusion, which is perfectly correct when suitably interpreted, appears to contradict a former statement; but further examination will show that they may be reconciled. The mistake made by Mr. Preece was in arguing from the particular to the general. If we wish to be accurate, we must go the other way to work, and branch out from the general to the particular. It is true, to answer a possible objection, that the want of omniscience prevents the literal carrying out of this process; we shall never know the most general theory of anything in Nature; but we may at least take the general theory so far as it is known, and work with that, finding out in special cases whether a more limited theory will not be sufficient, and keeping within bounds accordingly. In any case, the boundaries of the general theory are not unlimited themselves, as our knowledge of Nature only extends through a limited part of a much greater possible range.

Now a telegraph circuit, when reduced to its simplest elements, ignoring all interferences, and some corrections due to the diffusion of current in the wires in time, still has no less than four electrical constants, which may be most conveniently reckoned per unit length of circuit—viz., its resistance, inductance, permittance, or electrostatic capacity, and leakage conductance. These connect together the two electric variables, the potential difference and the current, in a certain way, so as to constitute a complete dynamical system, which is, be it remembered, not the real but a simpler one, copying the essential features of the real. The potential difference and the permittance settle the electric field, the current and the inductance settle the magnetic field, the current and resistance settle the dissipation of energy in, and the leakage conductance and

Silvanus P. Thompson (1851–1916)

A polished engineering professor in the modern mold, Thompson could speak the language of both the practical and the theoretical electrical engineer. His books *Calculus Made Easy* and *The Life of Kelvin* are still in print!

caused Heaviside later to write[51] of

> ...a certain peculiar concurrence and concatenation of circumstances last year rendering it impossible for me to communicate the practical applications of my theory...the resultant effect of which was to screen Mr. Preece from criticism...

and[52]

> But in the year 1887 I came, for a time, to a dead stop, exactly when I came to making practical applications in detail of my theory...in opposition to the views at that time officially advocated.

That same month of July, however, did see the *Philosophical Magazine* print the last installment (Part 7) of the discontinued series, with a *big* blast at Preece. First pointing out an erroneous statement made by Preece in his Royal Society paper[40] announcing the "*KR*-law," concerning the relative capacitances of earth-return and metallic circuits, he then wrote,[53]

> It is, however, a mere trifle in comparison with Mr. Preece's other errors; he does not fairly

appreciate the theory of the transmission of signals, even keeping to the quite special case of a long and slowly worked submarine cable, whose theory, or what he imagines it to be, he applies, in the most confident manner possible, universally!

All through the summer of 1887 Heaviside kept up the pressure on Biggs to allow him to "expose" Preece. By September 20 Biggs felt compelled to appeal to Oliver's primary desire to see his technical work published, and to ask him not to take up valuable journal space snapping after Preece's ankles:

[To] continue as fast as I can to insert your articles I must beg of you to leave the hot controversial discussions for a future opportunity. If Mr. Preece ultimately adopts your views as you seem to think he will, you can [then] justly point out his change... .

Biggs *would* dig in his heels if Oliver went too far. For example, the September publication by Preece of a paper[38] on induction moved Heaviside to write a sarcastic reply, but it didn't see the light of day until first printed[54] in *Electrical Papers*. Why Biggs rejected it for *The Electrician* is easily understood from this opening passage:

It [Preece's paper] contains an account of the latest researches of this scientist on [induction]. The fact that it emanates from one who is...one of the acknowledged masters of his subject would alone be sufficient to recommend this paper to the attention of all electricians. But there is an additional reason of even greater weight. The results and the reasoning are of so surprising a character that one of two things must follow. Either, firstly, the accepted theory of electromagnetism must be most profoundly modified; or, secondly, the views expressed by Mr. Preece in his paper are profoundly erroneous. Which of these alternatives to adopt has been to me a matter of the most serious and even anxious consideration. I have been forced finally to the conclusion that electromagnetic theory is right, and consequently, that Mr. Preece is wrong, not merely in some points of detail, but radically wrong, generally speaking, in methods, reasoning, results, and conclusions.

Biggs' professional interest in curbing Oliver's hot pursuit of Preece came to a sudden end less than a month later. *The Electrician* of October 14 ran just a short announcement, tucked up in a corner:

PERSONAL—Mr. C. H. W. Biggs has resigned the position connected with this Journal which he has held since its commencement in 1878.

The journal also sent out a form-letter to all its contributors:

MY DEAR SIR,

You will gather from an announcement in our columns this week that certain changes are taking place in connection with the Editorship of this Journal. As you have been a contributor to our columns, I think it desirable to inform you that the paper will continue to be conducted upon the same general lines as hitherto, and that we shall at all times, as heretofore, be happy to receive communications from you, whether in the shape of original Articles, Correspondence, or Notes on matters of passing interest.

I trust that we may continue to receive your co-operation and support.

I am,

Yours faithfully,

W. H. SNELL

Acting Editor

What a shock! Biggs was gone and, despite the encouraging tone of his letter, young Mr. Snell would prove to be of different mind concerning Oliver. William Henry Snell, who had been appointed Biggs' assistant[55] the previous year, wouldn't be editor for long, as he would die at the early age of 31 in March 1890, and in fact for most of that time he was editor in name only. As his assistant recalled[56] many years later,

> The Editor in those days was W. H. Snell, a little be-spectacled, Japanesy gentleman, very quiet and gentle in manner... . I had hardly got into my stride as a sort of glorified office-boy when poor Snell fell ill and after a protracted struggle, died... . During the year of the editor's battle with ill-health I was in sole charge of *The Electrician*... .

Still, Snell was editor long enough to have an impact on Oliver—on November 30, 1887 he wrote to terminate Heaviside's current series of articles:

> ...although I rate the intrinsic value of your papers very high indeed, I much regret I have to tell you that I have been unable to discover that they are appreciated by anything like a sufficient number of our readers to justify me in requesting you to continue them. I have taken especial pains to inform myself upon this point and after inquiring in the quarters where students might confidently have been expected I have not been able to discover *any* [see Note 57]... . Pray accept my expressions of sincere regret.

Despite the closing line, Heaviside probably didn't feel all that kindly (in an 1892 letter[58] to Oliver Lodge he called Snell "a smart young man with shortsighted views"), but it was too late for anybody to shut Heaviside up. Biggs had lasted just long enough to get Oliver's fundamental work on induction, transmission lines, and distortionless transmission into print.

But was Snell right? *Did* anybody other than a few professors read him? Snell was, perhaps, both right and wrong at the same time, as the following incident shows. When his last article of the discontinued series was run on December 20, 1887, Oliver added a final note (dated November 30, the date of Snell's letter): "The author much regrets to be unable to continue these articles...having been requested to discontinue them." Then, less than two months later, in February 1888, Oliver sneaked back into the literature with more on the seemingly taboo topic of induction with a very mathematical article with a title that didn't make one think of telephones ("On electromagnetic waves, especially in relation to the vorticity of the impressed forces; and the forced vibration of electromagnetic systems"). Appearing in the *Philosophical Magazine*,[59] it caught the eye of *The Electrician* which ran[60] the following editorial note:

> Speaking of long-distance telephony, there is a paper in the *Philosophical Magazine* this month which no one interested in this subject should omit to read. It is unhappily true that there are very few men engaged in practical work who will be able to follow Mr. Oliver Heaviside's abstruse exposition of the principles of electromagnetic induction; but everyone can recognise the importance of a statement like the following, coming as it does from one of the first living authorities in the mathematical theory of electricity... . Mr. Heaviside says, "It seems to be imagined that self-induction is harmful to long-distance telephony. The precise contrary is the case. It is the very life and soul of it... . I have proved this in considerable detail; but they will not believe it." The proof referred to may be studied in detail in *The Electrician,* Vols. 18 and 19; but, unfortunately, it is not in such a form that he who runs may read. It would, perhaps, be a good thing, if we were some day to publish a gloss upon these papers *adapted to the understanding of the ordinary electrical engineer* [my emphasis].

This, in turn, prompted a letter[61] to the journal from a reader who wrote of *The Electrician*'s quote:

> This appears to be so entirely contradicted by other authorities on the subject and by experiment that I feel sure all telephonists must—like myself—wish that Mr. Heaviside would give us an explanation of his reasons *which can be understood by us as well as by the talented few* [my emphasis] who can fully appreciate his article.

Snell felt differently, however, and ran no such explanation from Oliver.

Snell *was* right—most readers could *not* understand Heaviside's mathematics. And Snell was wrong, too—a lot of those same readers somehow knew Oliver Heaviside had very important things to say and they *did* try to read between his lines of equations. Still, the fact remained that apparently nothing was being *done* in response to Heaviside's theoretical analyses. How could he get people to pay attention to him and *force* them to read his work, even if (as *The Electrician* editorial put it) it "should be lost in a wilderness of quaternions"?

The events of the autumn of 1888 were just what Heaviside needed to make sure people did pay attention.

Oliver Lodge's Oscillating Leyden Jar

The other important train of experimental events that literally forced Oliver's writings on induction into prominence began, ironically enough, with Preece himself about six years before Hughes' Presidential talk, and came to a head in a stupendous squabble between Preece and Oliver Lodge two years after the Hughes–Heaviside affair. During the years 1878–1881 Preece was a member of a study group that examined in detail the nature of lightning rods, and which made recommendations on their proper construction and installation. This was a matter of great importance, one dear to Preece, as he had devoted much thought to the topic as a direct consequence of his Post Office duties.

As of 1880 many tens of thousands of lightning rods were in operation all over England, to offer protection to the enormous capital investment in telegraph lines and instruments. Preece was an influential force in the study group, and its report[62] accurately mirrored his personal beliefs—the ideal lightning rod is a thick, solid copper conductor with low resistance, mounted high above everything else in the area, and run to a solid earth ground. Ohm's law was all the theory involved, according to Preece, and so if you give a charged cloud a tempting target like a big, fat copper rod, why then where *else* would the next lightning bolt go! The rod was, in Preece's mind, only a "drainpipe" to siphon off a cloud's overflowing electrical charge. In this, I should add, Preece was merely agreeing with his hero, Faraday, who had held the same view.

To back up this position Preece did some "simulated lightning" experiments in 1880, using copper conductors of various lengths and shapes (solid cylinders, hollow tubes, and ribbons), with a 42.8-microfarad capacitor charged to 3317 volts (from 3240 batteries in series!). He discharged the capacitor through the various conductors, observing results he claimed[63] were "...very much of the character of lightning." His report concluded with:

> It therefore appears proved that the discharges of high potentials obey the laws of Ohm, and are not affected by change of form [of the conductor]. Hence, extent of surface [of the conductor] does not favour lightning discharges. No more efficient lightning conductor than a cylindrical rod or a wire rope can therefore be devised.

Later that year Preece wrote a remarkable paper[64] in which he presented an *electrostatic* analysis of a lightning discharge, based on the electric field between a charged cloud and the Earth, just before "...a rent or split occurs in the air along the line of least resistance—which is disruptive discharge, or lightning." It was a clever analysis, but a *static* one that ignored the fact that the actual discharge itself is anything *but* static! This would, in fact, be the major point of contention between Preece and Lodge, with Lodge asserting lightning is a high-frequency *oscillatory* process in which inductive, *not* ohmic, effects are all important. It was this battle, as we've seen in Chapter 5, that was the major "social" event of the 1888 B.A. meeting, with

Oliver J. Lodge (1851–1940)

Lodge was at the heartbeat of science and technology during a long life that saw more happen in those fields than had occurred in the previous 2000 years. An intelligent, able, and highly respected man, his intellect burned evenly and widely, but lacked Heaviside's intense brilliance of genius in any particular area.

its resulting debate concerning the relative merits of the "practical" and the "theoretical" man.

Preece's result was:

> Hence a lightning-rod protects a conic space whose height is the length of the rod, whose base is a circle having its radius equal to the height of the rod, and whose side is the quadrant of a circle whose radius is equal to the height of the rod.

That is, Preece claimed everything in his "cone of safety" would be immune to lightning strikes and, indeed, stated

> I have carefully examined every record of accident that I could examine, and I have not yet found one case where damage was inflicted inside this cone when the building was properly protected.

The last two words are crucial, because if damage *did* occur then Preece automatically concluded the building had not been "properly protected," e.g., the lightning rod had not been "properly" grounded. It was a statement that allowed Preece an escape-hatch no matter what happened! As he put it in graphic terms,[65]

> Lightning conductors, if properly erected, duly maintained, and periodically inspected, are an *absolute source of safety* [my emphasis]; but if erected by the village blacksmith, maintained by the economical churchwarden, and never inspected at all, a loud report will some day be heard, and the beautiful steeple will convert the churchyard into a new geological formation.

Lodge rejected all this, particularly Preece's "cone of safety" (see Tech Note 3) and based *his* ideas concerning lightning on the observation that a Leyden jar, when discharged, would often do so in an oscillatory[66] fashion. Lodge liked to express this analogy in a graphic manner by saying that people were going to have to get used to the idea of living inside a "Leyden jar" formed by clouds and the Earth.

Because of the high-frequency (one megahertz is mentioned often—but see Tech Note 3) oscillation of a lightning discharge, argued Lodge, a low-*resistance* path is *not* what is wanted at all, but rather an "alternate path" that offers a low *impedance* to the resulting high-frequency currents. Lodge wrote on these ideas in *Philosophical Magazine* in August 1888 in his paper "On the theory of lightning-conductors"; his opening paragraph cited Oliver and the skin effect:

> Quite recently it has been recognized, first quite explicitly perhaps [see Note 67], by Mr. Heaviside...that rapidly alternating currents confine themselves to the exterior of the conductor... .

A few months before this article appeared, in March, Lodge had given an invited demonstration lecture at the Society of Arts on lightning and lightning conductors. There he had presented his reasons for why Preece and the Lightning Rod Conference Report were in error. The discharge spark from a Leyden jar *looked* to the eye very much like lightning, and Lodge made the great leap to supposing lightning bolts also oscillated. In this he was wrong, but that wasn't understood until much later, and Lodge's arguments and his cleverness with demonstrations *looked* quite plausible. The importance of this is that Lodge, unlike Oliver, was a man with credentials who was actually citing Oliver in support of his position, and from this Heaviside could only gain stature. Indeed, Lodge's praise could hardly have been increased. At one point he said,[68]

> ...I must take this opportunity to remark what a singular insight into the intricacies of the subject, and what a masterly grasp of a most difficult theory, are to be found among the eccentric and in some ways repellent writings of Mr. Oliver Heaviside.

In any case, what Lodge concluded from his experiments was that it wasn't Preece's thick, solid copper rod that was wanted for a lightning rod, but rather one of low inductive impedance and great *surface* area. And, despite its magnetic properties allowing poorer surface penetration of alternating currents, cheap iron was actually to be *preferred* over expensive copper, according to Lodge. It was, suggested Lodge, perhaps the very rapidity of a lightning flash that rendered the skin effect unimportant in determining the exact *material* to be used in a protective rod, but of course it *was* important in selecting the rod's *shape*. Heaviside was very much aware of Lodge's work. In August he wrote[69] in *The Electrician* a note entitled "Lightning discharges, etc." and, in support of Lodge, he said of the initiation of a current in a conductor (i.e., the first rush of current in a lightning rod),

> ...at first the conduction-current is purely superficial. It is clear then at the very front of a wave, where conduction is just commencing on the surface, the conductor cannot be treated as if it had the same properties (conductivity, inductivity, permittivity) as if it were material in bulk, for only a thin layer of molecules is concerned... . Thus, *iron may behave, superficially, as if it were non-magnetic* [my emphasis].

Heaviside never missed a chance to promote his own work, and this essay was no exception. First pointing out how his earlier, now terminated series in *The Electrician* treated the theory of wave propagation on wires, the very area in which Lodge was doing his dramatic experiments, he launched a private shot at Snell:

> I was informed (substantially) that no one read my articles. Possibly some few may do so now, with Dr. Lodge's experiments in practical illustration of some of the matters considered.

Preece found this rejection of his position upsetting, and thus was set the stage for the 1888 "battle at Bath" during the British Association Meeting in September.

"Experience" versus "Theory"

Lodge's position, of course, disturbed Preece greatly. As President of Section G (the Mechanical Engineering Section) of the B.A. meeting, Preece included in his opening remarks a clear challenge to Lodge[65]:

> Some of our cherished principles have only recently received a rough shaking from the lips of Prof. Oliver Lodge, F.R.S., who, however, has supported his brilliant experiments by rather fanciful speculation, and whose revolutionary conclusions are scarcely the logical deduction from his novel premises.

Others knew there was controversy brewing, and FitzGerald (who was President of Section A, the Mathematical and Physical Sciences Section) said in *his* opening remarks[70]:

> With alternating currents we *do* propagate energy through non-conductors. It seems almost as if our future telegraph cables would be pipes [a nice prediction of wave-guides]. Just as the long sound waves in speaking tubes go round corners, so then electro-magnetic waves go round corners if they are not too sharp. Prof. Lodge will probably have something to tell us on this point with lightning conductors.

FitzGerald knew all about Lodge's Leyden jar experiments as he had been (by Lodge's own admission), the moving force behind them. As Lodge explained[71] it, the nature of lightning and of proper lightning rods had *not* been his real reason for the jar experiments at all, but rather he had been seeking an *experimental* demonstration of Maxwell's theory:

> I had been thinking...on the direction of how directly to manufacture light, i.e., how to construct an electrical oscillator of a given and sufficient frequency.

Optical frequencies were far beyond Lodge's capabilities, of course, but he had succeeded in creating waves *on wires* and it was that accomplishment he originally thought would be the hit of the 1888 B.A. meeting. In July, however, he found that Hertz had already generated waves *in space* and thus Lodge, like Hughes, missed getting credit for one of the great experimental discoveries of the 19th century.

The same experiments put him on a collision course with Preece, however, and in that sense Lodge still had a "great hit." The showdown came during a joint meeting of Sections A and G, with FitzGerald in the chair. Preece stood up and declared he had 500,000 lightning conductors

William Preece, carrying the banner of "experience" on a lightning rod, strides over the vanquished Oliver Lodge who lies in the mud (along with *his* rod and banner of "experiment"). This illustration (entitled "Ajax Defying the Lightning") appeared on the cover of the December 1888 issue of *Electrical Plant*.

under his personal supervision and asked "if that were not enough to give a man experience, then what was?" Preece was willing to concede "slow oscillations," but ridiculed Lodge's one megahertz number. Most importantly, he did not believe in the influence of self-induction (and the skin effect) in lightning conductors, and "looked upon the way which lately self-induction had been brought into account for every unknown phenomenon as being very much what the Americans call a species of bug."

And then, to quote *The Electrician*'s summary,[72] Preece went on to say that

> ...Prof. Lodge had made a discovery—he did not know what it was—but that Prof. Lodge, in being possessed by his mania, he would call it, for self-induction—he had self-induction before his eyes, and nothing else—had neglected to study Prof. Poynting's Paper to the advantage that one would expect. If he had studied Prof. Poynting's Paper, which showed that energy passed through the dielectric, and not through the conductor, he might have applied that principle to his experiments, and proved, with equal satisfaction to himself, that the peculiar effects he produced were due to something or other in the dielectric.

For Preece, of all people, to recommend Poynting to Lodge was absurd (as we saw in Chapter 7 Lodge was well versed on Poynting, while Preece himself couldn't have gotten past the first equation in Poynting's paper, and it is evident Preece was just "showing off"). The bizarre humor in this did not escape Heaviside's attention, as we'll soon see.

Lodge began his reply by admitting he did not have 500,000 conductors under his control, indeed he had none, not even on his house, and then he carefully but in no uncertain terms stated Preece didn't know what he was talking about. In fact, using the most civilized Victorian language possible, Lodge in effect called Preece "an ass"! Preece actually played Lodge's straight man in setting himself up for this incredible put-down. In his introductory remarks to Section G, Preece referred to the March Society of Arts lecture and said Lodge was selected to "eulogize" the Lightning Rod Conference Report and instead he had acted as a Balaam who rode upon the British Ass (the British Association) and cursed the work of the conference. Lodge replied that he had no idea he was suppose to bless the report—only to give a lecture on lightning rods; he then turned Preece's own words on him, saying that Balaam was himself (Lodge, the prophet who went and said what he ought to have said), Balak was the Society of Arts, and he didn't know who the third party (the ass) was, unless he was the party who spoke against the prophet (at this point *The Electrician* reported "great laughter" from the audience which obviously understood Lodge's drift).

And so the B.A. meeting of 1888, as discussed in Chapter 5, degenerated into a nasty squabble, not over such technical matters as induction, or lightning, or iron versus copper, but into one about who is "better," the *practical* man or the *theoretical* man. The strained relations between the two men, immediately after the B.A. meeting, are evident in a letter[73] Preece sent to Lodge (October 26, 1888) in which, after accusing Lodge of being the "principal culprit" in precipitating the whole unseemly business, he went on to write

> You rather accept the dictum that the practical man is behind the theoretical one. Now I have been as you know very actively engaged on the practical side for 35 years, and I cannot recall to mind one single instance when I have derived any benefit from pure theory.

Preece also stated his belief that William Thomson had not used theory to any gain during the Atlantic Cable Project but rather, all the work had been done by practical men "without any help from your so called Theorist." Lodge showed this bizarre letter around to friends (including Heaviside); Thomson called it "really quite too monstrous," and George Carey Foster (Professor of Physics at University College, London, and inventor of the Carey-Foster bridge for measuring resistance) declared Preece "a thorough-paced humbug" with much of the letter being nothing but "absolute rot and astounding self-conceit." All in all, an unpleasant affair, indeed, was the 1888 B.A. confrontation, right in the murderous spirit of Jack the Ripper who was at the height of his activities, in London's East End, at the time of the meeting.

But the Preece–Lodge confrontation had served Heaviside's purposes, too, by spotlighting his work. By October, with the end of the B.A. meeting, Oliver felt secure enough to write,[51]

> Is self-induction played out? I think not. What *is* played out is...the British engineer's self-induction, which stands still and won't go. But the other self-induction, in spite of strenuous efforts to stop it, goes on moving; nay, more, it is accumulating momentum rapidly, and will, I imagine, never be stopped again. It is, as Sir W. Thomson is reported to have remarked..."in the air". Then there are the electromagnetic waves. Not so long ago they were nowhere; now they are everywhere, even in the Post Office. Mr. Preece has been advising Prof. Lodge to read Prof. Poynting's paper on the transfer of energy. This is progress, indeed! Now these waves are also in the air, and it is the "great bug" [Preece's favorite sneer] self-induction that keeps them going.

Heaviside's Vindication

The real breakthrough for Heaviside came just months later, with Sir William Thomson's Presidency of the IEE. In his Presidential Address[25] of January 10, 1889, Thomson publicly anointed Heaviside as an authority, and such was Thomson's enormous influence that there was no one thereafter who doubted it (even Preece, in his soul, must have been shaken). Thomson's praise was effusive:

> ...it [the theory of wave propagation along wires] has been worked out in a very complete manner by Mr. Oliver Heaviside; and Mr. Heaviside has pointed out and accentuated this result of his mathematical theory—that electromagnetic induction is a positive benefit: it helps to carry the current. It is the same kind of benefit that mass is to a body shoved along a viscous resistance [see Note 74]... . Heaviside's way of looking at the submarine cable problem is just one instance of how the highest mathematical power of working and of judging as to physical applications helps on the doctrine, and directs it into a practical channel.

Heaviside, of course, was ecstatic at this and on January 18 he wrote[75] to Thomson:

> I have just read your Address to the new Inst. of E.E., and write to thank you for your most hearty appreciation of my work in connection with waves along wires. I am ashamed to see it occupy so much space in your Address, and think it must have crowded out your own matter which would have been of greater and more permanent scientific interest [this is about as humble as Oliver got in his life!]. Your appreciation is most welcome after the long-continued indifference and (sometimes) opposition I have met with.

C. H. W. Biggs was pleased, too, and from his new editorial chair at *The Electrical Engineer (London)* he used the occasion to unload[76] some of *his* frustrations about what clearly had not been an entirely voluntary departure from *The Electrician*:

> ...we would direct attention [to] the eulogy upon the work of Mr. Oliver Heaviside. This may not be the best place to record certain facts, in connection with Mr. Heaviside's writings, but we venture to do so for many reasons, not the least of which is the pleasure we feel in having our views corroborated by so eminent an authority as Sir W. Thomson. Putting aside the papers that appeared in the *Phil. Mag.*, articles that were given in the *Electrician* were given in spite of the most strenuous opposition by the proprietors and every member of the staff, except the late editor. Hence possibly the singularity of the sudden cessation of the articles which has so frequently been remarked upon that the reason may not be

> uninteresting. Previous to Sir W. Thomson's acknowledgment of the magnitude and value of Mr. Oliver Heaviside's work, Mr. Preece had called attention to certain parts of it as worthy of the closest attention and forestalling later workers in the same direction.

This, it seems to me, is as close as libel-suit-conscious Biggs dared get to declaring publicly that Preece had brought pressure to bear on him at *The Electrician*. Indeed, Preece still fought on in spite of Thomson's Address.

At the May 9, 1889 meeting of the IEE, for example, he dramatically declared[77] his unshaken faith in lightning rods as specified by the Lightning Rod Conference and, "for the good of science and the honor of the Institution of Electrical Engineers!", he offered to grasp a lightning rod in his hand during the next thunderstorm. And not only that, he would also "sit upon a barrel of gunpowder without fear of personal results." The next week, at the May 16 meeting, the editor of the Conference Report (G. J. Symons) rose to say,[78] first, "I think Dr. Lodge has not realized...the terrible mischief which he may be doing [in criticizing lightning rod practice]" and second, if Preece did in fact sit out in the rain on a barrel of gunpowder, clutching a lightning rod, he would not be in danger. But Symons *did* hasten to add the careful disclaimer "my opinion on purely electrical matters is worth nothing."

As an indication of the change in climate from 1887–88 to 1889, *The Electrician* now began to make fun of Preece in its reports of these meetings (perhaps because the editorial assistant, W. G. Bond, was running things, with Snell now caught up in his fatal illness). In editorial notes,[79] for example, the journal declared that the War Office had agreed to provide a barrel of gunpowder, and that "Mr. G. J. Symons thought that Mr. Preece, while grasping the rod, should be otherwise well-insulated from the earth. But on this point opinion seems to differ."

Such goings-on, of course, changed no minds among the main participants. As Preece himself put it in a letter[73] (May 17, 1889) to Lodge:

> You and your friends believe one thing—W.H.P. and the old boys believe another, and no one seems inclined to be convinced even against his will.

But *others* were now beginning to think about using Heaviside's revolutionary wave propagation theory of transmission lines. After Heaviside's death, for example, Alfred Rosling Bennett (1850–1928), a well- known early telephone engineer, recalled[80] his early experiments in increasing the inductance of telephone circuits:

> ...I endeavored in 1890 to test one of his [Heaviside's] theories by placing coils in weather-tight boxes along the recently erected Stirling-Dundee copper telephone trunk line... . Careful observation, however, revealed no variation in the normally good speaking, the reason no doubt being the distance—about 100 miles—was too short to enable any noticeable difference to be developed... . I recollect Lord Kelvin, then Sir William Thomson, telling me about Heaviside's Papers one afternoon in Glasgow, and he was considerably astonished and pleased to learn that one of the devices was already under trial. He asked for full particulars.

In an earlier letter[81] Bennett took exception to the generally held belief that the idea of raising the inductance of a circuit by using discrete (i.e., "lumped") "loading coils" was due to others in America, and implied that he had gotten the idea for coils from Heaviside, via Kelvin:

> It is very probable that the idea of such coils originated with Heaviside; but, as I have said on other occasions, Lord Kelvin, then Sir William Thomson, in discussing Heaviside's investigations with me in 1888, said that practical effect could probably be given to them by intercalating suitable coils at intervals in long telephone lines, and he spoke, as I then thought, as if the notion was his own. But he also led me to understand that he was, or had

> been, in correspondence with Heaviside—of whose labours he spoke in the highest terms—so that he may have been only repeating something communicated by the latter; although, so far as I can make out, the first public mention of coils by that pioneer was in 1893... .

Bennett was correct in his reference to Oliver's first *public* announcement of lumped loading in 1893, as it was then that he wrote,[82]

> Instead of trying to get large uniformly spread inductance, try to get a large average inductance... . This means the insertion of inductance coils at intervals in the main circuit.

Any earlier ideas Kelvin had on lumped loading almost surely *did* come from Heaviside. A letter (July 29, 1922) to the President of the IEE explains why Oliver kept this concept out of print for *six years.* He recalled when, in the *autumn of 1887* he and Arthur went to the *Bull and the Bush* teagarden in Hampstead Heath (near where Oliver was born), had a dinner of "stout porkchops," and afterward went for a walk

> ...where I explained everything ["loading in lumps and the true conditions of telephony"] to him. He was immensely struck by what I said, but that crafty old Preece was in the way. The P.O. wires would be wanted and he [Arthur] would have to get P's permission, and tell him everything [obviously, the Heaviside brothers wanted to test lumped loading coils in 1887!]. So it was agreed that I should not immediately publish the lumped loading I proposed. There was no money to patent it and I had tested P myself to prove his dishonesty.

But those were the bitter days of 1887, and now it was 1889 and Heaviside was at last a man to be taken seriously. Just how pleased he was with life, and his sweet triumph over Preece, is indicated by a poem in one of his research notebooks[83]:

> Self-induction's "in the air"
>   Everywhere, everywhere;
> Waves are running to and fro
>   Here they are, there they go.
> Try to stop 'em if you can,
>   You British Engineering man!
> Conceive him (if you can)
>   The engineering man,
> Docking and blocking and burking a Paper
>   Up in St. Martin's-le-Grand!

And we can only begin to imagine the pleasure he must have gotten from reading Oliver Lodge's words in *Nature*,[84] a second public endorsement (to the general world of science) of his talent. After first writing of "the genius of Maxwell and of a few other great theoretical physicists whose names are on everyone's lips," Lodge went on to inform readers

> ...of one whose name is not yet on everybody's lips, but whose profound researches into electro-magnetic waves have penetrated further than anybody yet understands into the depth of the subject, and whose papers have very likely contributed to the theoretical inspiration of Hertz—I mean that powerful mathematical physicist, Mr. Oliver Heaviside.

A Change of Scene—and Fame

The year 1889 brought another change to Oliver's life—he and his parents left London never to return. Nearly thirty years before his other older brother, Charles, had begun training as an apprentice in the music business, as an instrument maker. Later, after marrying Sarah Way (whose sister Mary would years later play an important role in Oliver's life), he accepted a job

Oliver's family at Berry Pomeroy Castle in the early 1890s. His father is the stern-faced, bearded man in the center, and his mother stands to her husband's right. Miss Mary Way peers over Mrs. Heaviside's right shoulder, and Oliver's brother, Arthur, stands hat-in-hand at the far right. His other brother, Charles, is the man kneeling between Thomas Heaviside and Arthur. And far in the back, with only his head visible next to an archway of the castle, is pipe-smoking Oliver himself.

in the music store of J. Reynolds, in the seacoast resort town of Torquay[85] in the southwest of England. Charles prospered, and by 1889 he was a partner in the business. Indeed, things were going well enough to allow the opening of a second store in nearby Paignton. That autumn then, with Rachel and Thomas Heaviside both in their seventies and in less than good health, Oliver and his parents accepted Charles' invitation to live in the house next to the Reynolds music store in Paignton, at 15 Palace Avenue.

A poem in one of Oliver's research notebooks,[86] "The Vision," written on his 39th birthday (May 18, 1889) some months before the move, shows that the event must have been considered for some time, with not a little concern for what the future might bring:

Coming events cast their shadows before,
  And darken the field of view;
I cannot, therefore, surely tell
  When I shall meet with you.
But nevertheless, and notwithstanding the universal gloom,
  Through which the view of futurity can only vaguely loom,
The clouds at times appear to break, and then I seem to see
  Illumined by the rising sun, a vision of Torquay.

Oliver's concerns were mostly unfounded, however, and his life at Paignton would prove to be a happy time (although, as we'll see in the following chapters, he did fight some more technical battles). It was at Paignton, it seems, that he took up cycling with enthusiasm, and enjoyed exploring the local eerie and romantic ruin, the 12th century Berry Pomeroy Castle. Occasionally, when Searle visited from Cambridge, he and Oliver would bicycle out to the castle for a picnic. Many years later Searle recalled[87] one such trip, as an example of his friend's ''biting sarcasm'':

> Of one room, which had not been a dungeon because there had been no bars to the window, Oliver said ''It is very damp; it must have been the servant's bedroom''.

Surrounded in Paignton not only by his parents, but now also by his brother's large family of five children, Heaviside's social horizons were somewhat broadened. One of them later wrote[88] to Searle about those years:

> I remember, in the big upper stock-room of my father's music saloons, how, with my father playing a march, Oliver, at the head of us, would march around, in and out among the pianos (perhaps a dozen or more), we hanging on to his coat tails in a row, one behind the other.

Those *were* happy days for him, and on the technical side, too, his affairs brightened even more. In March 1890 Snell finally died and *The Electrician* once again had a new editor. Bond stayed on as assistant, and Alexander Pelham Trotter (1857–1947) was selected for the job. Trotter was a technical man (whose specialty was illumination engineering, about which he authored two books) who, after having Maxwell as a teacher at Cambridge, took an Honors degree in the *Natural Science Tripos.* He shared Heaviside's love of cycling and, indeed, while at Cambridge became the first man to break the ''three-minute mile'' on a bicycle. In the mid-1880s he first met Preece, and worked as his assistant, conducting illumination tests of street lighting.

At the time of his editorial appointment Trotter was both in private practice as a London consulting engineer, and serving as Technical Secretary to the Electrical Trades' Section of the London Chamber of Commerce. He was clearly a man who knew and was known by all who mattered in the practical electrical world. And best of all (for Oliver), his opinion of Heaviside wasn't Snell's. Trotter wanted Heaviside back as a regular contributor. Even before the new editor approached him, however, Heaviside began to reap the fruits of his recent elevation to an acknowledged ''authority,'' from a different source at *The Electrician.* The journal's publisher, George Tucker (1852–1916), who was a great admirer of Oliver, tried to get a photograph and biographical sketch for publication. Heaviside refused, not out of some juvenile sense of ''getting even,'' but on general principles. He had a near obsession about photographs and self-promotion (except through his *technical* writing). In one obituary notice, for example, the author wrote,[89]

> At one time I tried to get a portrait of him for the Institution of Electrical Engineers, but failed; he did not wish to have his photograph exhibited, he thought that ''one of the worst results (of such exhibitions) was that it makes the public characters think they really are very important people, and that it is therefore a principle of their lives to stand upon doorsteps to be photographed''.

Reflecting the same attitude, another writer recorded[90] Oliver's irreverent reaction to a photograph of a group of members of the Institution of Electrical Engineers:

> Giants at the back. Pygmies at the front. I gave it, framed and glazed too, away to a Newton Abbot furniture dealer, for nothing, along with an old kitchen table.

Tucker never gave up asking for a photograph, however, but Oliver, in turn never gave up saying no. But he must have been pleased, secretly, to be asked.

In his unpublished reminiscences[91] Trotter wrote

> Silvanus Thompson and [John] Perry suggested that I should invite Oliver Heaviside to resume his articles. They admitted that his mathematics were peculiar, difficult and obscure, but that FitzGerald and a few others had recognized that his results were of great importance... . I turned over the pages of his first [terminated] series, and felt that to devote space to such difficult matter, to be read by so few, would not be fair to subscribers who had a right to expect the articles in *The Electrician* to be reasonably intelligible. And yet, here seemed to be an opportunity to save from oblivion something that might turn out to be a work of importance. [I was told] that Heaviside was a difficult man to deal with.

Just *how* difficult he quickly learned! On October 23, 1890 he wrote to Oliver:

> Shortly after I undertook the editorship of *The Electrician* I had the pleasure of meeting your brother from Newcastle [Arthur] where he told me that you were thinking of casting some of your recent unpublished writings into a somewhat popular form. While fully recognizing the great value of your work, I think you will understand that the proportion of space which can be allotted in *The Electrician* to different subjects must be considered with reference to the number of readers who are interested in them. I believe that the large majority of electrical engineers cannot read more advanced mathematics than those which have been employed by Fleming [see Note 92] in his articles on The Alternate Current Transformer. If you could manage to either bring down some of your researches to the level of such readers, among whom I must class myself, or else if you would select such portions as can be dealt with in an elementary form, I should be very glad indeed to see your articles resumed in *The Electrician*.

Heaviside replied by agreeing to submit new articles, in principle, but only under *seven conditions*! This astonishingly bold (even arrogant) stance shows how much Heaviside believed his fortunes had reversed since the gloomy days of 1887. And to his credit, Trotter's response was a good-natured acceptance. Some of the conditions were easy to meet. He agreed, for example, to accept and insert articles on a regular basis, "Fortnightly, if possible, but must yield to pressure of current news," and that there would be plenty of notice if another discontinuance should be necessary. On one issue Trotter stood firm, however, and that was Oliver's continuing obsession with Preece. In his reply (October 29, 1890) to Heaviside he was very explicit on this point:

> I object to contentious discussions, as they are apt to degenerate into a tone that is unsuitable to this paper. I reserve the right to cut out or to modify such matters. Such revision would of course appear on the proof to you.

Trotter *did* yield to Heaviside's demands, in general, though, and he did so because, as he wrote in his reminiscences,

> ...I was told by some of the leading scientific men of the day that Heaviside would [then] be encouraged to continue his work. [It was] left entirely to me, if I thought that the articles would be a credit to the paper.

Trotter did feel obliged, once again (November 18, 1890), to warn Heaviside of his dislike for personal arguments:

> I hope you will let this matter begin de novo [anew]... . I know nothing of any former controversy. We all know that W. Preece made a mistake...[and] the Nameless One [should] be left to another opportunity. I should very much prefer this alternative.

But beyond this concern, it is clear that Trotter wanted Heaviside back. Perhaps Trotter was also attracted to Oliver by a common thread of humor that ran through both of their personalities. Trotter seems to have been a man with a balanced view of himself who, while serious, was not overly so[93]; in any case Oliver was lucky a second time to find an editor who "understood" him—and on January 2, 1891 the new series of articles on electromagnetic theory began, with the understanding they would eventually be collected into book form. Because of the "continuing saga" nature of the writing, with each article beginning where the last one left off, it was also felt all of his earlier work should be reprinted as books. After a bit of correspondence about just who had to give whom permission to reprint, Macmillan and Company brought out all of Oliver's work, up to the end of 1890, as the two-volume *Electrical Papers,* in 1892.

Back in Print—in Style!

The reviews, on both sides of the Atlantic, were a not very consistent mixture of the good and the not-so-good:

> ...almost every page bears the impress of a vigorous and original mind... . [Note 94]

> The preface is characteristic of the author, and the body of the work would be more dignified if he had confined his personality to that alone. The singular propensity of the new school of English scientists, of which Doctors Heaviside [Heaviside had no doctorate!] and Lodge are representatives, to mar their scientific writings by a facetiousness entirely out of place is probably not appreciated beyond the circle of their immediate friends. [Note 95]

> The articles on electric and electromagnetic theory form the best commentary that has been written on Maxwell's "Electricity and Magnetism". [Note 96]

> Mr. Heaviside's work is often caviar to the general, and is not presented always in the best manner. We have heard of an editor who said he could reprint the same Heaviside article three times under different headings without anybody detecting the fraud, but perhaps that was a humorous exaggeration. Mr. Heaviside has never sought to be popular... . [Note 97]

> Mr. Heaviside's discussions are addressed only to the mathematical physicist, and are quite beyond the lay reader. They have been recognized as of a high order of merit by scientific men, and have taken their place as a valuable contribution to the scientific literature... . [Note 98]

> Thickly scattered among the severely technical matter which forms the bulk of the book, there is much vigorous writing which the non-mathematicians can appreciate... . [Note 99]

> Few persons will find the work easy reading, yet it is much less difficult than might at first appear, and well deserves attentive study... . The work is frequently interrupted by remarks which put author and reader upon good terms with each other, and which sometimes excite a smile, as when, for instance, a formula resulting from a long train of mathematical reasoning is pronounced "ridiculously simple". [Note 100]

It was his old friend at Trinity College, Dublin, George Francis FitzGerald, who perhaps wrote the best and most knowledgeable review.[101] While pointing out that it was Heaviside who had "cleared away the *debris*" of Maxwell's original work, and that inductive loading was *Oliver's* contribution (see Tech Note 4), he did not fail to be critical also as, for example, when he had sharp things to say about the writing style:

> Many complain that Browning is unintelligible. Some say that he writes nonsense. His

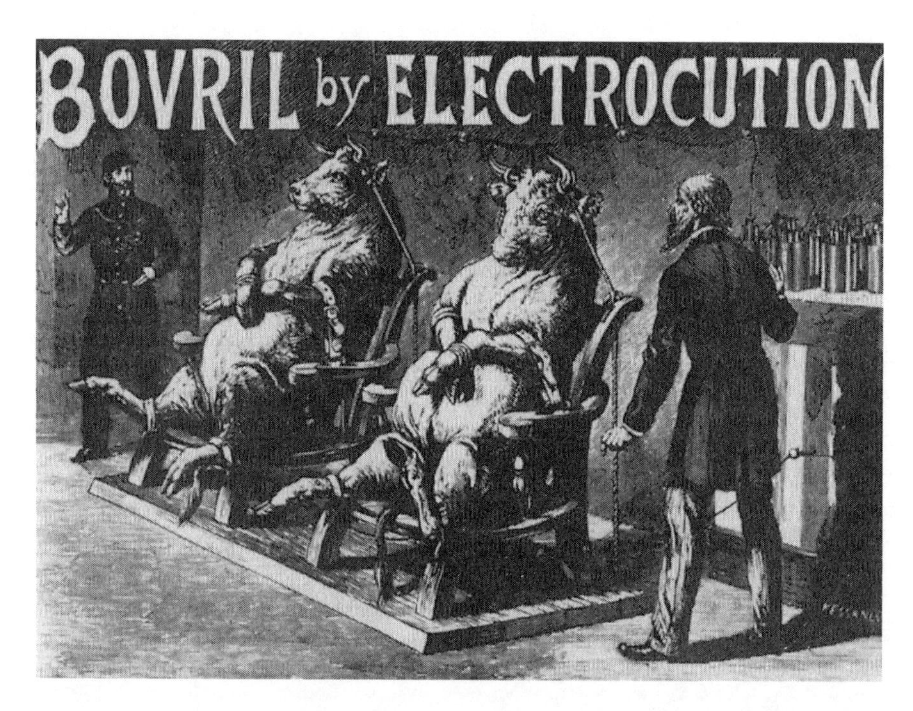

This gruesome advertisement appeared in the 1891 Christmas issue of *The Graphic,* showing the Victorian fascination for electricity had advanced considerably beyond that of the days of poor Bodger (who had been blown apart by lightning, in Chapter 6, thirty years before). Bovril was a beef bouillon hot drink, which another ad asserted to be one of "The Two Infallible Powers" (the other was the sitting Pope, Leo XIII)!

> admirers worship him. Oliver Heaviside has the faults of extreme condensation of thought and a peculiar facility for coining technical terms and expressions that are extremely puzzling to a reader of his Papers. So much so, that there seems very little hope that he will ever attain the clarity of some writers, and write a work that will be easy to read. In his most deliberate attempts at being elementary, he jumps deep double fences and introduces short-cut expressions that are woeful stumbling-blocks to the slow-paced mind of the average man... .

FitzGerald also expressed regret at Heaviside's throwing aside of quaternions in favor of vectors (discussed in Chapter 9), and at what he felt to be Oliver's futile battle[102] to reform the accepted system of units ("When people tolerate miles and furlongs and pecks and bushels and barrels and firkins and hogsheads, etc., etc., how can they be expected to get up any enthusiasm over the eviction of 4π?"). But there was something for everyone in his review, including Oliver, who must have loved reading (at least as much as Preece hated!):

> Mr. Heaviside is, by the practically unanimous opinion of scientific experts, and by his success as a prophet, pointed out as [an] authority...who should be consulted by all those who desire to improve telegraphic signaling. He is a scientific expert of the best kind in these subjects, and it is very much to be regretted that scientific experts are, in these countries, so highly paid to advance causes and so very seldom paid to advance practice. The telegraph department has, however, shown itself anxious to improve the means of communication by applications of the most recent scientific ideas, and it is inconceivable that any personal friction should really stand in the way of the utilisation of what the Royal Society has crowned with a Fellowship.

PRICE 21s., POST-FREE. WILL LAST FOR YEARS.

OVER A QUARTER OF A MILLION PATIENTS HAVE BEEN SUCCESSFULLY TREATED FOR RHEUMATISM, LUMBAGO, SCIATICA, GOUT, KIDNEY COMPLAINTS, EPILEPSY, PARALYSIS, INDIGESTION, CONSTIPATION, FEMALE COMPLAINTS, GENERAL AND LOCAL DEBILITY, FUNCTIONAL DISORDERS, &c.

The following are selected from thousands of reports received. The originals may be seen at the Company's Rooms.

GOUT. "Completely cured me."—Major PAKENHAM, Armagh.

HYSTERIA. "The Belt has done wonders."—Mrs. M. ORIEL, Dowlais.

LUMBAGO. "It has been an untold comfort to me."—Miss HUDSON, Twyford.

RHEUMATIC FEVER. "I can speak positively of its advantages."—Dr. C. LEMPRIERE, D.C.L., Senior Fellow St. John's College, Oxford.

SCIATICA. "Worth fifty pounds."—Mr. R. J. WATSON, Harwich.

KIDNEY DISORDERS. "I am like a new man."—Mr. W. P. POUGHER, Plymouth.

BRONCHITIS. "A wonderful change."—Rev. E. AUBREY, Ystrad, Pontypridd.

NEURALGIA. "A most complete success."—M. A. LANGDALE, Esq., Leacroft.

RHEUMATISM. "Completely cured me."—W. TURNER, Esq., Ashby-de-la-Zouch.

NERVOUS HEADACHE. "Derived much benefit."—Mr. W. HOLT, New Stafford.

INDIGESTION. "I wish I had purchased one before."—Mr. C. HITCHCOCK, Bicester.

NERVOUS DEBILITY. "My health has improved every day."—Mr. W. BLISSETT, Wallingford.

CAN YOU AFFORD TO DIE? Can you afford to drag on a miserable existence? Can you afford to pay doctor's bills? when all disorders of the Nervous System, Impaired Vitality, and Defective Organic Action can be speedily, effectually, and permanently cured by wearing the ELECTROPATHIC (Battery) BELT, which is guaranteed to restore impaired vital energy, to invigorate the debilitated constitution, stimulate the organic action, promote the circulation, assist digestion, and promptly renew that vital energy the loss of which is the first symptom of decay. Every man and woman suffering from Rheumatism, Lumbago, Sciatica, Indigestion, Constipation, or any Disorder of the Nervous System, should at once stop physicking, and send a P.O.O. or Cheque for 21s., for an ELECTROPATHIC BELT (post-free), and obtain immediate relief.

LIVER COMPLAINT. "I have no pain now."—Mr. T. SMITH, Hey, Lees.

EPILEPSY. "Wonderfully improved."—Mrs. GODWIN, Andoversford.

SHORTNESS OF BREATH. "Greatly relieved."—Mrs. C. FRAYER, 72, Elm Park-gardens, S.W.

ASTHMA. "Quite left me."—Mrs. J. LOW, Stirling, N.B.

CONSTIPATION. "A wonderful effect."—Mr. H. T. STOKES, Demerara.

WEAKNESS. "Still improving."—Mrs. MENHENICK, Colwyn Bay.

DYSPEPSIA. "I feel altogether stronger and better."—Miss WELLS, Scarboro'.

PILES. "Better in every way."—Mr. R. STRINGER, East Bradenham.

BAD CIRCULATION. "Feel like a different person."—JOHN ALGAR, Esq., Campsey Ashe, Wickham Market.

PALPITATION. "Quite free."—Mr. G. GATE, St. Helier, Jersey.

PARALYSIS. "Very great benefit."—Mrs. SHARP, Maidstone.

WRITER'S CRAMP. "A complete cure."—Miss E. HASTINGS, Eastbourne."

SPINAL WEAKNESS. "Rapid improvement."—Mr. W. WATSON, Newspynie, N.B.

MR. C. B. HARNESS, the Eminent Consulting Medical Electrician, has been specially retained by the Company, and may be consulted daily (without charge), on all matters relating to health, and the application of Curative Electricity. Residents at a distance, and those unable to avail themselves of a personal consultation, are invited to write for a PRIVATE ADVICE FORM, which will be forwarded POST-FREE on application to the MEDICAL BATTERY COMPANY, LTD., 52, OXFORD-STREET, LONDON, W.

On receipt of Post-Office Order or Cheque, value 21s., crossed "London and County Bank," the MEDICAL BATTERY COMPANY, LIMITED, 52, OXFORD-STREET, LONDON, W., undertake to forward the ELECTROPATHIC (Battery) BELT, as described above, for either Lady or Gentleman, free by post to any part of the United Kingdom.

"ELECTROPATHY; or, HARNESS' GUIDE TO HEALTH," will be sent, post-free, on application to the

Medical Battery Co., Ltd., 52, OXFORD-STREET, LONDON, W.

The Pope was not the only celebrity to appear without his permission in a Victorian ad pushing the wonders of electricity. Here we see, in the April 10, 1886 issue of *The Illustrated London News,* the crusty English Prime Minister William E. Gladstone, himself, wearing one of Mr. C. B. Harness' battery-powered "electropathic belts" while chopping down a tree (or perhaps beheading a member of his cabinet who had displeased him—the ad is unclear on this detail!). The fine print reads: "CAN YOU AFFORD TO DIE? Can you afford to drag on a miserable existence?" Such belts were especially recommended for "weak men suffering from the effect of youthful errors" (as well as from piles, epilepsy, writer's cramp, rheumatism, gout, bronchitis, and liver and kidney problems). The same company also sold an electric corset for women ("The VERY THING for Ladies") who were bedeviled by a "weak back." Mr. Harness also marketed "Dr. Scott's Electric Hair Brush," which once (it was claimed) grew hair on a man "whose head was as bald as a bladder of lard."

If the reviews were generally satisfactory, the sales were not. Macmillan printed 750 copies of *Electrical Papers,* initially priced at thirty shillings. After five years, they had sold only somewhat more than half. To get rid of the rest of their less than best-seller, the publisher steadily dropped the price: first to ten shillings, then two shillings and sixpence, then one shilling and sixpence, and finally, as Heaviside wrote[103] in 1901,

> They can be picked up cheap, because the remainder was sold off in quires for a few pence per volume, on account of the deficiency in storage room. So look in the four penny boxes. Though somewhat vexed at first by this disposal of my laboured lucubrations, it has, later, given me and others occasion for much laughter.

Despite these brave words, *at the time* the poor sales of *Electrical Papers* must have caused more than a little "vexation." Put simply, there was no money coming in, right when both Oliver and his long-suffering family must have thought, at last, all the years of work (for Oliver) and all the years of supporting a jobless relative (for the rest of the Heavisides) were about to pay off. It was Oliver's first (but not his last) big financial crisis.

His Friends Try to Help

At the beginning of 1894 FitzGerald, Lodge, and John Perry wrote and tried to offer a way out. This was the first of a series of letters which reveal much of Heaviside's touchy sensibilities, as well as FitzGerald's near saintly patience. In places Heaviside's replies are missing, but it is usually possible to figure out, in general, what he must have said from FitzGerald's response (all letters to Heaviside, after the first one, were over FitzGerald's signature alone).

> February 6, 1894
>
> Dear Heaviside
>
> As persons deeply interested in scientific research, like yourself, may we ask your kind attention to and acceptance of the following request.
>
> In further recognition of the acknowledged value of your contributions to science in carrying forward the work of Faraday and Maxwell, it has been proposed to grant you an honorarium by the Royal Society, inadequate though it will necessarily be.
>
> It has fallen to us as personal friends to have the honor, as well as the pleasure, of making this proposal known to you. As a matter of form and to avoid the humiliation of having to report a refusal, all we ask is an expression of your willingness to accept it if offered, and thus give the greatest gratification to your friends and put them under a lasting obligation to yourself.

> February 12, 1894
>
> My dear Heaviside
>
> I really don't know what funds the Royal Society has out of which it is expected that they would offer to pay you something on account of your distinguished work. All I know is that I am informed that there is an intention of offering to give you something, but that it was a condition precedent to ensure that you would not refuse and that for some reason, perhaps a bad one, I was pitched upon to concoct a letter to you, which I did with the assistance of others. I also know that the Royal Society does not throw money about to everybody, but only gives it to those who deserve it on account of their scientific work, so that I do not myself see any difference between your accepting this and any professor, Faraday, for instance, accepting payment for his work. That payment should be made in advance is, if anything, less defensible than payment afterwards...—well, I can see no possible objection to receiving it, I know I take it every quarter day.

I would be the very last person to wish to trap you into taking what afterwards would disgust you so that I do not think you should ask no questions and, if you desire me, I will write to Lord Rayleigh for information about how there are funds at the R.S. available for a payment such as I understand is proposed. I would like you however to formulate the questions so that I may not blame myself afterwards, if there is any misunderstanding.

Hoping that you will finally see your way to accepting as I am sure you deserve and would spend it better than anyone else I know... .

February 15, 1894

My dear Heaviside

Different people have such very different ideas as to what is humiliating etc. etc. that I have written to Lord Rayleigh for more information about the matter. In the meanwhile, from what Perry writes me, I fancy that the fund is not paid away generally to the Prince of Wales or the Duke of Westminster, even though they might make scientific advances, but to people who have not got much inherited wealth and are not able to earn much by their scientific work. I suspect, in fact, in making payments out of the fund the poverty and opportunities of the scientific person who is to get this payment for his work is taken into account.

If you consider it humiliating not to be well off and not to enjoy sufficiently good health to earn more than a hodman, though the scientific world with one consent says that your work is worth that of hundreds of hodmen, I dare say I ought to say millions, then I am afraid that you will deprive the scientific world of the satisfaction of endeavoring to emphasize this, their belief, in a tangible way by paying you for a small part of the value they have received from you.

As I said before I really want, before all things, to avoid leading you into a step that you would be sorry afterwards you had taken.

P.S. When I come to think of it, it seems to me that the reason I take money from Trinity College for my work is because I am poor.

This last letter from FitzGerald brought a quick response from Heaviside, which shows quite clearly his feelings of social inferiority:

February 16, 1894

My dear FitzGerald,

...I don't think you take money from T. C. ..."because you are poor", but because it is your proper payment for your work, and to which you have an equal right whether you are poor or rich. Again to emphasize this point, if the scientific world says my work is worth that of hundreds of hodmen or millions (I quote your words) and the scientific world chooses to pay me distinctly for that work according to their own (quoted) estimation, why then I should be proud to take it! It would be a pot of money.

However, even if I may have evidently emphasized in my previous a particular side of the question, still the fact remains that that side is a very real one. To give an extreme case, people sometimes starve rather than go to the workhouse. Why? The associations, I suppose, and then pride.

As for the power of money, I know more of that than most people. I had an example last evening. I went to a little musical party, my brother's family with a professional to assist, or lead rather. It was all very nice till a certain person came in, when a strange change occurred. All mutual intercourse ceased. Mr. Poker became the sun, and all revolved round him. I became a perfect nonentity—not even introduced, except later perfunctorily as an afterthought. So it continued all the while he remained, and when he left he (perhaps naturally) did not think it worthwhile including me in his farewells. Having gone, things righted themselves, and I received semiapologies. He was a *gentleman,* obliged to be very civil to him, important to be right with him. On inquiry I learned he moved in the best society, and was an absentee Irish landlord!

Of course you should understand that I entirely sink the fact that I have personal significance in other respects when I (very rarely) go out, but it is not pleasant to be entirely snuffed out by the arrival of a *gentleman*!

February 20, 1894

My dear Heaviside

You are right that the College does not pay me because I am poor but I work for the College because I am poor.

Anyway we can't expect to be paid for our work in every respect. We all, I hope, do a great deal of unpaid work. Scientific men have not enough resources at their command to pay at its value for *all* the work that is done for Science but they have some money to do so, and in choosing those whose work they will pay for, in part, they must choose according to some reasonable principle and as a matter of fact in England and by the Royal Society the choice is I expect partly determined by the pecuniary needs of those who have worked for Science.

The reason why the workhouse has a bad name is because most of those relieved there are paid, not because they have been useful members of the community, but because they have been useless members. I have never heard of an officer's widow who starved rather than receive a pension. Pensions, if deserved, bless him that gives and him that takes. The Duke of Wellington was endowed by the state for services rendered but I don't suppose that if he had been already a Rothschild he would have been given so much money.

I am afraid we have all to put up with a lot of social folly owing to inherited customs i.e. owing to the unreasonableness and social instincts of the race. We suffer for the folly as well as the wickedness of others and the suffering is none the less because it is unavoidable and not our fault.

I have not yet heard from Lord Rayleigh or would let you know what is the nature of this fund I hear may be available.

February 21, 1894

My dear Heaviside

Lord Rayleigh has sent me the information I asked for and I find that the Fund out of which it is hoped that you may be offered some remuneration is a Fund entitled The Scientific Relief Fund and that it was established "for the aid of such Scientific men or their Families as may from time to time require and deserve assistance." Lord Rayleigh says "I do not see why a note from the Scientific Relief Fund should be regarded as a compassionate allowance." I presume because it is not in the usual sense of the word "compassionate" i.e. the kindly overlooking of faults but is only given to those who deserve it for their work, not who claim it out of compassion for their errors.

I am afraid I can add no more to what I have said. It is I am afraid rather insulting your intelligence to be writing these long arguments. I can only add that having made as sure as I can that you are in possession of the facts as far as I know them and of the arguments that bear upon it, I will feel quite satisfied however you decide: though my satisfaction will be regretful satisfaction if you refuse and a pleased satisfaction if you will accept the remuneration if offered.

February 22, 1894

My dear FitzGerald,

Lord R. is a man of few words. Or perhaps you have left out his opinion of my audacity, daring to look a gift horse in the mouth! Anyhow, it would not be seemly for me to ask more. Will you, therefore, be so good as to convey to Lord R. my most respectful declination of the offer?

> I fear I have been the cause of a great waste of your valuable time, all about nothing too. I am very sorry.
>
> The time may come when I may be compelled to beg for assistance, *of any kind.* May it be far off!

On February 24 FitzGerald replied to call Heaviside's final words "pathetic," and also to suggest another way for Oliver to get money—write a popular-style book. Oliver answered on February 26 and said he preferred to use "self-assertive" to characterize his refusal of money obviously designed "for the assistance of decayed men who have fallen on evil days." Heaviside was, of course, not totally unappreciative of his friend's concern, and also wrote "...our recent correspondence has been very gratifying. The milk of human kindness is such a scarce commodity in general, that it should be treasured when found, and I greatly appreciate your desire to serve me." But as for FitzGerald's new suggestion, all he got back for his trouble was "I do not write for the masses..." and besides, so declared Oliver, it wasn't necessary because "If my income is small, so are my expenses, and I can scrape along." To explain these words, curiously upbeat for an unemployed author whose books weren't selling, Heaviside revealed to FitzGerald he had agreed to produce a second volume of new work, to follow the first volume of *Electromagnetic Theory*. The initial volume, released the previous year (1893) by *The Electrician* Printing and Publishing Company, would do better, he believed, than had *Electrical Papers*. As he forecast[104] things:

> I have an American public as well. I expect to sell at least 1000 copies of Vol. I. E.M.T., 500 in a few months, another 500 in a year or two, and then a slow steady sale.

More Battles

The years in Paignton would prove to be technically active ones, years during which Heaviside would participate in the beginning of the acceptance of vector analysis as a general tool for physical scientists, would experience success in attacking enormously difficult problems with his operational methods (as well as bitterness in a second period of suppression of his work similar to 1887), and would play an important side-role in a debate with Lord Kelvin on the age of the Earth. All of these episodes are treated in detail in the following three chapters.

These same years would take their emotional toll, too. His mother died in Paignton in October 1894 at age 75 and, only two years later, in November 1896, his 83-year-old father followed her. And while Heaviside's star had risen in the electrical world after 1889, the fact is Preece's never set. Oliver had to endure watching his "crafty" foe continue to play the role of official expert for a still enthusiastic audience. No matter Lodge at Bath in 1888 and Thomson at the IEE in 1889—Heaviside was the "Hermit of Paignton" (as *The Electrician* called him in 1896) while Preece was an honored man.

In 1893, for example, Preece was elected President of the IEE, and in his Inaugural Address[105] of January 26 he made some remarks that show he had not forgotten past battles, no matter how genial a face he put on for public display. On lightning he said, with some merit,

> Professor Lodge has...endeavored to modify our views as to the behaviour of lightning discharges and as to the form of protectors, but without much success. His views have not received general acceptance, for they are contrary to fact and to experiment.

With (much) less merit he defended the "*KR*-law" with

> It is very much the fashion to deny the accuracy of the *KR* law. This is probably the result of

ignorance of its meaning or of its interpretation. Some speak of it as empirical, others scoff at it as imaginary, and some sneer at it as an impossible law... .

And then Preece showed how smoothly he could adjust his historical perspective. *Now,* in 1893, inductance was *not* an "evil" but rather a "negative capacity" that could reduce the *effective KR* (this is a new concept!), and so President Preece was able to explain why some circuits with *apparent KR* values too big for good telephony could actually work. With this ingenious argument he was able simultaneously to admit induction wasn't all bad, retain the *KR*-law, and continue to ignore Heaviside and his work.

There can be little doubt about whom Preece had in mind as sneering—just three weeks before Preece's Address, Heaviside had written[106] in *The Electrician*:

> Like a kind of fly-wheel, the self-induction imparts inertia and stability, and keeps the waves going. It is the long-distance telephoner's best friend who was, not many years since, spurned with contempt from the door... . There was also some considerable fuss made about a supposed...*KR* law...according to which you could not telephone further than KR = such or such a number...in spite of repeated attempts to bolster up the *KR* law, the critical number has kept on steadily rising ever since... . Make your circuits longer, and it will go up a lot more.

Not only Oliver was "sneering" at the *KR*-law, as shown by an editorial[107] in the American journal *The Electrical World,* written in response to Preece's IEE Address. Calling the law Preece's "pet," the journal went on to joke that, for Preece to keep the *KR* of actual, in-service cables (known to provide good telephony) below his maximum of 15,000, "he will soon reach a point where he will have to give *K* a minus number." And as an indication that Heaviside's work, even in 1893, still had some way to go to be appreciated, the editorial concluded with

> It is quite evident that *KR* is not a formula for telephonists to be guided by. The principles underlying telephonic transmission have not yet been the subject of really painstaking original investigation...there is a splendid field here awaiting the discoverer, and doubtless some day the fundamental laws of telephonic transmission will be given to the world.

Returning to Preece and his Address; near the end of it he launched a general, all-purpose assault against Theory, stating

> It is a misfortune that a beautiful hypothesis like Maxwell's electro-magnetic theory of light has been discussed almost solely by mathematicians. Its consideration has been confined to a small and exclusive class...and this is to be regretted, for, after all, *it is the many, and not the few, that determine the acceptance or refusal of a theory* [my emphasis—this incredible statement indicates Preece thought a *vote,* not *facts,* should determine the fate of Maxwell!]... . I have no sympathy with the pure mathematician who scorns the practical man, scoffs at his experience, directs the universe from his couch, and invents laws to suit his fads.

Preece was certainly thinking of people like Heaviside and Lodge when he said this—certainly FitzGerald so believed, leading him in his review[108] of EMT 1 to attempt a rebuttal which had a curious turn to it. FitzGerald ended his review with words that must have puzzled nearly all who read them:

> ...telephoning through iron wires [is a question of great importance, because of self-induction effects] and it must ultimately be appreciated by practical telegraphists, even though they for a time sneer at those whose bodily infirmities prevent them from helping the advance of mankind in any other way than by using their powerful intellects to solve the problems that at present require solution, in order that the practical man may improve his practice.

When Heaviside read this, as part of a review of *his* book, he took offense and angrily wrote to FitzGerald:

May 26, 1894

My dear FitzGerald,

..."Bodily infirmities." You gave me a most disagreeable shock. To have one's bodily infirmities introduced to public notice is quite painful. Have I ever said anything to you to warrant the statement? I think not. My bodily infirmities should not be publicly advertised, even if they did prevent, etc. Some may pity me, which I don't want, and dislike; and my enemies, if I have any, will chuckle. And then the mischief may be multiplied by copying. And matters will be made worse by taking any notice of it, or contradicting it. The only thing, I suppose, is to ignore it, and let it be forgotten as soon as possible. But I am sure you did not mean any harm by it, though even in a little town like this it would be very unpleasant if it got circulated. People are horribly rude, and delight in giving offense...common people, of course.

To some who know me the "bodily infirmities" will cause laughter, as I am considered anything but infirm, perhaps strong above average. I never obtrude my infirmities upon them, and they know nothing or next to nothing of them. My principal infirmity is that I am a chronic dyspeptic, and it occasionally gets awful, and brings on nervous disturbances, which may culminate in epilepsy some day [his mother had it], or may not, according as I live it down, or not. But this sort of thing is strictly for home consumption; most certainly not for public entertainment or otherwise.

Lenard's experiment [see Note 109]. Very remarkable.

FitzGerald's reply explains the odd ending to his review of EMT 1:

May 29, 1894

My dear Heaviside,

I thought several times over and changed my mind half a dozen times about including that remark intended for a rap at Preece for his remark about people who lectured from a sofa. His remark has made me very angry whenever I thought of it and I thought it a shame that no one should have a rap at him on account of it, so that I am afraid now that in venting my spleen at Preece who, I fear, won't care, I have hurt you. I am very sorry for it. I suppose I should have quoted Preece so that my remark should have been unmistakably founded upon his and not appear, as I fear it does, like an independent remark in special connection with your work. I intended it to be in special connection with his remark but I am very much afraid, now that my attention is called to it, that other people will not remember his remark as intensely as I do and so will not put the two together. I must try and restrain my spleen in the future. I suppose no good is ever done by giving way. I am very sorry, but I am afraid I can do nothing to repair the mischief, for any notice of it would now only produce more attention.

This explanation satisfied Oliver and on May 30 he answered with "I didn't know you were thinking of one of the eminent electrician's funny remarks... . But I do not mind the eminent a bit, now that he is no longer able to sit upon me and suppress as he used to... ."

But as we'll see later, Preece was never far from his thoughts, even years after the man's death. Theirs was a struggle that would be finished only with the end of both of them.

Tech Note 1: The Skin Effect

In a *perfect* conductor, one with a conductivity $\sigma = \infty$, an incident electromagnetic wave *cannot* penetrate into the conductor *at all*. This is what Heaviside meant in 1893 when he wrote the apparently paradoxical statement[110]: "A perfect conductor is a perfect obstructor, but does not absorb the energy of electromagnetic waves." Even in real conductors, where $\sigma < \infty$, it has been found that σ is sufficiently large that up to extraordinarily high frequencies the penetration depth is small, on the order of a small fraction of a wavelength. That is, the conductor appears to have a "skin," beneath which wave penetration can be considered insignificant. The field equations in a conductor obeying Ohm's law can be solved and, as in the simpler case of $\sigma = 0$ (the vacuum of space), the result is ***wave propagation.*** Only now, with $\sigma > 0$, the wave amplitude decays with distance. This decay occurs as the electric field of the wave induces $\sigma\vec{E}$ currents which drain the energy out of the wave ("Ohmic Losses"). The wave amplitude decays exponentially, i.e., both $\vec{E}$ and $\vec{B}$ inside the conductor, at distance x from the surface, behave like

$$|\vec{E}| = |\vec{E}_s| e^{-Kx}$$

$$|\vec{B}| = |\vec{B}_s| e^{-Kx}$$

where $|\vec{E}_s|$ and $|\vec{B}_s|$ are the field amplitudes *at the surface.* As Heaviside showed in 1888[111]:

$$K = 2\pi f \left(\frac{\mu\epsilon}{2}\right)^{1/2} \left[\left\{1 + \left(\frac{\sigma}{2\pi f\epsilon}\right)^2\right\}^{1/2} - 1\right]^{1/2}$$

where f = frequency of the wave (in hertz). It is usual to define a "good" conductor to be one that has a conductivity large enough so that

$$\sigma \gg 2\pi f\epsilon.$$

Then we can write the approximation

$$K = (\pi\mu\sigma f)^{1/2}.$$

The "skin depth," d, is defined to be the distance at which the wave amplitude has decayed from its value at the surface by a factor of e, so that

$$d = \frac{1}{K} = (\pi\mu\sigma f)^{-1/2}.$$

This simple result explains many otherwise curious observations. For metals, for example, σ is sufficiently large that even well down into the visible portion of the electromagnetic spectrum the skin depth is only on the order of *billionths* of a meter, and this is why metals are reflective, not transparent. Even a very thin sheet of aluminum foil makes a most unsatisfactory window!

Another interesting example of this effect is that of radio communication through seawater with submarines (Heaviside discussed the problem of electrical waves in seawater in 1897,[112] although of course there was no mention by him of submarines!). For radio waves to penetrate to submarine depths, say on the order of at least ten meters (this allows the submarine to remain sufficiently deep to avoid creating a revealing surface wake), the required frequency is

$$f = 1/(\pi\mu\sigma d^2)$$

which, for values of μ and σ appropriate for seawater, works out to be about 3 KHz. And for land-based communication to distant, submerged submarines, the frequency must be much lower still, e.g., a few tens of hertz.

TECH NOTE 2: THE "*KR*-LAW"

A reading of Preece's original derivation[40] of the "*KR*-law" is confusing (at least *I* couldn't easily follow it) because he changes symbols in midstream and also because he uses misleading names for physical quantities. For example, he uses S to represent the "limiting distance of speech" when, in fact, the S he derives is *not* the length of the transmission circuit. The following is essentially what he meant.

For human speech to be recognizable it must have a spectrum with energy at frequencies that exceed some minimum value, e.g., 1000 Hz or more. This says the time-constant of the transmitting circuit must not exceed some *maximum* value. That is, there is some *upper* bound on the value of the circuit KR (the product of capacitance and resistance has units of time). If x is the length of the circuit that just reaches this maximum delay, and if k and r are the per unit length capacitance and resistance, respectively, then $K = kx$, $R = rx$, and $KR = krx^2$. As did Preece, set this maximum permissible total KR equal to the crucial value of A and arrive at

$$x = \sqrt{\frac{A}{kr}}.$$

Obviously the bigger A is, the longer is the circuit over which speech can be transmitted. But what *is* the value of A? This gets us into a very strange aspect of Preece's paper. He does give values for A (for overhead wires, $A = 15{,}000$ for copper and 10,000 for iron). But *how* he arrived at these numbers is a puzzle and, in fact, there was nothing constant about them. That is, if it was found that a circuit could transmit speech over a greater distance than predicted with the previous value of A, why then Preece would just declare a *new* (bigger) value of A!

An example of how Preece used this "law" is his treatment of the telephone circuit connecting Paris and Brussels. With a resistance and capacitance of 2.4 ohms and 0.012 microfarad per kilometer, respectively, the maximum distance for speech transmission is given by

$$x = 5892\sqrt{A} \text{ kilometers;}$$

this is so large (for either iron or copper) that it exceeds, by far, the actual circuit length of 320 kilometers and, as Preece wrote, "the speaking must be excellent."

TECH NOTE 3: PREECE AND LODGE ON LIGHTNING

Preece claimed[64] that a low-ohmic, well-grounded conductor, represented by the vertical line AB in Fig. 8.1, would protect everything inside the tent-shaped volume. The length of DB is equal to that of AB (as is the length of BE), and the curved sides are quarter-circular arcs of radii equal to the length of AB. The broken lines represent a building of height h and maximum width w, just contained in the "cone of protection."

Oliver Lodge ridiculed this claim during the 1888 British Association meeting, pointing out that even if true (which he didn't believe), it would require "gigantic" rods to protect a building. Lodge didn't back up this assertion with any calculations, but he was correct. It is not difficult to show that the length of AB, the length of the minimum sufficient lightning rod, is given by

$$AB = \frac{1}{2}(w + 2h) + \sqrt{wh}.$$

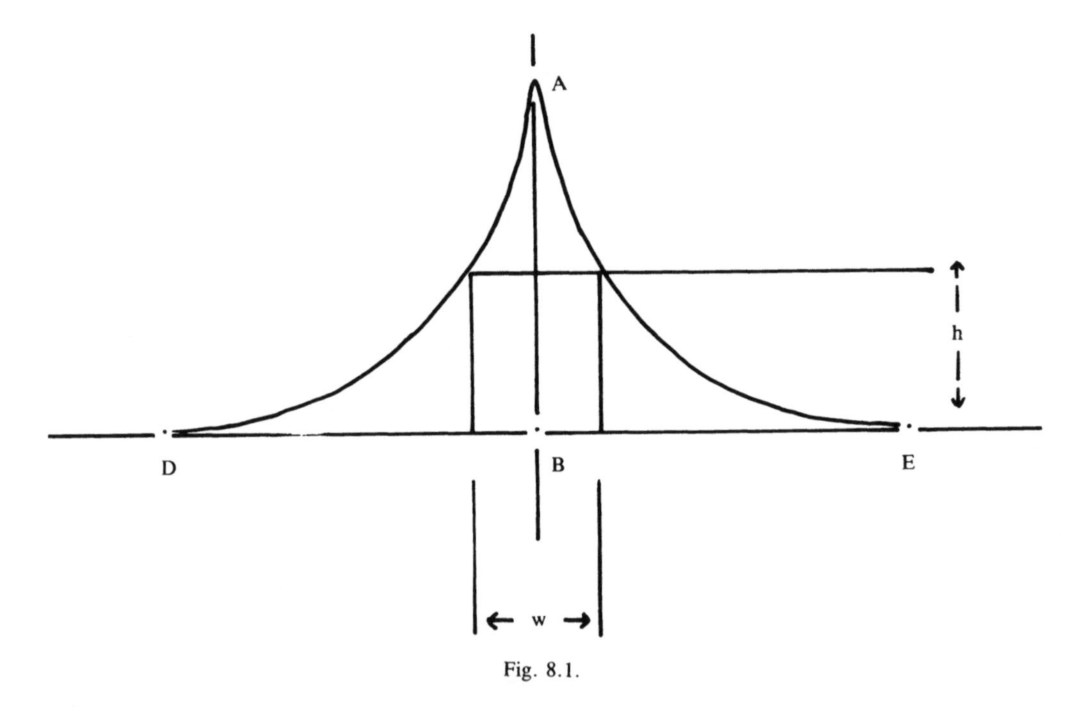

Fig. 8.1.

Just picking some "reasonable" numbers out of the air, say, for example, $h = 100$ feet and $w = 300$ feet for a government office building, gives a result of a thick copper rod 423 feet high! Needless to say, no such rods were ever installed, so by Preece's own theory no building ever *really* was protected. Preece made no reply, but later, in a letter[113] he referred to Lodge's ideas as based more on the presence of "impudence" than on impedance. Of course Lodge's objection can obviously be at least partly rebutted by using more than one rod. And, in fact, the Lightning Rod Conference Report recommends multiple rods be used, one at every elevated point of a building so as to have it literally bristling with protection, like a porcupine.

Both Lodge and Preece had serious misconceptions about the nature of lightning. Preece thought bolts are limited to 500 feet in length, while Lodge thought a mile to be possible (actually, up to *two*-mile-long bolts have been reliably and repeatedly measured). Lodge calculated the voltage of a mile-long bolt to be 5000 million volts by extrapolating from the 30,000 volts required to produce a one-centimeter spark in air, while Preece felt the voltage increased much slower than linearly with bolt length (he suggested it increased as the cube root of the length). In fact, Lodge was *far* off-base, with actual voltages rarely exceeding 100 million volts, while Preece's suggestion gave 1.6 million volts for a mile-long bolt—too low, but a lot closer than Lodge's value. Preece thought a bolt *wouldn't* strike a lightning rod if it was "properly" installed and even Lodge, who must have been prepared for almost anything from Preece, was taken aback at the 1888 B.A. meeting when he heard *that*! To balance the ledger, Lodge was wrong about the oscillatory behavior of lightning. By experimenting with various Leyden jars Lodge found, for example, that a gallon-sized jar discharged through a wire suspended around a room gave one-megahertz oscillations, while a pint-sized jar (discharged through tongs) gave fifteen-megahertz oscillations. Why he settled on one megahertz is not clear, but in any case when Lodge calculated the inductive impedance of various rods at a one-megahertz frequency he was totally beside the point. And as for Preece's "cone of safety" which Lodge made so much fun of, the fact is the idea is *not* dead. In a recent book[114] on lightning you can find the "cone of protection" concept discussed, along with some interesting photographs in support of the idea, and accompanied by the blanket statement "When properly installed, lightning rods can provide virtual immunity from direct lightning strokes."

Tech Note 4: Heaviside and S. P. Thompson on the Distortionless Circuit

In late August 1893 Silvanus P. Thompson delivered a paper, "Ocean telephony," at the International Electrical Congress at the World's Fair in Chicago. In this paper[115] he described his inductively loaded "distortionless" cable, patented in England in 1891 (he obtained U.S. patents in 1893). The conference opened on August 21 and, by amazing timing, Heaviside appeared in *The Electrician* issues of August 25, September 15, and October 6 (reprinted in EMT 1, pp. 403–428) with detailed prose discussions of the distortionless circuit conditions he had announced over six years earlier, in June 1887. Oliver's brother, Arthur, was also in attendance at the Congress (along with Preece), and it is certain that Oliver was privy to all advance program information and knew ahead of time of the nature of Thompson's paper.

In his 1893 review[101] of *Electrical Papers,* FitzGerald specifically mentioned Thompson's cable and pointed out Heaviside's priority. Unlike Heaviside's concept of adding *series* inductance to the cable, Thompson had inductive coils connected in *shunt,* i.e., across the two wires of a metallic circuit. The advantage of this approach is the inductance coils must also be of high resistance (to avoid shorting the cable) which is obviously very easy to arrange, as opposed to Oliver's requirement of very low resistance inductance coils. The big disadvantage of Thompson's approach, a fatal one, is that as a matter of practical engineering it simply doesn't work (as Heaviside said, "Thompson got the inductances in the wrong place").

While it is as clear as new window glass that Thompson got the basic idea for his cable from Heaviside, he made no attempt to bring Oliver in with him in seeking the patent. Heaviside never complained in public about this intellectual appropriation by Thompson, but he must have in private because, in a letter (August 16, 1893) to him, FitzGerald wrote: "S.P.T. and his patent:—I expect he did get the inspiration from you... . I do hope S.P.T., if he makes anything of his patent will pay you well for your inspiration and appoint you electrical advisor to his company at a large salary." Nowhere in his paper to the Electrical Congress, however, does Thompson mention, or even hint at, Oliver's work, so it is unlikely he ever intended to share anything (least of all, money) with Heaviside.

Thompson's paper received a good deal of visibility, and continued to receive it for some years. For example, the *Pall Mall Gazette* of January 16, 1897 ran a delightfully Victorian interview with him. Beginning by reporting how Sir Henry Mance's recent Presidential Address to the IEE had "let drop an allusion to a novel method of approaching" the problem of sending messages over an insulated line, the newspaper went on to declare proudly

> A representative of the *Pall Mall,* who was present, promptly sought out the inventor, S. P. Thompson, F.R.S., and buttonholed him in a corner about the details of his system.

After a fair amount of technical detail, the interview concluded with the following words from Thompson (nowhere does Heaviside's name appear):

> I have been at work a good many years on this matter. I saw what was coming [the need for long-distance telephony], and I have been over all the ground, I believe, that it is possible to cover. Yes, I have patents, of course, and as I do not believe that there is any other satisfactory way of overcoming the difficulty, I intend to wait and see what happens when the practical problem arises.

I find the entire affair extraordinary, because of both Thompson's ignoring of Heaviside and Heaviside's lack of apparent anger at Thompson. I think we have here a puzzle for the historians—at least one reader of *The Electrician* shared my bewilderment at this event, and wrote[116] to ask "Cannot Mr. Heaviside be persuaded to give his views on the method proposed by Prof. Thompson?" As close as Heaviside came to answering this plea was in the above-mentioned articles in *The Electrician,* in which he just once mentioned Thompson in passing. These articles are particularly interesting as they are almost totally prose (with the little mathematics that does occur so elementary that even Preece could have had no trouble with it). His discussion is based on the two fundamental results that express the magnitudes of the signal attenuation and distortion on a circuit, in terms of the four circuit parameters:

$$\sigma=\frac{R}{2L}-\frac{K}{2S}>0 \qquad \text{(distortion)}$$

$$\rho=\frac{R}{2L}+\frac{K}{2S} \qquad \text{(attenuation)}$$

where, as Oliver observed, "Both σ and ρ should be as small as possible, of course." R, L, K, and S are the per unit length distributed circuit resistance, inductance, leakage conductance, and capacitance, respectively. It is very important to note that Heaviside's K is *not* Preece's K, i.e., Preece (and everybody else) used K for capacitance in those days. Oliver's articles give a nice tutorial on how to get small values of σ and ρ by varying the four parameters (but most important is the L).

There *is* one portion of Heaviside's discussion that he almost surely wrote with Thompson in mind. One advantage to Thompson's high-resistance inductive shunts is that they automatically introduce artificial faults into the circuit (a value of 3000 ohms per shunt each ten miles is mentioned in his Electrical Congress paper), and such faults can often be helpful (increasing K reduces σ and, while it also increases ρ, the trade-off is not *necessarily* bad). Thompson wrote as though this idea of artificial faults was something original with him, i.e., after mentioning the "sheer horror with which all good telegraph engineers regard the idea of having a fault on their line," he said "Deliberately to insert, as the author proposes, a series of faults, as shunts across the cable from the going wire to the return wire, will, to many of them, appear sheer madness." Oliver, of course, had made this suggestion *years* earlier (see Chapter 4), and to make this clear to readers he quoted his own paper.[117]

Notes and References

1. "The telegraph systems of the world," *McClure's Magazine,* vol. 5, pp. 99–112, July 1895.
2. One reason for the near monopoly by England in cable manufacturing was its control of Malaysia, the primary source of gutta-percha (see Chapter 3) for cable insulation.
3. G. Herrle, "The submarine cables of the world," *National Geographic,* vol. 7, pp. 102–107, March 1896.
4. Such speeds are not possible *by hand* transmission, of course, and certainly could not be decoded by ear! They were achieved by prepunching messages on a paper tape and then processing them with the *automatic transmitter/receiver apparatus developed by Wheatstone.* This very high speed transmission was of major importance where multiple transmissions of the same, often quite long, message were required to ensure accuracy in the final decoded version, such as newspaper stories. A million news words a *night* out of the London central telegraph office, alone, was not uncommon.
5. In 1885 one could send a twelve-word (including address and signature) telegram anywhere in England for sixpence. Each additional word was a halfpenny. By comparison with telephony, the fee for three minutes of talk in 1893 between Paris and London was eight shillings (the price of a 192-word in-country telegram).
6. Some ingenious, if farfetched, schemes to get around the slow transmission speeds of existing cables were developed by inventors. John V. Gibboney, for example, of Lynn, Massachusetts received a U.S. Patent for his

system that used a phonograph to prerecord a telephone message, then played the record slowly over a long-distance cable (where another phonograph at the receiving end rerecorded the message). Upon playing back the second record at high speed the original frequencies were restored to produce understandable speech! What could reasonably be called a *conversation* was out of the question, of course—see *The Electrical Engineer (New York)*, vol. 12, p. 576, November 25, 1891.

7. *The Electrician,* vol. 33, p. 461, August 17, 1894.
8. Quoted from D. G. Tucker, "Beginnings of long-distance telephony, 1882–1887," *Electronics & Power,* vol. 20, pp. 825–827, October 17, 1974.
9. "Electric induction between wire and wire," *The Electrician,* vol. 17, pp. 410–412, September 24, 1886 (including discussion among Preece, William Thomson, S. P. Thompson, and J. H. Poynting). Oliver mentioned these experiments, attributing the work to Arthur, in EP 2, pp. 185–186, 237–239. And in his address as the new President of the Newcastle-on-Tyne section of the IEE in 1900, Arthur specifically took credit for this work (*The Electrician,* vol. 45, pp. 247–250, June 8, 1900). Preece reported on a continuation of these experiments at the next B.A. meeting at Manchester, in "On induction between wires and wires," *The Electrician,* vol. 19, pp. 461–464, October 7, 1887 (including discussion between Preece and William Thomson).
10. "Recent progress in telephony," *The Electrician,* vol. 9, pp. 389–393, September 9, 1882.
11. D. G. Tucker, "Francois van Rysselberghe: Pioneer of long-distance telephony," *Technology and Culture,* vol. 19, pp. 650–674, October 1978.
12. Preece used this term during the lightning rod debate with Oliver Lodge at the contentious (recall Chapter 5) 1888 B.A. meeting at Bath—see *The Electrician,* vol. 21, p. 646, September 21, 1888. Details of this debate are discussed later in this chapter.
13. Early telegraph circuits were formed using single lines of iron, with the loop completed through an earth-return. Inductive effects could be tremendously reduced by eliminating the earth-return, replacing it with another wire, and twisting the wire-pair together. It was also found to be helpful to use copper rather than iron. As an added benefit, copper doesn't rust! Such all-metal circuits with no earth-return were called "metallic circuits."
14. See Chapter 3, Note 5.
15. "On electrical motions in a spherical conductor," *Philosophical Transactions of the Royal Society,* vol. 174 (part II), pp. 519–549, 1883. Charles Niven (1845–1923) ran an essentially parallel and nearly simultaneous course (in some respects even anticipating Lamb). A remarkably able applied mathematician who was Senior Wrangler in 1867, his work appeared in "On the induction of electric currents in infinite plates and spherical shells," *Philosophical Transactions of the Royal Society,* vol. 172 (part 2), pp. 307–353, 1881. Niven, as did Lamb, followed Maxwell's theory and was explicit in pointing out to his readers the theory's central (and then revolutionary) concept—"...energy is supposed to be seated everywhere in the surrounding medium... ."
16. One interesting calculation Lamb made was the determination of the time constants for electric currents (however started) in a sphere. For an iron ball one foot in diameter, even the slowest mode (the one taking the longest to decay by a factor of e) has a time constant of just six seconds. For copper and iron spheres the size of the Earth, however, Lamb calculated the astonishing time constants, for the longest lasting mode, of 10 and 330 million years, respectively!
17. EP 1, pp. 439–440.
18. The story of Hughes' near miss didn't come out until twenty years later, when John Fahie, as part of his *History of Wireless* (1901), wrote to Hughes to ask about talk he had heard concerning the 1879 experiments. Hughes related how he had invited several prominent men (including Preece) to view his work, all of whom agreed that "all the results could be explained by known electromagnetic *induction* [my emphasis] effects" and therefore "would not accept my view of actual aerial electric waves."
19. D. W. Jordan, "D. E. Hughes, self-induction, and the skin effect," *Centaurus,* vol. 26, pp. 123–153, 1982.
20. Ibid, p. 144.
21. R. Appleyard, *The History of the Institution of Electrical Engineers,* London: IEE, 1939, pp. 61–62.
22. EP 1, pp. 95–112. There is a note in NB 3A:17 that this paper was "in Preece's hands."
23. EP 1, pp. 181–190.
24. NB 2A:10–11. The "absolute zero" Heaviside was referring to is the graduated sound intensity scale Hughes used, about which *The Electrician* (vol. 7, p. 21, May 31, 1879) reported: "The range of sound is sufficient at the maximum [setting]—200—for everyone who is not absolutely deaf; 0 or zero is a point of positive silence from the instrument... ." As late as 1894 Preece was still talking (Note 7) of this as an "absolute zero of silence." In EP 2, p. 266 Heaviside mentioned "Prof. Hughes' oddly named Sonometer..." in an essay on induction balances.
25. *Journal of the Institution of Electrical Engineers,* vol. 18, pp. 4–35 (in particular, p. 9), 1889. Kelvin very quickly appreciated the physical significance of the skin effect himself, and was just as quick to introduce it to his students, as shown in the memoir by his long-time assistant, Magnus Maclean: "Kelvin as a teacher," *Journal of the Institution of Electrical Engineers,* vol. 56, pp. 169–192, March 1918. It is not clear, however, whether Kelvin appreciated the theoretical basis (Maxwell's theory) that underlies the skin effect.
26. EP 2, pp. 28–32.
27. It was common knowledge that Hughes was often hard to understand. One commentary (*Nature,* vol. 29, p. 459, March 13, 1884) described this unfortunate characteristic as follows: "If Prof. Hughes were as great a

master of writing English as he is of experimenting, his views on magnetism would receive speedier acceptation, for they would then probably be understood without that close study which his involved sentences and heterogeneous paragraphs now demand. It is very remarkable that such an ardent worker, such a deep thinker, and such a clear and simple experimenter should have such difficulty in expounding his views on paper."

28. EP 2, pp. 33–38.
29. *The Electrician,* vol. 16, p. 495, April 30, 1886.
30. *The Electrician,* vol. 16, p. 510, May 7, 1886.
31. The first instance that I have found of Heaviside using the modern term of *skin effect* was dated November 1893 (EMT 1, p. 443), although he did use the word *skin* in January 1891 (EMT 1, p. 14) in connection with surface conduction.
32. EP 2, pp. 168–170.
33. EP 1, pp. viii–ix (Preface).
34. But he did consider nonuniform cases, too. For example, in 1884 (EP 1, pp. 399–400) he discussed an induction-free transmission line with distributed capacitance and leakage *varying* directly with the distance from one end. This is the so-called Bessel cable, named after the Bessel functions that occur in the solution.
35. This sequence of articles is in EP 2, pp. 39–323; they originally appeared in both *The Electrician* and *Philosophical Magazine.*
36. Quoted from D. W. Jordan, "The adoption of self-induction by telephony, 1886-1889," *Annals of Science,* vol. 39, pp. 433–461, September 1982.
37. "The relative merits of iron and copper wire for telegraph lines," *The Electrician,* vol. 15, pp. 348–351, September 18, 1885.
38. "On the coefficient of self-induction in telegraph wires," *The Electrician,* vol. 19, pp. 400–401, September 16, 1887.
39. Quoted from Jordan (Note 36).
40. "On the limiting distance of speech by telephone," *Proceedings of the Royal Society,* vol. 42, pp. 152–158, March 3, 1887.
41. William Edward Ayrton (1847–1908), Professor, from 1884 until his death, of Applied Physics and Electrical Engineering at the City Guilds Central Technical College in Kensington. For many years Ayrton was a close technical collaborator with Heaviside's friend, John Perry.
42. Quoted from Jordan (Note 36).
43. There is some mention of the "Bridge System" in EP 2, pp. 250–251, and EMT 1, pp. 433–437.
44. EP 2, pp. 323–354.
45. *The Electrician,* vol. 19, pp. 79–81, June 3, 1887; reprinted in EP 2, pp. 119–124. The "experimental evidence" Heaviside was thinking of, on the merit of copper-covered steel wires, was the work of van Rysselberghe (Note 11)—see EP 2, p. 399.
46. This operational derivation of the distortionless circuit is *not,* historically, the original derivation. Originally Heaviside thought in terms of how to achieve a transmission circuit with no reflected waves from the receiving end back to the transmitting end (see EP 2, pp. 119–124) in Arthur's "Bridge System" (see EP 2, p. 402).
47. EP 2, p. 119. In the original publication in *The Electrician* (see Note 45) the final word *practician* was *scienticulist.*
48. This was common 19th century slang for "murdering by strangulation," as did the infamous William Burke who sold fresh and undamaged (except for their throats) corpses to medical students who wished to study anatomy. Burke went to the gallows in 1829 for his deeds, but Preece was luckier.
49. Years later, in a letter to the President of the IEE (July 29, 1922), Heaviside wrote of his brother as being "very afraid of P."
50. This seems to indicate Arthur was dressed down *in person* by Preece, who showed him the paper with the marginal comments, but did not return it.
51. EP 2, p. 489.
52. EP 1, p. x (Preface).
53. EP 2, p. 305.
54. "Mr. W. H. Preece on the self-induction of wires," EP 2, pp. 160–165, dated September 24, 1887.
55. Biggs and Snell also were *technical* collaborators, giving a joint paper to the 1887 B.A. meeting on transformers and alternating current distribution.
56. W. G. Bond, *The Electrician,* vol. 87, p. 603, November 11, 1921. Bond would himself eventually serve as editor (1895–1897) and was an admirer of Oliver's talents.
57. A story about this incident has grown over the years, beginning with the belief that Snell had a loose insert put into each copy of an issue of *The Electrician,* asking for those who read Heaviside's articles to write and say so. There is reason to believe Heaviside believed this (see E. J. Berg, "Oliver Heaviside," *Journal of the Maryland Academy of Sciences,* vol. 1, pp. 105–114, 1930), but Snell's letter seems to indicate he merely asked around. Another version, explaining why no such loose insert has ever been found, is that due to a printer's error the inserts were left out, and that's why Snell got no replies!
58. Quoted from Jordan (Note 36).
59. EP 2, pp. 375–396.
60. *The Electrician,* vol. 20, p. 309, February 3, 1888.

61. *The Electrician,* vol. 20, pp. 452–453, March 2, 1888.
62. G. J. Symons (Ed.), *Lightning Rod Conference Report*, London: E.&F.N. Spon, 1882. The group, while not an official body, was nevertheless impressive, with members from the Institute of British Architects, the Physical Society, and the Meteorological Society, as well as the Society of Telegraph Engineers and Electricians.
63. "On the proper form of lightning conductors," *The Electrician,* vol. 5, pp. 199–200, September 11, 1880.
64. "On the space protected by a lightning-conductor," *Philosophical Magazine,* 5th Series, vol. 10, pp. 427–430, December 1880. See also the critical reply to Preece's analysis in *Nature,* vol. 23, p. 386, February 24, 1881.
65. *The Electrician,* vol. 21, p. 565, September 7, 1888.
66. That Leyden jar discharges could oscillate had been known for decades before Lodge. By lowering the oscillation frequency sufficiently, for example, one could *hear* the audio tone of the spark, an experiment Lodge loved to perform at demonstrations. Even so, it was still an exciting event when high-speed photography allowed recording of the oscillations, as described in the paper by C. V. Boys, "Notes on photographs of rapidly moving objects, and on the oscillating electric spark," *Proceedings of the Physical Society,* vol. 11, pp. 1–15, November 1890.
67. Lodge was trying here to be careful in his history. As he wrote in a footnote, "It is not possible, I think, to give Mr. Heaviside the credit of the original discovery of this theorem (though, doubtless, he discovered it for himself)... ." Lodge went on to mention others who discovered the skin effect, including Lamb, but then he added that even so, of all the others "none of the *actual* statements were quite so explicit as that of Mr. Heaviside in 1885." This actually was a weakening of the position he had taken just a month earlier in a letter to *The Electrician* (vol. 21, p. 303, July 13, 1888), in which he wrote "I am now informed [by Oliver!] that Mr. Oliver Heaviside published this fact [skin effect], as deduced mathematically by himself, a year before Prof. Hughes' experiments which illustrated it and brought it into prominence were published...and to Mr. Heaviside, therefore, the priority is distinctly due. I have no opportunity just now of verifying this statement, but I have not the slightest reason to doubt it; and probably to readers of *The Electrician* it is a familiar fact."
68. Later, when Lodge collected all of his work in this area into book form (*Lightning Conductors and Lightning Guards,* London: Whittaker, 1892), he deleted the "eccentric" and "repellent." See pp. 46–47.
69. EP 2, pp. 486–488.
70. *The Electrician,* vol. 21, p. 570, September 7, 1888.
71. *The Electrician,* vol. 21, pp. 607–608, September 14, 1888.
72. *The Electrician,* vol. 21, p. 646, September 21, 1888.
73. Quoted from B. J. Hunt, " 'Practice vs. theory': The British electrical debate, 1888-1891," *Isis,* vol. 74, pp. 341-355, September 1983.
74. This was, indeed, a very difficult concept in the late 1880s. In a marginal note (see Chapter 7, Note 14) Heaviside wrote that when he first suggested inductive loading a well-known physicist, Thomas Blakesley, said "it would be like making humps on a road to increase the speed of vehicles." Blakesley was no fool—he laid much of the groundwork for modern, routine mathematical analysis of a-c circuits (in *The Electrician*). To Blakesley we owe the word *quadrature* to denote a 90° phase shift.
75. Quoted from B. R. Gossick, "Heaviside and Kelvin: A study in contrasts," *Annals of Science,* vol. 33, pp. 275-287, 1976.
76. "The Presidential Address" (editorial), *The Electrical Engineer (London),* vol. 3, pp. 52–53, January 18, 1889. Later in the year Biggs wrote a vigorous *defense* of Preece when he was misquoted by the popular press in his critical remarks about using electricity for criminal executions—see *The Electrical Engineer (London),* vol. 4, pp. 250–251, September 27, 1889. The precise relationship between Biggs and Preece is still unclear.
77. *Journal of the IEE,* vol. 18, pp. 457–458, 1889.
78. Ibid, pp. 500–503.
79. *The Electrician,* vol. 23, p. 23, May 17, 1889, and p. 51, May 24, 1889. I could find nothing in the recorded discussion of the IEE meetings to support *The Electrician*'s remarks, and I suspect it was all just a bit of editorial fun at Preece's expense.
80. *The Electrician,* vol. 94, p. 276, March 6, 1925.
81. *The Electrician,* vol. 73, p. 66, April 17, 1914. Kelvin mentioned such talks with Bennett in his IEE Presidential Address (Note 25).
82. EMT 1, pp. 441–446, in an essay entitled "Various ways, good and bad, of increasing the inductance of circuits." It is here that I've found the earliest use by Heaviside of the term *loading*, in connection with *uniform circuits,* i.e., *"...load the dielectric itself with finely-divided iron, and plenty of it."*
83. NB 7:94. The initials "W.H.P." written in the margin leave no doubt as to the identity of the "British Engineering man."
84. *Nature,* vol. 39, p. 473, March 14, 1889. Perhaps I am reading too much into Lodge's words, but his particular turn of phrase makes me think he is having a bit of fun here with a play on the skin effect itself!
85. The *quay,* for non-British readers, is pronounced *key.* In a letter to Hertz (November 12, 1889) Heaviside called Torquay "the Montpellier of England."
86. NB 10:74.
87. *The Heaviside Centenary Volume,* London: IEE, 1950, p. 93.

88. Ibid, p. 96.
89. F. Gill, "Oliver Heaviside," *Electrical Communication,* vol. 4, pp. 3–6, July 1925. Frank Gill was President of the IEE in 1923.
90. R. Appleyard, *Pioneers of Electrical Communication,* London: Macmillan, 1930, p. 224. How such a picture came to be in Oliver's possession is a bit of a mystery, but one possibility is that Arthur gave it to him. Just before Preece delivered his inaugural address as IEE President in 1893, Arthur received an IEE prize of money and just such a photograph as Oliver described.
91. In the possession of the IEE Archives, London.
92. Trotter was referring to John Ambrose Fleming (1849–1945) who, in 1904, built the first vacuum-tube (or "valve," as he called it) diode. In a letter (May 10, 1922) to the President of the IEE Heaviside showed he respected Fleming: "He was the first man to bring my theory in mathematical form before the S.T.E. after Sir W.T. had introduced it to them. But of course there was no one to understand it. Fleming later on made very ingenious apparatus to show wave propagation. I have often wondered why he didn't go further, and do the discontinuous loading trick." See also Fleming's interesting recollections, *Fifty Years of Electricity: The Memories of an Electrical Engineer,* New York, NY: D. Van Nostrand, 1921.
93. After Trotter left *The Electrician* in 1895 he spent twenty years in government service—and in his later life he liked to list his favorite recreation as "remembering I am no longer a Government official"!
94. *Nature,* vol. 47, pp. 505–506, March 30, 1893.
95. *The Electrical World,* vol. 21, p. 250, April 1, 1893.
96. *Physical Review,* vol. 1, pp. 152-156, 1892-93.
97. *The Electrical Engineer (New York),* vol. 18, p. 357, October 31, 1894.
98. *The Popular Science Monthly,* vol. 43, p. 560, August 1893.
99. *The Athenaeum,* vol. 102, p. 67, July 8, 1893.
100. *The Nation,* vol. 56, p. 199, March 16, 1893.
101. *The Electrician,* vol. 31, pp. 389–390, August 11, 1893.
102. Heaviside's was a loud and continuous voice arguing for the removal of the "4π" numerical factor which shows up ("erupts" was his word) so often in the formulas of physics. The problem as Oliver saw it (EMT 1, pp. 116–127) was the illogical definition of the Coulomb force between two electric charges, q_1 and q_2, separated by distance r, as $F = kq_1q_2/r^2$, with the constant k chosen to be unity. This causes 4π factors to occur elsewhere (an "obnoxious effect") because we live in a three-dimensional space and all phenomena that "spread" in such a space do so through spherical surfaces with areas that are 4π times the radius squared. In Oliver's metaphor of the plague, "...the unnatural suppression of the 4π in the formulae of the central force, where it has a right to be, drives it into the blood, there to multiply itself, and afterwards break out all over the body of electromagnetic theory." If, instead, the constant $k = 4\pi$, then the 4π does not "erupt" in other, more commonly used formulas, and such a choice for k is the "rational" one, as Heaviside put it. This new value for k, however, introduces changes into the sizes of the electrical units, and this has political problems similar to that of getting Americans to give up quarts for liters and miles for kilometers. These new units are the "Heaviside rational units." In the Preface to EMT 1 (dated December 16, 1893) Oliver expressed admiration for how chemists had given up HO as the chemical formula for water when they found out it was H_2O. Why, he asked, couldn't physicists be as rational? Obviously privy to this argument, FitzGerald wrote on December 22 in an attempt to persuade his friend to fight other, more important wars: "You won't be able to move physicists as the H_2O question moved chemists. Besides HO had very little money at its back but that wretched 4π is a millionaire and you might as well try to get a patent against Rothschild as break its vested interests."
103. EMT 3, p. 93.
104. Heaviside was wrong—five years later (May 1899) he wrote to FitzGerald that in that time not quite 600 copies of EMT 1 had been sold.
105. *Journal of the Institution of Electrical Engineers,* vol. 22, pp. 36-74, 1893.
106. EMT 1, pp. 320–321.
107. *The Electrical World,* vol. 21, p. 117, February 18, 1893. The editorial observed that a telephone circuit between Boston and Chicago had a *KR* of 54,000 and it worked just fine. Since Preece had set an upper *KR* limit of 15,000 for good telephony, something was obviously not right: this problem was due, as the journal pointed out, to the fact that induction had "emerged from the English Channel and set to work demolishing" capacitance!
108. *The Electrician,* vol. 33, pp. 105–106, May 25, 1894.
109. Heaviside was almost certainly referring to a report, in the same issue of *The Electrician* carrying FitzGerald's review, of Philipp Lenard's study of the deflection of cathode rays (not yet known to be electrons) by an externally applied magnetic field.
110. EMT 1, p. 328.
111. EP 2, p. 422.
112. EMT 2, pp. 536–537.
113. *The Electrician,* vol. 21, p. 712, October 5, 1888.
114. P. E. Viemeister, *The Lightning Book,* Cambridge, MA: MIT Press, 1972, pp. 164, 194, 202–205.
115. Thompson's paper was reprinted by *The Electrician* (vol. 31, pp. 439–440, August 25, 1893, and pp. 473–475, September 1, 1893) and abstracted versions also appeared in *The Electrical Engineer (New York)* (vol. 16, pp.

196–198 with discussion on p. 207, August 30, 1893) and in *The Electrical World* (vol. 22, pp. 177–178, September 2, 1893). All three journals had reporting problems. *The Electrician* got the two Olivers mixed up, and cited Lodge as the discoverer of the distortionless circuit, while *The Electrical World* quoted Oliver (rather than Arthur) in the discussion as saying the key idea is to use "plenty of copper," which is something *Preece* would have said! *The Electrical Engineer* did mention Arthur in the discussion, paraphrasing him as saying "His experience in the British postal telegraph service confirmed the results predicted by Prof. Thompson. In a set of experiments on underground telephone lines they found beneficial results followed the insertion of electromagnets at certain points."

116. *The Electrician,* vol. 31, pp. 507–508, September 8, 1893. A few years later a very nice explanation was given in a *student* paper delivered at the IEE ("The working of long submarine cables," *The Electrician,* vol. 39, pp. 740–746, October 1, 1897), which began by referring to Heaviside as a "high authority." The younger generation was beginning to pay attention to Oliver.

117. EP 1, p. 77.

9
The Great Quaternionic War

... there will be a splendid row, which is some consolation.

— Peter Tait to John Tyndall, in an 1872 letter declaring Tait's joyful anticipation of a scientific controversy

I never met Willard Gibbs; perhaps, had I done so, I might have placed him beside Lorentz.

— Albert Einstein, in a 1954 interview

A text-book of vector algebra...is much needed, as many physicists are becoming interested in the new algebra, owing in great measure to Mr. O. Heaviside's able exposition of its principles and applications in the Electrician *and elsewhere.*

— Commentary in *Nature*, 1892

... no figure, nor even model, can be more expressive or intelligible than a quaternion equation.

— Peter Tait, 1890

Maxwell, I fear, is responsible to a large extent for the discredit into which quaternions have fallen among physicists.

— Alexander McAulay, 1892

... the invention of quaternions must be regarded as a most remarkable feat of human ingenuity. Vector analysis, without quaternions, could have been found by any mathematician...but to find out quaternions required a genius.

— Oliver Heaviside, 1892, in a subtle hint that quaternions aren't as easy as vectors

It is as unfair to call a vector a quaternion as to call a man a quadruped.

&

Quaternions furnish a uniquely simple and natural way of treating quaternions. *Observe the emphasis.*

— Oliver Heaviside, 1892, in not so subtle comments

More Debates

The great debate with Preece was the central one of Heaviside's life, but in this chapter (and the next two) we'll take a look at three others. The common thread that runs through all these confrontations is *not* that Heaviside was *always* right (but I believe that to be nearly so and, of course, obviously so in the debate with Preece), but rather the insight they give us into Oliver's workings as we watch him in controversy. With Preece we saw how Heaviside handled (often ineptly) a *politically* powerful but technically weak adversary who was demonstrably wrong. With the war of words over quaternions, however, Heaviside went against a man, Professor Peter Tait, technically as gifted as himself. But now he also had an equally talented ally, Yale's Professor J. Willard Gibbs, a man of genius sometimes called the "Maxwell of America."

Together Gibbs and Heaviside would create and actively promote a new mathematical tool, one that today every scientist and engineer takes for granted—modern vector algebra and calculus. At first each worked independently and unaware of the other, separated by the Atlantic. Later they worked together in spirit, still apart physically and never meeting, but linked by a mutual respect for the other's contributions.

Vectorial systems can be traced back to considerably before both men, certainly to the early 1840s, but those were preliminary, tentative attempts that were still groping for just the right physical insights and mathematical notation. As men with one foot in each arena, physical science *and* mathematics, both Gibbs and Heaviside brought the critical mix of physical intuition and mathematical insight to bear as no pure mathematician or experimental physical scientist could. But this doesn't mean they convinced their colleagues overnight. On the contrary, they had to rebut an earlier system, the noncommutative quaternions of the Irish

Oliver Heaviside, about the time of the quaternion debate.

mathematician Hamilton, a system that had its share of devoted supporters, some of whom felt *very* intense, indeed, about the matter. In America, for example, one of them wrote[1] in 1857,

> It is confidently predicted, by those best qualified to judge, that in the coming centuries Hamilton's Quaternions will stand out as the great discovery of our nineteenth century... . And if the world should stand for twenty-three hundred years longer, the name of Hamilton will be found, like that of Pythagoras, made immortal by its connection with the eternal truth first revealed through him.

As the rest of this chapter will show, it was no easy matter for Gibbs and Heaviside to overcome such passion. And among all the English-speaking scientists, none was more passionate in his devotion to quaternions (see Tech Notes 1 and 2 at the end of this chapter) than the Scottish mathematical physicist, Peter Tait.

Peter Tait, the Warrior of Victorian Science

Peter Guthrie Tait (1831–1901) was lucky in his friends, counting Maxwell as a schoolboy chum from their days as teenagers at Edinburgh Academy, and Lord Kelvin as an idolized intimate till the day he (Tait) died. Tait was, like Heaviside, a man also blessed with great intellectual ability (he was Senior Wrangler in 1852; was appointed to his first academic post, Professor of Mathematics at Queen's College, Belfast, at age twenty-three; and at his death was the author of over 350 papers and the author or co-author of nearly two dozen books!). He also shared Heaviside's trait of a sharp tongue that often pushed his pen. Tait's willingness to enter into controversy went beyond merely a natural desire to defend his views, however. He loved an argument for its *own sake.*

Almost nothing, it seems, was too trivial for Tait to battle about in print. In 1884, for example, he had harsh words in *Nature* for R. T. Glazebrook,[2] then Secretary of the Cambridge Philosophical Society, over his (Tait's) claimed difficulty in obtaining the Society's publications!

But this was nothing compared to the hot debate Tait got into with John Tyndall.[3] Tyndall championed Julius Robert Mayer (1814–1878) as having originated the idea of conservation of energy, as well as having determined the quantitative relationship between heat and work, and as being at least as deserving of recognition for these contributions as were the far better known Joule and Helmholtz.[4] William Thomson, who was Joule's champion (and Tait, whose attitude was that Thomson could make no mistake), quickly attacked Tyndall in 1863, and the controversy sporadically burst into print over a period of *years.* Thomson had the good sense, after stating his position in the matter, to drop it. Not so Tait, who raged on alone against Tyndall for more than a decade. In person Tait could be a charming man (Tyndall himself recorded[5] having met Tait in 1867, when they shook hands and had a "very cordial" exchange), but in print he could be as aggressive and as nasty as a bull elephant with an infected trunk.

Even while the Mayer dispute festered, Tait attacked Tyndall on a different front. This time the issue was Tyndall's support of the work of another on how glaciers move, as opposed to the work of James David Forbes (late Professor of Natural Philosophy at the University of Edinburgh). As one of Forbes' biographers (and by now at Edinburgh[6] himself), Tait roared in on the attack once again and this affair became even more bitter than the one over Mayer. A hint of just *how* bitter can be gotten from a note published by the editor of *Nature* (Norman Lockyer, who knew both men). Lockyer wrote[7] that he was unhappy at the personal nature of Tait's comments, and then he revealed that Tyndall (usually a most gentle soul) had recently

Peter Tait (1831–1901)

A man whose creative abilities were matched by a strong sense of national pride, he refused fellowship in the Royal Society of London (his with a nod of the head if he wanted it), declaring fellowship in the Royal Society of Edinburgh was quite good enough for him. He was a close friend of the father of Robert Louis Stevenson (*Treasure Island*), did important work on the "four-color" map problem of mathematics, and wrote some fascinating analyses on the physics of golf balls in flight!

replied in kind. After thinking things over, however, Tyndall had quickly sent Lockyer a follow-up note admitting he had written in haste and that

> ...I find two passages...I think it desirable to cancel. The first is that in which I speak of lowering myself to the level of Professor Tait; the second is that in which I reflect on his manhood. These passages I wish to retract.

This caused Tait to reply the very next day with a bitter personal letter to Lockyer[8]:

> The fact is that your impartiality as Editor has all along *told against me*... . You allow Tyndall, under pretext of withdrawing them [the insults] to *reprint* two of the low things he said. (Enough, however, remains unretracted to make it impossible for me to meet him except with the tip of my toe.)

My assessment of Tait as an aggressive man who played *very* rough, indeed, isn't just a modern one based on a speculative reconstruction of long-ago events, as one 1874 newspaper comment shows[9]:

> The Professor must be a very lovable or a remarkably lucky man—for he seems to love nobody and nothing, and everybody seems to love him, at least enough never to be angry with him. He is...an academical Ishmael; his hand is against every man, but no man's hand is against him.

And years later, even the author of an admiring obituary[10] felt obliged to write

> ...the sense of bereavement is too near to us to permit of the necessary historical abstraction. Nor is this the time to enlarge on the polemical discussions in which Tait took part. Ready to take a blow, he did not always spare his strength in giving one, and his opponents did not always relish his rough play. ...undoubtedly some of them were led, temporarily at least, greatly to mistake his character.

The writer then went on to declare of Tait that

> Personal contact with him at once dissipated any such misconception. To feel the magic of his personality to the full it was necessary to visit him in the little room at the back of his house... . Ten minutes in that sanctum would have made a friend of his bitterest foe, and the conquest would have been mutual and permanent, for it seemed to be an axiom of Tait's that a man who had become his friend could sin no more.

Heaviside never met Tait, so it's anybody's guess if the "magic" of Tait's personality would have been sufficient to soothe their differences. What happened between the two men is that Heaviside made the mistake (in Tait's view) of writing disparaging things about quaternions, the creation of a man Tait had long revered (since Oliver was *three*!) and to whom he had attached himself as chief disciple for spreading the good word concerning quaternions ("4nions," as Maxwell called them).

Heaviside, however, wouldn't listen, and even rebutted the value of quaternions in physical theory, and for that Tait had *no* forgiveness. His wrath for the anti-quaternion infidels was comparable to his wrath in the Tyndall controversy.

William Hamilton and Quaternions

William Rowan Hamilton (1805–1865) was a brilliant Irish mathematician[11] whose genius for languages was evident at a very early age. He could read at three, by four had started on Latin, Greek, and Hebrew, and by ten had become familiar with Sanskrit. At seventeen his powers in mathematics, too, became obvious; overall Hamilton was so impressive that he was knighted at age thirty. His work has been given every conceivable level of praise and criticism. One writer[12] has said he was "After Isaac Newton the greatest mathematician of the English-speaking peoples... ," while another[13] has called him "An Irish Tragedy" for having wasted the rest of his life on the quaternions he discovered in 1843. This latter assessment is no doubt a bit extreme, as every physicist in the world today knows of the "Hamiltonian" formulation in mechanics. The ghost of Hamilton need not fear obscurity, although Hamilton himself most likely would have been happier if, instead of this fame among physicists, he were remembered for his quaternions. Ironically, it seemed for a while, at first, that he might be.

It was Tait, in fact, who introduced[14] his boyhood friend, Maxwell, to quaternion concepts, and Maxwell later used them in the *Treatise*. Ironically (from Tait's point of view) it is clear from their own words that it was from the *Treatise* that such quaternion foes as Heaviside and

William Rowan Hamilton (1805–1865)

A great mathematician, but he was frustrated in love. His love for another man's wife continued, even as he remained devoted to his own wife, even after her death, until the day *he* died.

Gibbs first developed their dislike of them! Maxwell had his doubts about quaternions, too, and after 1873 they disappeared from his work. Maxwell liked quaternions for their aid in *thinking* about physical quantities, but he remained unconvinced of their value in actual calculation—as he put it, [15]

> Now, Quaternions, or the doctrine of Vectors, is a mathematical method, but it is a method of thinking, and not, at least for the present generation, a method of saving thought. It does not, like some more popular mathematical methods, encourage the hope that mathematicians may give their minds a holiday, by transferring all their work to their pens.

In the same essay we also find Maxwell's gloomy (but certainly correct) evaluation of the reaction of most 19th century mathematicians and physicists to quaternions:

> Hamilton himself, the great master of the spell, when addressing mathematicians of established reputation, found, for his Quaternions, but few to praise and fewer still to love.

Still, Maxwell was at least sympathetic to the *spirit* of quaternions. Tait's other great friend, William Thomson (Lord Kelvin), would have nothing to do with them at all. After Tait moved from Belfast to Edinburgh he met Thomson and sometime later the two men began their joint authorship of the famous *Treatise on Natural Philosophy*. When this book[16] appeared in 1867 it had a tremendous impact, but not because of any endorsement of quaternions. They appear, in fact nowhere in the book, and after Tait's death Kelvin explained[10] why to the writer of the obituary notice in *Nature*:

> We have had a thirty-eight years' war over quaternions. He had been captivated by the

> originality and extraordinary beauty of Hamilton's genius in this respect, and had accepted, I believe, definitely from Hamilton to take charge of quaternions after his death, which he has most loyally executed. Times without number I offered to let quaternions into Thomson and Tait, if he could only show that in any case our work would be helped by their use. You will see that from beginning to end they were never introduced.

Kelvin's rejection of *any* system but that of cartesian components for writing the equations of physics was, in fact, across the board. As he wrote in 1892 to the author of a book of vectors

> I do think...you would find it would lose nothing by omitting the word "vector" throughout. It adds nothing to the clearness or simplicity of the geometry, whether of two or three dimensions. Quaternions came from Hamilton after his really good work had been done; and though beautifully ingenious, have been an unmixed evil to those who have touched them in any way, including Clerk Maxwell.

And in 1889 he wrote to Heaviside

> ...it was, as you say, Maxwell that first gave *curl*, as he in fact tells us himself in the first volume of his [*Treatise*]. It is rather the symbolic system connected with it [i.e., the vector formulation of *curl*] in your own and Maxwell's papers that I object to, than the word itself, and I cannot agree with any attack on Cartesian coordinates.

This last quote is from a letter (dated April 27, 1889) prompted by letters Kelvin (then still William Thomson) had received from Heaviside months before. In his Presidential Address to the IEE on the evening of January 10 Thomson had attributed the word *curl* to the mathematician William Kingdon Clifford (1845–1879), and then he said the word had "...I grieve to say, been adopted by Mr. Oliver Heaviside." After reading this, Oliver wrote to Thomson on January 18 to correct him, i.e., to tell him he had taken *curl* from Maxwell, not Clifford. Later, on February 27, Heaviside wrote again and said "There is an intolerable waste of space in mathematical papers (physical) in printing in Cartesian formulae for perfectly well-known functions." Heaviside's remarks[17] did nothing to change Thomson's mind, however, and in 1896 he wrote to FitzGerald

> ..."vector" is a useless survival, or offshoot, from quaternions, and has never been of the slightest use to any creature. Hertz wisely shunned it, but unwisely he adopted temporarily Heaviside's nihilism [Note 18]. He even tended to nihilism in dynamics [i.e., an earlier use by Hertz of vector ideas], as I warned you soon after his death. He would have grown out of all this, I believe, if he had lived.

Time has shown Kelvin to have been wrong about vectors, but the modern assessment of quaternions is almost as negative. This is the *modern* reaction, however. When Tait first learned of quaternions in 1853 by reading Hamilton's just published book *Lectures on Quaternions*,[19] he fell in love with them and immediately put Hamilton on an idol's pedestal. Almost literally until he lay on his deathbed he proselytized for quaternions with an ardor and faith that came dangerously close to fanaticism. And like the Inquisition of a different time, he could be ruthless in dealing with those who treated quaternion dogma lightly.

Before 1890—The Calm before the Storm

Hamilton and Tait carried on a voluminous correspondence, initiated by Tait in 1858, five years after his excited reading of *Lectures*. Something like fifty letters were exchanged, with one of them reaching the astonishing length of ninety-six (yes, 96!) pages. Tait made swift progress in mastering quaternions, especially in their application to physics, an area mostly ignored by the mathematical Hamilton. So rapid was his progress, in fact, that in 1859 he began writing his own book, *An Elementary Treatise on Quaternions.* It did not appear until 1867, however, as Tait willingly withheld it—by Hamilton's request—until after Hamilton's own second attempt to explain the awkward *Lectures* could be published. Tait literally had to wait until Hamilton died, and even then it wasn't until the next year, 1866, that *Elements of Quaternions* finally came out. Released at last from his odd[20] promise, Tait sent his *Treatise* to the printer the year after that.

With the passing of the master, Tait inherited the crown of "Lord of Quaternions," a position that was actually "officially" endorsed by Hamilton in *Elements*[21]:

> ...Professor Tait...appears to the writer eminently fitted to carry on, happily and usefully, this new branch of mathematical science: and likely to become in it, if the expression be allowed, one of the chief successors to its inventor.

From the moment of the appearance of Tait's *Treatise* until his death, he considered the advancement of quaternions to be a major goal of his life. At first it seemed he might succeed in spreading their fame and use and, in the Preface to the Second Edition of his *Treatise,* he wrote (in 1873, the same year Maxwell's *Treatise* made prominent use of quaternions),

> ...I have been...surprised and delighted by so speedy a demand for a second edition... . There seems now at last to be a reasonable hope that Hamilton's grand invention will soon find its way into the working world of science, in which it is certain to render enormous services, and not be laid aside to be unearthed some centuries hence by some grubbing antiquary.

Beginning in the 1870s, however, developments were brewing that would eventually cause Tait to swallow these happy words bitterly.

The Vector Analysis of Josiah Willard Gibbs

Today we remember the great American scientist J. Willard Gibbs (1839–1903), who spent his entire career at Yale University, mostly for his work in thermodynamics, but he was in fact a mathematical physicist with interests that roamed everywhere.[22] One of his "minor" contributions to mathematics, for example, is the famous *Gibbs phenomenon,* treating the behavior of the Fourier series expansion of a *discontinuous,* periodic function at the point of discontinuity. This analysis[23] is now studied by every undergraduate student in science, mathematics, and engineering in the world. Ironically, however, his greatest contribution to mathematics, vector analysis as we use it today, is not attributed to him by most mathematicians (and the number of modern electrical engineers and physicists who associate Gibbs with vectors is almost certainly not much larger).

Gibbs came to vectors by way of Maxwell. As he related in an 1888 letter to the German mathematician Victor Schlegel,[24]

> My first acquaintance with quaternions was in reading Maxwell's [*Treatise*] where Quaternion notations are considerably used. I became convinced that to master those subjects [electricity and magnetism], it was necessary for me to commence by mastering those methods. At the same time I saw, that although the methods were called quaternionic the idea of the quaternion was quite foreign to the subject... . I therefore began to work out *ab initio,* the algebra of [vectors, including the scalar dot and the vector cross products], the three differential operations $\vec{\nabla}$ [curl, divergence, and gradient]... .

By 1879 Gibbs was sufficiently well along in his work that he offered a course at Yale in

J. Willard Gibbs (1839–1903)

A man of great talent, he once modestly declared "If I have had any success in mathematical physics, it is, I think, because I have been able to dodge mathematical difficulties."

electricity and magnetism using vectors. In 1881 he had a private printing of the first part of his *Elements of Vector Analysis* done, with the concluding half similarly done in 1884.[25] Gibbs sent copies of it to many of the great scientists of the day, including Rayleigh, Kirchhoff, FitzGerald, and the two Thomsons (William and J.J.). He sent a copy to Heaviside, too (in 1888). And he also sent a copy, certainly never suspecting what an uproar it was to incite, to Tait.

Tait Throws Down the Gauntlet

Parallel with Gibbs, Heaviside had gone through much of the same evolution of thought concerning quaternions and their shortcomings when applied to physical theory. Like Gibbs, Heaviside had begun by reading Maxwell. In a 1902 review of a book on vector analysis by one of Gibbs' students, he wrote[26]

> Maxwell exhibited his main results in quaternionic form in his treatise. I went to Prof. Tait's treatise to get information and to learn how to work them... . On proceeding to apply quaternions to the development of electrical theory, I found it very inconvenient. Quaternionics was in its vectorial aspects antiphysical and unnatural, and did not harmonize

> with common scalar mathematics. So I dropped out the quaternion altogether, and kept to pure scalars and vectors, using a very simple vectorial algebra in my papers from 1883 onward. The paper at the beginning of vol. 2 of my *Electrical Papers* may be taken as a developed specimen... .

The paper[27] Heaviside was referring to is "On the electromagnetic wave-surface," which appeared in the *Philosophical Magazine* in 1885, but in fact he had been using vector methods in print since 1882. At the end of that year he wrote[28] for *The Electrician* "The relations between magnetic force and electric current," in which he mentioned Hamilton's quaternions and made plain his preference for more straightforward vector ideas in applications to physical theory. In the 1885 paper he was even more emphatic in stating his position:

> Owing to the extraordinary complexity of the investigation when written out in Cartesian form (which I began doing, but gave up aghast), some abbreviated method of expression becomes desirable...I therefore adopt, with some simplication, the method of vectors, which seems indeed the only proper method.

Following this was a brief, yet self-contained introduction to vector algebra and calculus (which was hardly anything Hamilton, by then twenty years in his grave, or Tait, either, would have recognized). Heaviside then promptly used his "new math" to answer real, physical electromagnetic questions. It is nothing less than a *tour de force*.

Tait had received Gibbs' privately printed vector papers directly from their author, and he also knew what Heaviside had been writing. This is why, when the third edition of his *Treatise* appeared in 1890, the tone of the new Preface was in marked contrast to that of the second edition of 1873:

> It is disappointing to find how little progress has recently been made with the development of Quaternions. One cause, which has been specially active in France, is that workers at the subject have been more intent on modifying notation, or the mode of presentation of the fundamental principles, than on extending the applications of the calculus... . Even Prof. Willard Gibbs must be ranked as one of the retarders of Quaternion progress, in virtue of his pamphlet on *Vector Analysis;* a sort of hermaphrodite monster, compounded of the notations of Hamilton and of Grassmann.

Tait expressed other complaints in his Preface, such as what he felt to be the "failure in catching the 'spirit' of the [quaternion] method" by most analysts. He went so far as to assert

> To try to patch up a quaternion investigation by having recourse to quasi-Cartesian processes is fatal to progress. A quaternion student loses his self-respect, so to speak, when he thus violates the principles of his Order.

For Tait, to "properly practice" quaternions required the adherence to tradition at the level of taking Holy Orders!

All this might be dismissed as "that's just the way Victorians wrote," but his remark about Gibbs did not go quietly into musty library stacks to be forever forgotten. It precipitated, in fact, what one writer has called[29] a "quarrel at a violent level," one which resulted in both Tait and Heaviside unleashing their considerable abilities in sarcasm and name-calling upon the other, like two sorcerers of ancient times in a test of their powers.

The Battle

Gibbs responded to Tait in a gracious but forceful defense of his ideas, in a long letter[30] to *Nature* in April 1891:

> It seems to be assumed that a departure from quaternionic usage in the treatment of vectors is an enormity. If this assumption is true, it is an important truth; if not, it would be unfortunate if it should remain unchallenged, especially when supported by so high an authority.

Gibbs went on to argue in support of his "notions and notations," and introduced for the first time the interesting observation that his approach to vector analysis could easily be extended "to space of four or more dimensions," while "the notions of quaternions will [not] apply to such a space."

Tait quickly replied,[31] and in passing made the following astonishing statement:

> It is singular that one of Prof. Gibbs' objections to Quaternions should be precisely what I have always considered...their chief merit: —that they are uniquely adapted to Euclidean space, and therefore specially useful in some of the most important branches of physical science. *What have students of physics...to do with space of more than three dimensions* [my emphasis]*?* [Note 32]

Two more letters,[33] one from Gibbs and one from Tait, continued the discussion in *Nature,* with neither one budging an inch, and after that the issue appeared to be losing its momentum. But at virtually the same time, Heaviside completed his very important paper, "On the forces, stresses, and fluxes of energy in the electromagnetic field" (but it didn't appear in print[34] until the next year, 1892, because of, as Heaviside explained it, "typographical troubles"), in which he blew fresh air on the dying embers of the debate. It is easy to imagine Tait's reaction to such Heavisidean pronouncements in this Royal Society paper as

> As Electromagnetism swarms with vectors, the proper language for its expression is the Algebra of Vectors... . The quaternionic basis is rejected... . It [electromagnetism] has, in my opinion, been retarded by the want of special treatises on vector-analysis adapted for use in mathematical physics, Professor Tait's well-known profound treatise being, as its name indicates, a treatise on Quaternions... . I reject the quaternionic basis of vector analysis.

It should be pointed out that none of this could have come as a surprise to Tait when he read it in 1892. As far back as November 1891 Heaviside had written for *The Electrician,* as the first installment of what became the book-length Chapter 3 of EMT 1 on vector analysis[35] (it was here he called Tait a "consummately profound metaphysicomathematician" in a section provocatively entitled: "Abstrusity of Quaternions and Comparative Simplicity Gained by Ignoring Them"):

> "Quaternion" was, I think, defined by an American school girl to be "an ancient religious ceremony". This was, however, a complete mistake. The ancients—unlike Prof. Tait—knew not, and did not worship Quaternions.

Returning to his 1892 paper, Heaviside went on to refer to the Gibbs–Tait letters in *Nature* (which he called a "rather one-sided discussion," giving the better of the exchange to Gibbs), and, in reference to Tait's accusation of Gibbs being a "retarder of quaternionic progress," he replied,

> This may be very true, but Professor Gibbs is anything but a retarder of progress in vector analysis and its application to physics.

Finally, later in the paper, he threw one last gibe at Tait:

> It rarely occurs that any advantage is gained by the use of quaternions, in saying which I merely repeat what Professor Willard Gibbs has been lately telling us... .

Heaviside sent a copy of this paper to Tait, as well as to Alexander McAulay (an Australian college lecturer in mathematics who had been 49th Wrangler in 1886)—he most likely sent McAulay a copy in response to reading his pro-quaternionic paper, "On the mathematical theory of electromagnetism," which appeared[36] in the same volume of the *Philosophical Transactions* as did his own paper. Certainly one of McAulay's comments could not have escaped Oliver's attention: "And I may remark in passing that what Professor Tait persistently and with complete justice emphasizes as one of the greatest boons that Quaternions grant to ungrateful physicists, viz., their *perfect naturalness,* seems to me to receive illustration in the methods about to be described." Perhaps he thought to show McAulay the error of his ways by sending him his paper! In any case, McAulay responded with a new letter[37] in *Nature*:

> The band of physicists who use and urge the use on others of vector analysis is woefully small [he called Gibbs and Heaviside "two of the justly best known of that band," making them sound something like the leaders of a gang of suspicious characters!]... . Prof. Gibbs and Mr. Heaviside have not yet convinced the rest of the small band—not to say each other—of the merits of their algorithms... .

Then came the insulting conclusion

> ...Maxwell, Clifford, Gibbs, FitzGerald, Heaviside prescribe a course of spoon-feeding... . Hamilton and Tait recommend and provide strong meat... . Let the spoon-feeders provide spoon-meat of the same *kind* as the other physicians.

In McAulay's eyes, apparently, being difficult was a virtue and one shouldn't make things *too* easy for students. At least he put Gibbs and Heaviside in the good company of Maxwell.

Tait's response to Heaviside's paper came just two weeks after McAulay's, also to *Nature,* in the form of a very mean-spirited (even nasty) letter[38]:

> I fancied that, in reply to the voluminous letters of Prof. Willard Gibbs, I had said in a few words [Note 39] all that is requisite (if indeed anything *be* requisite) to show the necessary impotence, as well as the inevitable unwieldiness of every system of (so-called) *Vector Analysis*... . A recent perusal of the first four pages of a memoir by Mr. O. Heaviside—for so far only could I go—has dispelled the illusion... . I particularly desired to read the memoir...as I hoped to learn from it something new in Electrodynamics. But, on the fifth page I met the check-taker as it were—and found that I must pay before I could go further. I found that I should not only have to relearn Quaternions (in whose disfavour much is said) but also to learn a new and *most uncouth parody of notations long familiar to me;* so I had to relinquish the attempt... . There I was content to leave the matter. But Mr. Heaviside has just published an elaborate attack [in *The Electrician*] on Quaternions... . In answer to his remarks [Note 40], in which he continues to point to me as the persistent advocate of a system which all right-minded physicists should avoid, I would simply refer him (and his readers, *if there be such* [my emphasis]) to a brief Address which I gave a short time ago... .

Most likely what upset Tait were Heaviside's blunt words in *The Electrician*:

> ...when Prof. Tait vaunts the perfect fitness and naturalness of quaternions for use by the physicist in his inquiries, I think he is quite wrong... . And yet this topsy-turvy system is earnestly and seriously recommended to physicists as being precisely what they want. Not a bit of it. They don't want it. They have said so by their silence. Common sense of the fitness of things revolts against quaternionic doctrines about vectors. Nothing could be more unnatural.

These were fighting words from Heaviside, and they stung Tait enough to prompt him into continuing his letter by telling *Nature*'s readers that a colleague, unlike himself, had actually read Heaviside:

> Dr. Knott [Note 41] has actually had the courage to *read* the pamphlets of Gibbs and Heaviside; and, after an arduous journey through these trackless jungles, has emerged a more resolute supporter of Quaternions than when he entered. He has revealed the (from me at least), hitherto hidden mysteries...of Prof. Gibbs' strange symbols... . And when, at my request Dr. Knott translated into intelligible form the various terms of one of the less formidable formulae of Mr. Heaviside's memoir, I was surprised to find two old and very unpretending friends masquerading in one person like a pantomime Blunderbore. In one of his Avatars the monster contains, besides the enclosing brackets, no fewer than 24 letters, 12 suffixes, 3 points, and 5 signs! When he next appears he has still the brackets to hold him together, but although he has now only 18 letters, he makes up his full tale of 44 (or 46) symbols; for he has 9 suffixes, 3 indices, 3 points, 5 signs, and 3 *pairs* of parentheses!

The idea of evaluating competing mathematical systems by counting symbols must have *seemed* odd to astute readers; the absurdity of it didn't escape Heaviside's notice, as we'll soon see. But first let me let Tait finish his angry letter:

> Dr. Knott's paper is, throughout, interesting and instructive; —it is a complete exposure of the pretensions and defects of the (so-called) Vector Systems... . I find it difficult to decide whether the impression its revelations have left on me is that of mere amused disappointment, or of mingled astonishment and pity.

Like Preece, Tait believed bluster and insults would carry the day for him, rather than logical arguments and facts. Perhaps he was misled by the unperturbed tone of Gibbs' earlier replies, but he got back far more than he bargained for from Heaviside. The next letter to *Nature* was from Gibbs[42] and it essentially ignored Tait and was a reply to McAulay. As was Gibbs' style, it is almost impossible to tell from it that a serious dispute was involved.

Then came Heaviside's answer[43] to Tait and there was no doubt a disagreement was in progress! Used to bullying (or at least *trying* to) "gentlemen" like Tyndall, Glazebrook, and Gibbs, this reply from a "roughneck" like Oliver must have been a bit of a shock. Heaviside dealt with McAulay with relative mildness,[44] but with Tait he took a different approach. In response to his arrogant, disdainful slap on Heaviside's cheek, what Tait got back was this solid right cross to the jaw:

> I have been, until lately, very tender and merciful towards quaternionic fads, thinking it possible that Prof. Tait might modify his obstructive attitude. But there is seemingly no chance of that. Whether this be so or not, I think it is practically certain that there is no chance whatever for Quaternions as a practical system of mathematics for the use of physicists. How is it possible, when it is so utterly discordant with physical notions... . A vector is not a quaternion; it never was, and never will be...the supposed proofs are perfectly rotten at the core. *It is to Prof. Tait's devotion to his master* [Hamilton] *that we should look for the reason of the little progress made in the last 20 years in spreading vector-analysis* [my emphasis].

While Tait must have been staggered by that, the knock-out blow came when Heaviside went on to write,

> ...Prof. Tait's letter...seems to be very significant. The quaternionic calm and peace have been disturbed. There is confusion in the quaternionic citadel; alarms and excursions, and hurling of stones and pouring of boiling water upon the invading host. What else is the meaning of his letter, and more especially of the concluding paragraph? ...It would appear that Prof. Tait, being unable to bring his massive intellect to understand my vectors, or Gibbs'...has delegated to Prof. Knott the task of examining them, apparently just upon the

> remote chance that there might possibly be something in them that was not utterly despicable... . He [Knott] has counted up the number of symbols in certain equations. Admirable critic!

The "symbol-count" criterion put forth by Tait in support of his claim for quaternions as "compact and elegant," and other notations as "artificial and clumsy," was directly addressed (and refuted) in the summer of the following year, by Professor Arthur Cayley in his paper[45] "Coordinates versus quaternions." Cayley[46] related a couple of illustrative and poetic counter-examples that he had previously presented to Tait, the second one[47] being more to the point:

> As another illustration which I gave him, I compare a quaternion formula to a pocket-map—a capital thing to put in one's pocket, but which for use must be unfolded; the formula, to be understood, must be translated into coordinates. [Note 48]

Shortly after the appearance of Heaviside's letter, the abstract of a paper by Knott, "Recent innovations in vector theory," was published[49] in *Nature.* Knott's tone was highly aggressive and sarcastic, and he referred to Gibbs as the "high priest" of vector analysts, and as "the prince of vector purists." Knott also referred to Gibbs' comment about spaces with dimension beyond the third, and showed he shared Tait's disbelief in their value:

> To elucidate the "nature of *things*" [Note 50] by an appeal to the fourth dimension—to solve the Irish Question by a discussion of social life in Mars—it is a grand conception, worthy of the scorner of the trivial and artificial quaternion of three dimensions.

Heaviside came in for his share of sneers, too, with Knott writing of "Heaviside's tall talk in the *Electrician* [showing] that, on the most lenient hypothesis available, our self-appointed critic of Tait's methods has never really read Tait's *Quaternions*." The "tall talk" Knott was upset about was Heaviside's remark[51] that "The reader of Prof. Tait's profound treatise on Quaternions will probably stick at three places..." and the first place specified is "difficult mathematics...combined with versors, quaternions, and metaphysics."

After this there were more letters to *Nature* from Knott, Gibbs, and others,[29] but from Heaviside and Tait there was silence—until the very end of 1893. Then Tait published[52] a review of McAulay's book *Utility of Quaternions in Physics.* This review was generally favorable, but Tait couldn't resist wrapping his praise for the book around another bomb for Gibbs and Heaviside:

> It is positively exhilarating to dip into the pages of a book like this after toiling through the arid wastes presented to us as wholesome pasture in the writings of Prof. Willard Gibbs, Dr. Oliver Heaviside, and others of similar complexion.

It is odd that Tait conferred the title of *Dr.* on Heaviside, as he never used it before and Heaviside wouldn't receive his honorary doctorate for more than a decade. It most likely was an attempt to embarrass Heaviside by drawing attention to his lack of formal credentials, but this just shows how little Tait could have understood a man like Heaviside. In any case, Heaviside let it pass (but he wrote in the margin of his copy of *Nature,* "Mud throwing again").

In his review, Tait went on to describe the proper use of quaternions as one who

> Intuitively recognizing its power, he snatches up the magnificent weapon which Hamilton tenders to all, and at once dashes off to the jungle on the quest of big game.

After reading this it is little wonder that Tait has been called[53] the "Rudyard Kipling of English Physics"!

For others (like Gibbs and Heaviside) who dared to tinker with Hamilton's original ideas,

Tait continued with his "big gun" metaphor and wrote of their efforts

> ...to convince a bewildered public...that, like the Highlander's musket, it [quaternions] requires to be treated to a brand-new stock, lock, and barrel, *of their own devising,* before it can be safely regarded as fit for service.

And then, in a softening of his hard-line position against *all* tinkering with Hamilton, Tait wrote

> Mr. McAulay himself has introduced one or two rather startling innovations. But, unlike the would-be patchers...he retains all the exquisitely-designed Hamiltonian machinery... .

This was all too tempting for Heaviside to pass up, and he literally got in the last word, in the last letter[54] of the debate. Oliver was just as tough and aggressive in this letter as he was in his first reply to Tait. Perhaps Tait was losing interest in continuing with an opponent as able as himself at verbal warfare, because he declined to reply, even when Heaviside wrote,

> That Prof. Tait should not be able to do justice to those who prefer to treat vectors as vectors and quaternions as quaternions...is naturally to be expected. He does not know their ways, either of thinking or working, as is abundantly evident in all that he has written adversely to Prof. Willard Gibbs and others. It is, however, a little strange, in view of Prof. Tait's often expressed conservatism regarding Quaternionics, that he should tolerate *any* innovations therein, such as Mr. McAulay has introduced. The latter may perhaps take this as a compliment to his analytical powers, which compel the former's admiration, and toleration of his departures from quaternionic usage.

Then, in an obvious rejoinder to Tait's sneer about Heaviside's readers, "if there be such," Oliver wrote of *his* work,

> I employ my usual notation for the benefit of readers (now becoming numerous) who, though they cannot follow the obscure quaternionic processes [see Tech Note 3 at the end of this chapter], can understand the plainer ones of vector algebra.

And then, finally,

> There is, I know, much more in Mr. McAulay's mathematics than Prof. Tait has yet fathomed... . I should not be writing this note were it not for the misconceptions that Prof. Tait indulges in about what he does not know, viz. vector algebra apart from quaternions.

The Aftermath

And so ended the public battle over quaternions versus vectors. Neither side had done much to "educate" the other, but the readers of *Nature* certainly had had an exciting spectacle put on for them. Rayleigh's comment[55] probably sums up what must have been the typical conclusion of those who had watched from the sidelines: "Behold how these vectorists love one another." Gibbs and Heaviside went right on using *their* vectors while Tait and Knott continued to the end to push for *their* quaternions. Indeed, in 1895 Tait replied to Cayley's likening of quaternions to a pocket-map and, sounding once again like Rudyard Kipling, wrote,[56]

> A much more natural and adequate comparison would, it seems to me, liken Co-ordinate Geometry...to a steam-hammer, which an expert may employ on any destructive or constructive work *of one general kind,* say the cracking of an egg-shell, or the welding of an anchor. But you must have your expert to manage it, for without him it is useless. He has to toil amid the heat, smoke, grime, grease, and perpetual din of the suffocating engine-room. The work has to be brought to the hammer, for it cannot usually be taken to its work. And it is not in general, transferable; for each expert, as a rule, knows, fully and confidently, the working details of his own weapon only. Quaternions, on the other hand, are like the

> elephant's trunk, ready at *any* moment for *anything,* be it to pick up a crumb or a field-gun, to strangle a tiger, or to uproot a tree. Portable in the extreme, applicable anywhere:—alike in the trackless jungle and in the barrack square:—directed by a little native who requires no special skill or training, and who can be transferred from one elephant to another without much hesitation. Surely this, which adapts itself to its work, is the grander instrument! But then, *it* [quaternions, aka elephant trunks] is the natural, the other [steam-hammers, aka co-ordinates, i.e., vectors] the artificial, one.

And some years later, in a review[57] of Wilson's book,[25] Knott said much the same thing: "...we can find no satisfactory reason for a man of Professor Gibbs' great powers leaving quaternionic paths... ."

Near the end of 1895 one last, curious letter[58] appeared in *Nature,* calling for the formation of "The International Association for Promoting the Study of Quaternions and Allied Systems of Mathematics." This letter, coauthored by Shunkichi Kimura[59] who was then also at Yale with Gibbs, suggested that in reply to the "many who are prejudiced against the calculus of quaternions," "quaternionists need only say that if the objectors approach the calculus of quaternions with proper care and meekness, they will...rejoice in having at their disposal an instrument of research mightier far than they had the slightest notion of so long as they were in the domain of cartesian coordinates." Tait was elected the first President of the Association (an honor which he declined), and its *Bulletin* appeared from 1900 until 1913. In 1899 Gibbs, himself, was invited to join, but it appears (not surprisingly!) that he decided to pass on the offer.

As the years passed, in fact, others never got on Knott's "quaternionic path" at all, and quaternions faded away while the Gibbs–Heaviside vector algebra became the everyday work tool of all modern engineers and scientists. But it didn't happen overnight and even FitzGerald felt that on this issue, Heaviside perhaps had gone too far. As he wrote to Oliver on September 26, 1892,

> I hope you will succeed in making the ordinary mathematical physicist think in vectors though I do not think your notation an improvement. You see I was "riz" on Tait and get very much muddled by your omission of *S* [Hamilton's symbol for the scalar part of a quaternion—see Tech Note 2 at the end of this chapter] and when one gets bothered every time one naturally takes a dislike to the botheration.

More sympathetic was *The Electrician,* which ran the following editorial comment[60] with the initial essay of Heaviside's vector analysis series:

> Mr. Heaviside's article in our present issue is devoted to a somewhat abstruse but highly important subject—the use of vectors in analysis. The writer discusses in his usual lively and combative manner the great advantages of vector methods over the ordinary Cartesian or quaternion processes... . Comparatively few are familiar with the habits of vectors, mainly owing to the absence of an elementary text-book treating of their properties... . We feel sure that a compact summary of the chief facts about them...will prove of great value. Electrical engineers are constantly dealing with vector quantities, although they may not be aware of the fact...Mr. Heaviside condemns the use of quaternions on account of their complication. The use of vector analysis appears hard merely because it is unfamiliar.

It is no doubt difficult for a modern engineer or scientist to appreciate, as this quote shows, that it was just a single lifetime ago that even the greatest scientists in the world were without vectors. In a letter to Oliver dated October 12, 1893, not really so long ago, we find Joseph Larmor, one of the great Victorian physicists, writing to say

> I am practically a convert to vector analysis and *I mean to learn up the machinery immediately I have time* [my emphasis].

How things have changed! But this doesn't mean quaternions are now *totally* dead. Every now and then they still pop up[61] and it may yet be proved that Hamilton was right when, in 1859, he wrote[62] to Tait about quaternions:

> *Could* anything be simpler or more satisfactory? Don't you *feel,* as well as think, that we are on a *right track,* and shall be *thanked* hereafter. Never mind when.

On the other hand, if the fate of Heaviside with respect to being remembered for his vector work is any guide, Hamilton and Tait may have a long wait for their thanks, no matter how useful quaternions may eventually become. It didn't take long, at all, for historical confusion to set in. At the end of 1901 the American Mathematical Society held its annual meeting in New York City. As *Nature* reported[63] the event,

> Prof. Gibbs's "Elements of Vector Analysis" (1881–4) attracted wide attention, though it was only a pamphlet (83 pp.) printed for the use of his students. *This Mr. O. Heaviside adopted, with slight modifications and expounded fully in his "Electromagnetic Theory"* (1893) [my emphasis].

To this "demotion" as a mere copier of Gibbs, Heaviside limited himself to just a short note to history in the margin of his copy of the journal:

> Nearly all my vector work was done before I received Gibbs' pamphlet in 1888. I had then done nearly all the work in the two volumes of *Electrical Papers*.

This is not to say Gibbs' importance isn't the equal of Heaviside's. While Oliver was ahead of Gibbs in actually *applying* vectors to electrodynamics, even here Gibbs contributed, too. A significant example of this, of particular interest to electrical engineers, is the reply from Gibbs to Kelvin's assertion that Röntgen's newly discovered rays (x-rays) were "condensational [i.e., longitudinal] waves in the luminiferous ether." Röntgen himself had early on speculated this was the case, and this prompted Kelvin's nephew J. T. Bottomley to write to *Nature* to remind[64] people that Kelvin had priority in the matter, having argued for their existence as long ago as 1884. Further, Kelvin had also stated then his belief that "the propagation of these waves would be enormously faster than the propagation of ordinary [i.e., transverse] light waves." Soon after Bottomley's letter Kelvin himself wrote[65] to elaborate, as well as to present a suggestion on how to *experimentally* test his ideas on generating waves of hyper-light speed. As he put it,

> It is not easy to see how this question could be answered experimentally; but remembering the wonderful ingenuity shown by Hertz in finding how to answer questions related to it [wave propagation], we need not perhaps despair to see *it* also answered by experiment.

Two months later Gibbs responded[66] to the challenge by solving the electromagnetic field equations for the particular details of Kelvin's proposed experiment, and showed how *only* light-speed transverse waves result. Not only did he thus refute Kelvin's pet idea, but he used *vector analysis* to do it (the hated *curl* is all through Gibbs' analysis). As both a rejecter of vectors and one who remained unconvinced[67] of Maxwell's theory, it must have been a doubly unhappy Kelvin who read his *Nature* that April day in 1896.

Off to War—Again

In his review[26] of Wilson's book Heaviside presented the author's proof of what Oliver thought a near-trivial statement ("Mr. Wilson says it is tolerably obvious, and then proceeds to *prove* it, presumably for the benefit of unhumorous people"). Declaring this proof to be inappropriate (he quoted FitzGerald's pet word "hugger-mugger" for something confused and muddled), he said *he* could do it "just by inspection." Heaviside then continued with the passage that gives us the introduction to the next chapter (and the next war, that arrived even as the smoke was still swirling about the quaternionic battlefield):

> It seems to me that the demonstration I have poked fun at is typical of a lot of work made up by the brain-torturers who write books for young people and college students who are going to be Senior Wranglers, perhaps. Let mathematics be humanized if possible. The best of all proofs is to set out the fact descriptively, so that it can be seen to be a fact.

This ungrateful attitude toward detailed mathematical proof was one that was certain to attract the (unfavorable) attention of mathematicians. This often expressed opinion was to be like erecting a lightning rod on top of an explosives factory and Oliver, seemingly to his surprise, attracted a fair number of bolts. To continue the analogy to its bitter end, his casual approach to proofs was eventually to get his mathematical factory blown up!

Tech Note 1: Numbers and Vectors—Real, Complex, and Hypercomplex

When mathematics broke free of the constraints of the infinite real line, and it was realized that a vastly richer system of numbers, the so-called *complex numbers* (a term due to the great German mathematician, C. F. Gauss), could be associated with the points of the infinite plane, an enormous intellectual step was taken. It is virtually impossible to exaggerate the importance of this step. Today's electrical engineers and physicists would be paralyzed if their beloved square root of minus one were to be taken from them, and many other scientists would be reduced to a nearly equal miserable state, as well.

Think geometrically. If we associate numbers with the points along a horizontal line (the *real axis*), then, even though this line goes to infinity in both directions, we still imagine we know its "middle," which we agree to call the *origin,* and define a *vector* as the directed line segment from the origin to one of the points. We can transform any such vector into another by multiplying by the appropriate number, e.g., $+2$ transforms into -2 when multiplied by -1. Multiplication by a positive number can be thought of as merely a contraction (or expansion). Multiplication by a negative number, however, has a more exotic interpretation, that of *rotation.* When we multiply $+2$ by -1 we rotate the $+2$ vector through $180°$ so that it is then pointing down the negative real axis toward -2. This idea of rotation is the breakthrough concept, because it gives us a *geometrical* interpretation of the invaluable $\sqrt{-1}$.

Let $i^2 = -1$ (and thus $i = \sqrt{-1}$). Geometrically we already know that multiplying by i^2 is equivalent to a $180°$ rotation, and since $i^2 = i \cdot i$, i.e., two successive applications of i, and since each i must have equal impact then each i must cause a $90°$ rotation! Thus is born the idea of drawing a vertical line, $90°$ from the horizontal real axis, and creating the pair of axes that define the coordinates of the *complex plane*.

The vertical axis is often called the *imaginary* axis, but in fact there is nothing imaginary about it at all (it has been drawn in Fig. 9.1). Although this idea seems to have been around since the late 1600s, it wasn't until 1799 that the Norwegian Caspar Wessel specifically called the vertical axis the "axis of imaginaries." The word *imaginary* is a holdover from olden times when mathematicians first stumbled upon the square root of minus one in solutions to certain *algebraic equations.* Not yet having a geometric interpretation for such solutions, they called them imaginary and then swept them under the rug. Even *with* the rotation concept, however, i has not had an easy road until comparatively recent times. As the Senior Wrangler of 1881 recalled,[68] "...it was an age when the use of $\sqrt{-1}$ was suspect at Cambridge even in

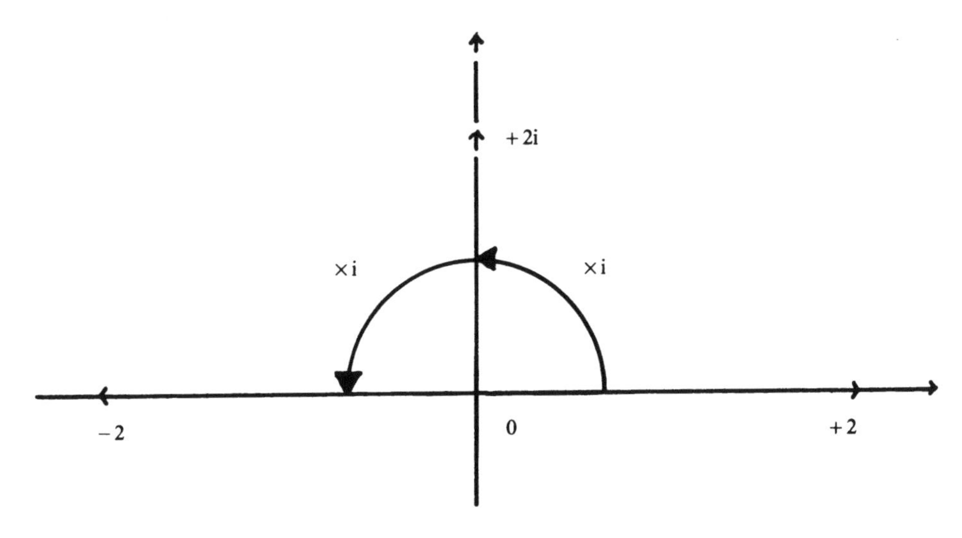

Fig. 9.1.

trigonometrical formulae... . The imaginary *i* was suspiciously regarded as an untrustworthy intruder.''

To every point in the complex plane we can associate a two-dimensional vector drawn from the origin; see Fig. 9.2. Each such vector has a real part, A, and an imaginary part, B, written as $A + iB$, and which we see, geometrically, make an angle θ with the real axis. We can think of there being a *unit* vector pointing along the positive real axis, and another *unit* vector (i.e., the imaginary i) pointing along the positive imaginary axis, and that an arbitrary vector can be written as the sum of multiples of these two basic unit vectors.

The length of the vector is, from the Pythagorean theorem, $\rho = (A^2 + B^2)^{1/2}$, and thus $A = \rho \cos \theta$ and $B = \rho \sin \theta$, and the vector itself is

$$A + iB = \rho\ (\cos \theta + i \sin \theta).$$

The expression in the parentheses is known, by Euler's identity, to be $e^{i\theta}$. Thus, any complex two-dimensional vector can be represented by the concise expression

$$\rho e^{i\theta}$$

where ρ is the length of the vector and θ is the angle, *in radians,* the vector makes with the real axis. This interpretation of complex numbers has been fruitful nearly beyond words. For example, we immediately have from all this that

$$-1 = e^{i\pi}$$

and from *this,* in turn,

$$\sqrt{-1} = i = (-1)^{1/2} = (e^{i\pi})^{1/2} = e^{i\pi/2}.$$

From these two results we can calculate such astonishing results as

$$\ln\ (-1) = i\pi$$

which just goes to show that you *can* calculate the logarithms of negative numbers (when the

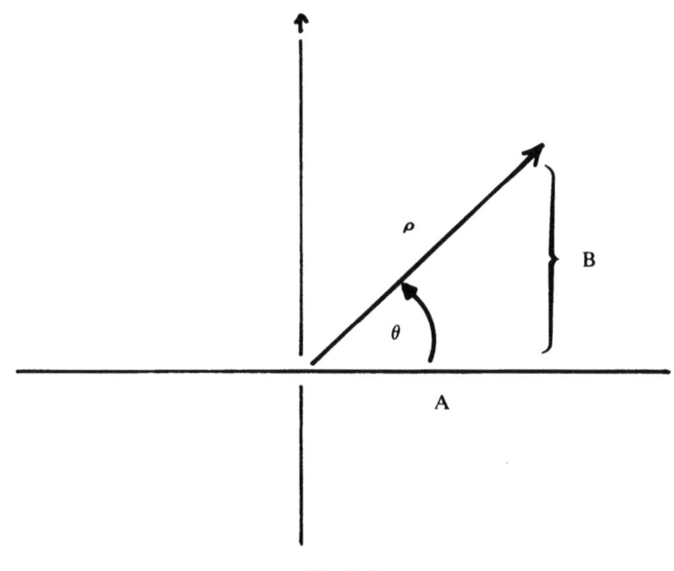

Fig. 9.2.

readout on your electronic calculator starts blinking when you try it, that's because the designers didn't do *everything* possible when they put their circuits and algorithms inside the black box). And how about the perhaps even more astounding conclusion that

$$(\sqrt{-1})^{\sqrt{-1}} = (e^{i\pi/2})^i = e^{-\pi/2} = 0.2078796.$$

Who would have suspected, at the beginning, that such wonderful knowledge could come from the "simple" idea of rotation!

The idea of rotating out of a space into a new one of higher dimensionality is one that science fiction writers and their readers have dearly loved since the last century. If only, they speculate, "we could rotate out of our three dimensional space of everyday life, why then we would find ourselves in the fourth (or fifth, or sixth, etc.) dimension!" This wonderfully imaginative idea was dramatically used by H. G. Wells (who argued that *time* is the fourth dimension) in his 1895 masterpiece, *The Time Machine,* in his passage describing the Time Traveler's demonstration of a miniature time machine to his disbelieving friends:

> We all saw the lever turn. I am absolutely certain there was no trickery. There was a breath of wind, and the lamp flame jumped. One of the candles on the mantel was blown out, and the little machine suddenly swung around, became indistinct, was seen as a ghost for a second...and it was gone... . Then Filby said he was damned.

In his highly entertaining book *Man and Time* (Doubleday, 1964, pp. 122–123), J. B. Priestly gave a sequence of photographs "showing" this demonstration, faithful even to the rotation. But both Wells' prose and Priestly's images are not really correct, of course, because they describe and show the rotation taking place in three-dimensional space, itself. The actual rotation would be four dimensional, and who among the readers of this book (none more than the author) wouldn't trade the contents of a well-stuffed safe-deposit box to learn the secret of how to perform *that* rotation!? Recall that in the *one*-dimensional space of the line we can transform a vector into another by multiplying by *one* number. It is obvious now that in the *two*-dimensional space of the plane we can transform a vector into another by multiplying by *two* numbers, i.e., to rotate a vector by angle θ *and* multiply its length by a factor of ρ, multiply its complex number representation by the complex number $\rho e^{i\theta}$. And then, once loose of the line and set free in the plane what, we may wonder (along with Hamilton), is to prevent us from rotating *again* into *three*-dimensional space? And shouldn't that, conjectured Hamilton, require multiplication of a vector by *three* numbers to transform it into another? As it turns out, it takes *four*.

Tech Note 2: Hamilton's Insight at the Brougham Bridge

In a letter[69] to Tait, dated October 15, 1858, Hamilton wrote

> ...I then and there felt the galvanic circuit of thought *close*; and the sparks which fell from it were the *fundamental equations between i, j, k...* . I felt a *problem* to have been at that moment *solved*—an intellectual want relieved—which had *haunted* me for at least *fifteen years before*.

What Hamilton was referring to was the solution that came to him in a flash of inspiration on Monday, October 16, 1843, while walking with his wife along the Royal Canal into Dublin to attend a Council Meeting of the Royal Irish Academy. It was then he realized that it takes *four* (not three) numbers to accomplish a three-dimensional transformation of one vector into another. In that instant Hamilton saw that it takes one number to adjust the length, another one to specify the amount of rotation, and *two* more to specify the *plane* in which the rotation takes place.

This *physical* insight led Hamilton to study hypercomplex numbers with *four* components (or *quaternions*[70]) of the form

$$q = w + ix + jy + kz$$

where w, x, y, and z are ordinary real numbers and where i, j, and k are each an imaginary unit vector pointing in the three mutually perpendicular directions of space, in a simple extension of the "ordinary" complex numbers of two-dimensional space. The imaginaries, besides their basic property of $i^2 = j^2 = k^2 = -1$, were *defined* to interact with one another (and the real number, one) as dictated by the multiplication table, in Fig. 9.3, which for the first time in mathematics postulated noncommutative multiplication, e.g., $ij = k$ while $ji = -k$. Matrix algebra shares this property, but it came after Hamilton's quaternions.

This sort of behavior (strange, even today, to students when they first encounter a mathematical system in which commutivity fails) was motivated in Hamilton's mind by the special requirements of rotations in three-dimensional space (as we'll see in Tech Note 3), but he wasn't unaware of other possible interpretations. As he wrote[71] in 1845,

> Is there not an analogy between the fundamental pair of equations $ij = k$ and $ji = -k$, and the facts of opposite currents of electricity corresponding to opposite rotations? [I presume Hamilton was thinking here of the *sense* of direction of the circular magnetic field around the currents.]

and later, in an 1854 letter[72] he wrote,

> ...Faraday may possibly remember my chat with him at Cambridge, in 1845, upon the subject of the analogy with the products of *ijk*, to the laws of electrical currents... . I sought to realize my old expectation [of developing his "electromagnetic quaternion"] to-day, after having translated Ampere's law into the notations of the Calculus of Quaternions.

In a letter[73] to one of his sons (dated August 5, 1865) Hamilton wrote of that magic moment of

	1	i	j	k
1	1	i	j	k
i	i	−1	k	−j
j	j	−k	−1	i
k	k	j	−i	−1

Fig. 9.3.

insight when he saw all this:

> I [could not] resist the impulse—unphilosophical as it may have been — to cut with a knife on a stone of the Brougham Bridge, as we passed it, the fundamental formula with the symbols i, j, k; namely,
>
> $$i^2 = j^2 = k^2 = ijk = -1$$
>
> which contains the *Solution* of the *Problem,* but of course, as an inscription, has long since moldered away.

Hamilton's quaternions, to use his own then newly invented terms, are the sum of a *scalar* (w) and a *vector* ($ix + jy + kz$). Hamilton wrote this sum as

$$q = Sq + Vq.$$

He very quickly, the very day of his discovery of quaternions, formed[74] one possible physical interpretation of what it means to add a scalar to a three-dimensional vector:

> In the quaternion (v, x, y, z), xyz [the vector] may determine *direction and intensity;* while v [the scalar] may determine the *quantity* of some agent such as electricity. x, y, z are *electrically polarized, v electrically unpolarized*... . The Calculus of Quaternions may turn out to be a Calculus of Polarities.

In his review[26] of Wilson's vector book Heaviside had a humorous way of addressing the question of adding a scalar to a vector:

> It is really quite legitimate to add together all sorts of different things. Everybody does it. My washerwoman is always doing it. She adds and subtracts all sorts of things, and performs various operations upon them (including linear operations), and at the end of the week this poor ignorant woman does an equation in multiplex algebra by equating the sum of a number of different things in the basket at the beginning of the week to a number of things she puts in the basket at the end of the week. Sometimes she makes mistakes in her operations. So do mathematicians.

TECH NOTE 3: QUATERNIONS ARE COMPLEX!

If we multiply two quaternions together (forming the *quaternion product*) using the noncommutative multiplication table, we have

$$q = w + ix + jy + kz = w + \vec{q}$$

$$q' = w' + ix' + jy' + ky' = w' + \vec{q}'$$

$$qq' = q'' = w'' + ix'' + jy'' + kz'' = w'' + \vec{q}''$$

where

$$w'' = ww' - xx' - yy' - zz'$$

$$x'' = wx' + xw' + yz' - zy'$$

$$y'' = wy' + yw' + zx' - xz'$$

$$z'' = wz' + zw' + xy' - yx'.$$

Note that if the quaternions q and q' both have zero scalar parts, i.e., if $w = w' = 0$, then both quaternions reduce to just vectors and their product is

$$\vec{q}\vec{q}' = -(xx' + yy' + zz') + i(yz' - zy') + j(zx' - xz') + k(xy' - yx').$$

Thus, for Hamilton, the product of two vectors is again a full quaternion with both a scalar and a vector part. This last expression contains two historically important results. First, for Hamilton, there is just *the* product of quaternions. For the modern Gibbs-Heaviside system, however, there are two distinct vector products, the *scalar* product and the *cross* product, and if one writes out these two products and compares them to the *quaternion* product of the two vectors, the result is the *difference* of the two vector products, i.e.,

$$\vec{q}\vec{q}' = \vec{q} \times \vec{q}' - \vec{q} \cdot \vec{q}'.$$

Modern vector analysis individually defines the two vector products and avoids Hamilton's mixture of them.

Second, if $\vec{q} = \vec{q}'$, then the quaternion product of a vector with itself is negative, i.e., since the scalar product of a vector with itself produces a positive number, and as the vector product of any vector with itself is zero, then

$$\vec{q}\vec{q} = -\vec{q} \cdot \vec{q} = \textit{a negative number},$$

a result that Heaviside found intolerable, as it leads to such *physically* strange conclusions as imaginary *lengths* for vectors and negative *kinetic energies* for moving masses. He called[43] this "the inscrutable negativity of the square of a vector in Quaternions." For Oliver, this was "the root of the evil" in quaternions.

Similar difficulties arise with the all-important differential vector operator, $\vec{\nabla}$, which was first defined by Hamilton. Tait wrote[75] of its origin as if Hamilton had found it inside a clam:

> No doubt, it was originally defined in the cumbrous and unnatural form

$$i\frac{d}{dx} + j\frac{d}{dy} + k\frac{d}{dz}$$

> But that was in the infancy of the new calculus, before its inventor had succeeded in completely removing from its formulae the fragments of their Cartesian shell, which were still persistently clinging about them.

Forming the quaternion product of $\vec{\nabla}$ with itself, in fact, leads to

$$\nabla^2 = -\left(\frac{d^2}{dx^2}+\frac{d^2}{dy^2}+\frac{d^2}{dz^2}\right),$$

the *negative* of the way we use ∇^2 today (and the way Heaviside used it a century ago). And of all things, it was even written in *Natural Philosophy* (Tait was coauthor) with the *positive* sign! This prompted Alexander Macfarlane (1851–1913), formerly one of Tait's students at Edinburgh and once an advocate of quaternions (who emigrated to America to become a professor of physics at the University of Texas, Austin), to contribute to the *Nature* debate with a letter[76] calling the quaternion version of $\vec{\nabla}$ "unreal." He asserted the burden of showing the *physical* meaning of the negative sign "lies on the minus men." The minus sign appears, as Macfarlane knew, from the fact that the imaginaries *ijk* are doing double-duty in quaternion notation. On the one hand, they are the unit space vectors, and on the other, they have the property that their squares are -1. The same symbol is used both for a vector and a "versor" (Hamilton's word for *rotator*), about which, in a second letter,[77] Macfarlane wrote, "I have said that some of the fundamental principles of quaternions require to be corrected, especially the one which identifies versors with vectors."

As an alternative to writing a quaternion as the *sum* of a scalar and a vector, Hamilton also wrote it as a *product*, i.e.,

$$q = TqUq$$

where the factor Uq (the *versor* of q) performs rotation, and the factor Tq (the *tensor* of q) adjusts the length during a vector transformation. And thus we come at last to Hamilton's original motivation—how to *rotate* vectors in three-dimensional space. To keep this from becoming a treatise on quaternions (but to satisfy the curiosity of those who have hung on to the end of a fairly "complex" Tech Note), I'll just give the final result.

Let $\vec{a}$ be the vector to be rotated, and q be the quaternion that accomplishes the job. That is,

$$q = w + ix + jy + kz = w + \vec{q}.$$

Define the *inverse* of q to be q^{-1} (i.e., $qq^{-1} = 1$) which, as can be verified by direct calculation, is

$$q^{-1} = \frac{w - ix - jy - kz}{w^2 + x^2 + y^2 + z^2}.$$

It is also verifiable by direct calculation that $q^{-1}\vec{a}q$ is itself a vector; call it $\vec{b}$. Then, it can be shown that $\vec{b}$ is $\vec{a}$ rotated in a plane perpendicular to $\vec{q}$, through the angle θ, where

$$\cos\left(\frac{1}{2}\theta\right) = \frac{w}{(w^2 + x^2 + y^2 + z^2)^{1/2}}.$$

Does this seem just a bit complicated to you? Well, Heaviside didn't think much of it, either! As he wrote,[43] in no uncertain terms:

> Quaternionists, I believe, rather pride themselves upon their power of representing a rotation

> by means of a quaternion... . The continued product of a quaternion q, a vector $\vec{a}$, and another quaternion q^{-1}, produces a vector $\vec{b}$, which is $\vec{a}$ turned around a certain axis through a certain angle. It is striking that it would turn out so; *but is it not also a very clumsy way of representing a rotation* [my emphasis], to have to use two quaternions, one to pull and the other to push, in order to turn round the vector lodged between them?

After that it should come as no surprise to read Heaviside's following words, from his review[26] of Wilson's book, in his tale of the curious youth who got hold of Hamilton's books and asked himself, "What could quaternions be?":

> He took those books home and tried to find out. He succeeded after some trouble, but found some of the properties of vectors...wholly incomprehensible. How could the square of a vector be negative? And Hamilton was so positive [a Heaviside pun?] about it. After the deepest research, the youth gave it up, and returned the books. He then died and was never seen again. He had begun the study of Quaternions too soon.

Others came to accept this point of view, and even Alexander McAulay, the quaternion advocate Heaviside admired, had to admit (in an essay[78] on how *good* quaternions are!) that things looked gloomy, indeed. As he put it:

> In writing the history of Mathematics of the 19th century the historian will be brought face to face with a phenomenon hard to account for... . Can any cause be assigned for [the] extraordinary case of arrested development [of quaternions]? The answer that by far the majority of physicists would give...is that [quaternions are] admirable...for expressing the results obtained by the old-fashioned and artificial methods...[but only after] the country has been reduced to order and civilization [should] quaternions be introduced as a luxury [for] the exhausted toilers... . Can the apathy of physicists with regard to quaternions be accounted for... ? ...Reasons can be given for this almost criminal negligence... . *Not much advance in Physics has been made by the aid of Quaternions.*

In the century that has passed since McAulay wrote, nothing has happened to make his words any less true. As Heaviside put it, in a letter to FitzGerald (dated February 26, 1894): "It is impossible to think in quaternions, you can only pretend to do it." And as the Senior Wrangler of 1905 wrote many years later[79] on how he had wasted time while preparing for his *Tripos* examination: "I spent a long time reading Tait's book on the futile subject of Quaternions."

Notes and References

1. *North American Review,* vol. 85, pp. 223–237, July 1857.
2. Glazebrook, passed over in 1884 for the Cavendish Professorship in favor of J. J. Thomson (despite the support of the incumbent, Lord Rayleigh), later became Director of the newly created National Physical Laboratory (again with Rayleigh's support). He was also one of the signers of Heaviside's nomination to Fellowship in the Royal Society. His quite reasonable, final response to Tait is in *Nature,* vol. 29, p. 335, February 7, 1884.
3. Tyndall (1820–1893) was Faraday's successor at the Royal Institution upon the great man's death in 1867. Tyndall's own death was truly tragic, caused by his devoted wife who accidentally overdosed him with the hypnotic drug, chloral, during an illness.
4. Helmholtz defined energy conservation more clearly than did Mayer, but under more limited conditions. Mayer imposed *no* limitations, and extended the law to every aspect of the universe, including living organisms. This was a daring act during the middle third of the 19th century when *life processes* were considered somehow exempt from natural laws. Mayer also originated the idea of the sun (and its heat) resulting from the continual amalgamation of colliding meteors, another idea that was "appropriated" by someone else (William Thomson, as we'll see in Chapter 11). Mayer it seems, had the fate of "losing" all his ideas, right *and* wrong, to others!
5. A. S. Eve and C. H. Creasey, *Life and Work of John Tyndall,* London: Macmillan, 1945, p. 124.
6. Tait became Forbes' successor as Professor of Natural Philosophy at Edinburgh in 1860 (he was selected over Maxwell), a post he retained until just before his death.
7. *Nature,* vol. 8, p. 431, September 25, 1873.
8. A. J. Meadows, *Science and Controversy,* Cambridge, MA: MIT Press, 1972, p. 35. Meadows also related how

Helmholtz declined to review some of Tyndall's work (which he admired) for *Nature*, for the most part because he did not want to become a new target for Tait's angry pen (and perhaps also his toe).

9. Eve and Creasey (Note 5), p. 178.
10. *Nature*, vol. 64, pp. 305–307, July 25, 1901.
11. The definitive modern biography of Hamilton is T. L. Hankins, *Sir William Rowan Hamilton,* Baltimore, MD: The Johns Hopkins University Press, 1980. Still *very* impressive, however, is the massive work (three volumes and well over 2000 pages!) by R. P. Graves, *Life of Sir William Rowan Hamilton,* New York, NY: Arno Press, 1975 (originally published 1882–89).
12. E. T. Whittaker, in *Lives in Science,* New York, NY: Simon and Schuster, 1957, p. 61.
13. E. T. Bell, *Men of Mathematics,* New York, NY: Simon and Schuster, 1937, p. 340. Bell also, just to keep the record straight, called Hamilton "by long odds the greatest man of science that Ireland has produced."
14. On December 13, 1867, Tait wrote to Maxwell "...I think you will see that 4nions are worth getting up for...they go into [physical mathematics] like greased lightning." Quoted from M. J. Crowe, *A History of Vector Analysis,* Notre Dame, IN: University of Notre Dame Press, 1967, p. 132.
15. This quote is from an unsigned review of the book *Introduction to Quaternions* (jointly authored by Tait and his former teacher, Philip Kelland) in *Nature,* vol. 9, pp. 137–138, December 25, 1873. Although unsigned, and not included in Maxwell's *Scientific Papers,* historians are nearly certain Maxwell was the author—for why, see Crowe (Note 14), p. 148.
16. The book was popularly known as T and T', the T' denoting Tait as the junior (or "derivative") author. This is particularly interesting as it explains the meaning behind Tait's private joke of referring to Tyndall as T'', i.e., as a scientist of the "second order." Tait may have shaken Tyndall's hand in public, but in private he did not forget those who had crossed him. Tyndall would not have agreed with the last line in *Nature*'s obituary notice for Tait (Note 10): "he was loved so well because he loved so much."
17. These quotes are from B. R. Gossick, "Heaviside and Kelvin: A study in contrasts," *Annals of Science,* vol. 33, pp. 275–287, 1976.
18. Kelvin was referring to Heaviside's well-known rejection of established authority and a marked tendency to "do it his way." Thomson's remarks appeared in S. P. Thompson's *Life of William Thomson,* London: Macmillan, 1910; when Heaviside read of his "nihilism" in a complimentary copy Thompson sent him, Oliver wrote back to say, "Found that I was considered to be a Nihilist. Never heard of that before. Not altogether wrong though. I have thrown bombs occasionally, and they exploded too, and did some damage to the grand old man's views... . The vector...never any use to anybody! What a story!"—quoted from Gossick (Note 17).
19. This was a notoriously difficult and awkwardly written book and we learn something of Tait's abilities from the fact that he was able to master much of it. The astronomer John Herschel, himself a Senior Wrangler, wrote to Hamilton in 1853 that he thought the book would "take any man a twelvemonth to read, and near a lifetime to digest... ."—quoted from Crowe (Note 14), p. 36.
20. Historians of science generally argue that by Victorian standards of conduct Hamilton's extraction of such a promise from Tait was not at all out of line. Still, it seems (to me) that such a request must have been at least a *little* unusual, even in those days, and it tells us something of the spell Hamilton had over Tait. A similar request to a modern scientist, of course, would almost certainly excite an obscene reply. In an 1891 letter to Oliver Lodge, Heaviside declared Tait's book to be "without exception the hardest book to read I ever saw."—see B. J. Hunt, *The Maxwellians* (Ph.D. dissertation), Baltimore, MD: The John Hopkins University, 1984, pp. 104–105.
21. Crowe (Note 14), p. 41.
22. The standard biography is L. P. Wheeler, *Josiah Willard Gibbs,* New Haven, CT: Yale University Press, 1951.
23. Gibbs' analysis is in the form of just two short letters to *Nature,* vol. 59 (December 29, 1898, p. 200, and April 27, 1899, p. 606). They are reprinted in *The Collected Works of J. Willard Gibbs,* vol. 2, New York, NY: Longmans, Green, 1931, pp. 258–260 (part 2).
24. The quote from Gibbs' letter is taken from Crowe (Note 14), pp. 152–153. Schlegel was a proponent of the early vector ideas of the German schoolteacher Hermann Gunther Grassmann (1809–1877), whose work Gibbs himself believed actually predated that of Hamilton (as it did, but Hamilton published first): Grassmann published his work, *Lineale Ausdehnungslehre,* in 1844. Later in his letter to Schlegel, however, Gibbs wrote "I am not, however, conscious that Grassmann's writing exerted any particular influence on my V-A [vector analysis]... ."
25. This work was never formally published, but is reprinted in *The Collected Works of J. Willard Gibbs* (Note 23), pp. 17–90 (part 2). It is helpful to read it along with the essay by Gibbs' student, Edwin B. Wilson, in *Commentary on the Scientific Writings of J. Willard Gibbs,* vol. 2, New Haven, CT: Yale University Press, 1936, pp. 127–160. Wilson published his own book on modern vector analysis shortly before Gibbs' death: *Vector Analysis,* New Haven, CT: Yale University Press, 1901. Heaviside reviewed this book favorably—see Notes 26 and 35.
26. *The Electrician,* vol. 48, pp. 861–863, March 21, 1902; reprinted in EMT 3, pp. 135–143.
27. EP 2, pp. 1–23.
28. EP 1, pp. 195–231.
29. A. M. Bork, "Vectors versus quaternions—The letters in *Nature,*" *American Journal of Physics,* vol. 34, pp. 202–211, March 1966.

30. *Nature,* vol. 43, pp. 511-513, April 2, 1891.
31. *Nature,* vol. 43, p. 608, April 30, 1891.
32. Today, of course, it is virtually impossible for physicists and electrical engineers to *avoid* spaces of high dimensionality, some even of *infinite* dimension. An *n*-dimensional space, in which the *metric* (the definition of the distance between two points) is the Pythagorean theorem, is called *Euclidean,* and if *n* is infinity then such a space is called a *Hilbert* space. The case of the continuous, distributed parameter transmission line analyzed by Heaviside can be given a modern mathematical treatment in Hilbert space, by considering the voltage at each point along the line as a vector along a different dimension axis. Thus, even a transmission line of finite length requires an infinite dimensional space for its total mathematical description.
33. *Nature,* vol. 44, pp. 79-82, May 28, 1891 (from Gibbs) and p. 105, June 4, 1891 (from Tait).
34. *Philosophical Transactions of the Royal Society,* vol. 183, pp. 423-475, 1892 (and reprinted in EP 2, pp. 521-574).
35. EMT 1, pp. 132-305. Heaviside called Chapter 3 "a stopgap till regular vectorial treatises come to be written suitable for physicists." It was here, for example, that the vector form of the *divergence* theorem (relating the *volume* integral of the divergence of a vector field, $\vec{F}$, to the integral of $\vec{F}$ over the *surface* of the volume) was first explicitly stated—see C. H. Stolze, "A history of the divergence theorem," *Historia Mathematica,* vol. 5, pp. 437-442, November 1978. Here we see Heaviside recognized a need, but failed to do the extra work to put Chapter 3 into book form. Ten years later when Gibbs' student E. B. Wilson (see Notes 25 and 26) finally did this, his book was *very* successful, with multiple printings and a paperback edition as recently as 1960. In his Preface Wilson wrote "...use...has been made of the chapters on Vector Analysis in Mr. Oliver Heaviside's *Electromagnetic Theory...* ," use a wiser Oliver would have made himself.
36. Note 34, pp. 685-779.
37. *Nature,* vol. 47, p. 151, December 15, 1892.
38. *Nature,* vol. 47, p. 225, January 5, 1893.
39. Tait seemed at last to be acutely aware that his relatively brief, "loud" letters made a strong contrast with Gibbs' long, detailed, *calm* rebuttals.
40. Tait was referring to Heaviside's article in *The Electrician,* vol. 30, pp. 149-151, December 9, 1892; reprinted in EMT 1, pp. 297-305.
41. Cargill Gilston Knott (1856-1922), formerly Tait's assistant and, by the time of Tait's letter, his colleague at Edinburgh as a lecturer on applied mathematics. Heaviside came to lump Tait and Knott together, calling the pair the "Edinburgh School of Scorners." After Tait's death, Knott wrote the very readable biography of his friend, *Life and Scientific Work of Peter Guthrie Tait,* Cambridge: Cambridge University Press, 1911. This book raises, I believe, the interesting possibility that Tait was ill-disposed toward Oliver long *before* the quaternion debate erupted. As Kelvin's intimate, if not from his own reading, it is certain that Tait knew of Heaviside's irreverent treatment of religion in general, and of the Archbishop incident in particular (see Chapter 7), and this no doubt distressed Tait. As Knott wrote (pp. 36-37): "Tait was indeed a close student of the sacred records. The Revised Version of the New Testament always lay conveniently to hand on his study table...and nothing pained him so much as a flippant use of a quotation from the Gospel writings. I have heard him reduce to astonished silence one guilty of this lack of good taste with the remark, 'Come now, that won't do; that kind of thing is taboo'." For Heaviside, of course, very little, if anything, was taboo.
42. *Nature,* vol. 47, pp. 463-464, March 16, 1893.
43. *Nature,* vol. 47, pp. 533-534, April 6, 1893.
44. Heaviside, in fact, felt there "was hope" for McAulay, and said "I have been much interested in his R.S. paper [see Note 36]. As the heart knoweth its own wickedness, he will not be surprised when I say that I seem to see in his mathematical powers the 'promise and potency' of much future valuable work of a hard-headed kind. This being so, I think it a great pity that he should waste his talents on such an anomaly as the quaternionic system of vector analysis." And in a letter to Gibbs (dated April 6, 1894) he wrote, "He seems to me to be a very clever fellow, and he knows it, and shows that he knows it a little too much sometimes."—quoted from N. Reingold, *Science in Nineteenth-Century America,* New York, NY: Hill and Wang, 1964, pp. 321-322.
45. *Proceedings of the Royal Society of Edinburgh,* vol. 20, pp. 271-275, 1894.
46. Cayley (1821-1895) was one of the truly great men of 19th century mathematics. His career started well as Senior Wrangler in 1842, and just got better and better. *One* of his contributions was the invention of matrix algebra, the second noncommutative algebra, that has had *much* better success with physicists and electrical engineers than have Hamilton's quaternions. So great were Cayley's fame and reputation that Lord Kelvin served as a pallbearer at his funeral, as did Stokes and J. J. Thomson.
47. Cayley's first example was: "...as I consider the full moon far more beautiful than any moonlit view, so I regard the notion of a quaternion as far more beautiful than any of its applications."
48. The lure of a highly compact notation *is* a seductive one. A modern example, from computer science, is the 1970s fad for APL ("A Programming Language"), in which one can write extraordinarily dense code. Unfortunately, the result was often a program that was dense because it crammed vast amounts of information into just a few lines (the "ultimate goal" for dedicated APL users was to do it all in a *single* line!), and was also *mentally* dense. These programs were for the brain like rocks are for the stomach. So tightly written were such programs that nobody but the authors could decipher their operation—and after a day or two even the authors were often seen

scratching their heads! This sort of thing is fun for crossword puzzles, but not for working engineers trying to solve real problems. I find it significant that the language of choice, in Professor R. N. Bracewell's new book *The Hartley Transform* (Oxford: Oxford University Press, 1986), is none other than good old BASIC! One doesn't hear much about APL anymore. Or quaternions.

49. *Nature,* vol. 47, pp. 590–593, April 20, 1893. A somewhat less aggressive version appeared under the title "The quaternion and its depreciators," *Proceedings of the Edinburgh Mathematical Society,* vol. 11, pp. 62–80, 1893.
50. Gibbs used this phrase in his original reply (Note 30) to Tait.
51. EMT 1, p. 289.
52. *Nature,* vol. 49, pp. 193–194, December 28, 1893. The Preface to McAulay's book was written in such extreme, overdone terms (referring, for example, to the "delirious pleasures" of quaternions) that even Tait had misgivings. As he wrote in his review, "It is much to be regretted that Mr. McAulay has not...let his Essay speak for itself. His Preface [is] the perfervid outburst of an enthusiast...and in some passages it runs a-muck at Institutions, Customs and Dignities... . We gladly pass from it to the main contents of the book."
53. Crowe (Note 14), p. 214.
54. *Nature,* vol. 49, p. 246, January 11, 1894.
55. Crowe (Note 14), p. 182.
56. "On the intrinsic nature of the quaternion method," *Proceedings of the Royal Society of Edinburgh,* vol. 20, pp. 276–284, 1895.
57. *Philosophical Magazine,* Series 6, vol. 4, pp. 614–622, November 1902.
58. *Nature,* vol. 52, pp. 545–546, October 3, 1895.
59. Two years later *Nature* (vol. 57, p. 7, November 4, 1897) ran a review of *Lectures on Quaternions,* Part i, a book by S. Kimura, then of Sendai, Japan. It seems likely that this is the same Kimura. The review itself was fairly negative, opening with "We are unable to read this treatise, because it is printed in Japanese." and closing with "When Part ii is published, and the author introduces his quaternions, he may be glad that the old scholars who protect the morals of his country are unable to understand what he is writing about." The review was signed only with the initials "J. P." and the staff at *Nature* has informed me that there is no longer any record of his identity. However, the reviewer revealed himself as one quite familiar with Japanese culture, leading me to believe he was Heaviside's friend, John Perry, who lived in Japan for several years (1875–1879) while teaching at the Japanese Imperial College of Engineering.
60. *The Electrician,* vol. 28, p. 23, November 13, 1891.
61. J. D. Edmonds, Jr., "Quaternion quantum theory: New physics or number mysticism?" *American Journal of Physics,* vol. 42, pp. 220–223, March 1974. Edmonds wrote of the vector–quaternion battle as a "death struggle that was fiery but rather final," but added that he hoped "quaternions may yet 'rise again'." His paper presented the case for how "quaternions would have greatly accelerated the advance of relativity and relativistic quantum theory if they had not fallen from grace with the physicists." See also his later paper, "Maxwell's eight equations as one quaternion equation," *American Journal of Physics,* vol. 46, pp. 430–431, April 1978.
62. As quoted by Tait in the Preface to his quaternion *Treatise,* 3rd ed., Cambridge: Cambridge University Press, 1890.
63. *Nature,* vol. 65, p. 546, April 10, 1902.
64. *Nature,* vol. 53, pp. 268–269, January 23, 1896. For Röntgen's speculation, see the English translation of his paper "On a new form of radiation," *The Electrician,* vol. 36, pp. 415–417, January 24, 1896; reprinted in Alembic Club Reprint No. 22, Edinburgh: E.&S. Livingstone, 1958, pp. 28–40. Professors Boltzmann and Lodge joined Kelvin in hoping the "new radiation" was longitudinal, too, because it would make the concept of the ether more acceptable. Contrary (as usual) was Heaviside who, although an "ether man" himself, declared longitudinal waves to have no place in Maxwell's theory.
65. *Nature,* vol. 53, p. 316, February 6, 1896.
66. *Nature,* vol. 53, p. 509, April 2, 1896.
67. Soon after Kelvin's letter to *Nature,* FitzGerald wrote to Heaviside (June 11, 1896) to say "...he [Kelvin] is not yet convinced nor does he, I think, even yet, understand Maxwell's notion of displacement currents being accompanied by magnetic force. I tried to get him to see that his own investigation of the penetration of alternating currents into conductors [the skin effect] was only the viscous motion analogue of light propagation but he shied at it like a horse at a heap of stones... ."
68. A. R. Forsyth, "Old *Tripos* days at Cambridge," *The Mathematical Gazette,* vol. 19, pp. 162-179, 1935.
69. Quoted from Graves (Note 11), vol. 2, pp. 435–436.
70. Whittaker (Note 12) tells us that the origin of this name, in Hamilton's own words, is that the term "occurs, for example, in our version of the Bible, where the Apostle Peter is described as having been delivered by Herod to the charge of four quaternions of soldiers... . And to take a lighter and more modern instance from the pages of *Guy Mannering,* Scott represents Sir Robert Hazelwood of Hazelwood as loading his long sentences with 'triads and quaternions'."
71. Quoted from Graves (Note 11), vol. 2, p. 489.
72. Quoted from Graves (Note 11), vol. 3, p. 482.
73. Quoted from Graves (Note 11), vol. 2, pp. 434–435.

74. Quoted from Graves (Note 11), vol. 2, pp. 439-440.
75. "On the importance of quaternions in physics," *Philosophical Magazine,* Series 5, vol. 29, pp. 84-97, January 1890.
76. *Nature,* vol. 48, pp. 75-76, May 25, 1893.
77. *Nature,* vol. 48, pp. 540-541, October 5, 1893. In modern vector analysis this problem is avoided by writing the unit vectors as $\vec{i}$, $\vec{j}$, and $\vec{k}$, with the scalar products of each one with itself as unity (and with the others as zero), e.g., $\vec{i}\cdot\vec{i} = 1$, and $\vec{i}\cdot\vec{j} = \vec{i}\cdot\vec{k} = 0$. The cross products are as Hamilton originally defined them.
78. "Quaternions as a practical instrument of physical research," *Philosophical Magazine,* Series 5, vol. 33, pp. 477-495, June 1892.
79. J. E. Littlewood, *Littlewood's Miscellany,* Cambridge: Cambridge University Press, 1986, p. 82.

10
Strange Mathematics

He introduced a new and radical mathematical attack on physical problems which was very powerful but also very obviously full of holes.

— Vannevar Bush, writing on Heaviside's operational calculus, 1929

Heaviside's methods seemed a kind of mathematical blasphemy, a willful sinning against the light. Yet Heaviside's results were always correct! Could a tree be really corrupt if it always brought forth good fruit?

— Prof. H. T. H. Piaggio, enthusiastically stating his position in *Nature,* 1943

Many readers of the Gazette *must have heard of Heaviside's operational method of solving the equations of dynamics and mathematical physics. If they have tried to learn about them from Heaviside's own works, they have attempted a difficult task. Nothing more obscure than his mathematical writings is known to me.*

— Prof. H. S. Carlslaw, expressing a somewhat different opinion in *The Mathematical Gazette,* 1928

I think I have given sufficient information to enable any competent person to follow up the matter in more detail if it is thought to be desirable. It is obvious that the methods of the professedly rigorous mathematicians are sadly lacking in demonstrativeness as well as in comprehensiveness.

— Oliver Heaviside, lashing out in bitterness over the Royal Society's rejection of one of his mathematical papers

An eminent authority once remarked that there is a lot of humbug in mathematical papers. He knew, having done it himself several times.

— Oliver Heaviside, EMT 3, 1900

"Rigorous Mathematics Is Narrow, Physical Mathematics Bold and Broad"

These words, in the opening pages of the second volume of *Electromagnetic Theory,* sum up Heaviside's philosophy and approach to mathematical analysis. They were written in bitterness, the result of what he felt to be ignorant treatment at the hands of mathematicians who were unwilling to step out of their rigid mold of *pureness* in order to appreciate what he had achieved. For that he never forgot or forgave, calling the mathematicians of the Royal Society "wooden-headed" in a 1922 letter written[1] nearly thirty years after the fact.

What precipitated this bitterness was the suppression of a major portion of his work, the operational calculus, from the pages of the *Proceedings of the Royal Society*. How hurtful and painful this was for Heaviside will soon be clear, but the irony of it is immediately evident. Election to Royal Society Fellowship had come to him, at last, after public acclaim by Lord Kelvin. All the years of rejection and dismissal by men of reputation, if not washed away, had at least been diluted (and perhaps even made sweeter) by his being brought into the fold of "certified authorities." He'd even played a bit hard to get at the end, and they had responded by making allowance for the eccentric nature of their newest colleague. They had allowed him his coyness and he had loved it.

And *then* they had reverted to form and *again* closed their eyes and ears to the new message he was bringing them. At least when Preece had denied the importance of inductance, Oliver could publicly (if cautiously) brand him a fool. Now it was a secret (in his mind, a star-chamber) process that censored his work. Nothing had changed after all, and rage as he might Heaviside could do nothing about it. The affair of the operational calculus burned in his throat to the day he died.

The Operator Concept

The use of operators in mathematics has a long and colorful history. One of Heaviside's great contributions to electrical theory was his application of operators to communication problems, but he certainly did *not* teach the mathematicians anything new. This point deserves to be strongly emphasized because the myth that Heaviside was somehow "done wrong" by stupid mathematicians, too rigid and ossified in their thinking to understand something new, is widespread (particularly among electrical engineers).

What Heaviside *did* do, and for which he truly deserves credit, was to show *how to apply to real, physical problems of technological importance* analytical techniques that had up till then been symbolic abstracts.

Operators, as they were used by Heaviside, allow the reduction of the *differential* equations of a physical system to equivalent (in some sense) *algebraic* equations. This is, of course, just what the Laplace transform does for the modern engineer and, in fact, Heaviside's operational calculus is just the Laplace transform in heavy disguise. The Laplace transform is a technique that has a fully developed, mathematically rigorous foundation. Heaviside's writings, however, swarm with unsupported, unproved, even contradictory statements. It should be no surprise that such goings-on made the hair stand up on the backs of mathematicians' necks, and encouraged many of them to dismiss Heaviside as a misguided symbol manipulator. The one aspect of the rejection of Heaviside's work that works against the mathematicians' response, however, is a simple one—they should have asked themselves *why,* if Heaviside was the trickster they thought he was, he often *did* arrive at answers that could be *verified* as correct?

Two early writers on Heaviside's use of operators put it this way:

> Heaviside was not only a great mathematician, he was also a great physicist; and *it is the knowledge of the physics of the problems* [my emphasis] which guided him correctly in many instances to the development of suitable mathematical processes. He concerned himself little with formal proofs or rigorous demonstrations. [Note 2]

and

> He was convinced about results as soon as he could verify them by severe experimental tests, and passed on without waiting to find formal proofs. He was a wanderer in the wilds and loved country far beyond railhead. [Note 3]

Heaviside's own words[4] show the truth of these assessments, at least insofar as his attitude toward proofs:

> To have to stop to formulate rigorous demonstrations would put a stop to most physico-mathematical inquiries. There is no end to the subtleties involved in rigorous demonstration, especially, of course, when you go off the beaten track. And the most rigorous demonstration may be found later to contain some flaw, so that exceptions and reservations have to be added.

and

> Now, in working out physical problems there should be, in the first place, no pretence of rigorous formalism. The physics will guide the physicist along somehow to useful and important results, by the constant union of physical and geometrical or analytical ideas. The practice of eliminating the physics by reducing a problem to a purely mathematical exercise should be avoided as much as possible. The physics should be carried on right through, to give life and reality to the problem, and to obtain the great assistance which physics gives to mathematics.

Both of these passages were written in November 1894, just months after the refusal of the Royal Society to continue to publish Heaviside's work on operators. He had, however, been using operators in his electrical analyses for many years.[5] Indeed, *mathematicians* had been using operators for *two centuries*! One writer,[6] in fact, has traced the origins of operational methods back to John Bernoulli in 1695. Fourier, too, as well as such superstars as Lagrange, Laplace, Cauchy, and Boole,[7] had freely taken advantage of the power of operators. So here we have an interesting question—how did Heaviside manage to take a subject that had been around for a very long time and turn it into a controversy?

Heaviside's Operators

Heaviside's systematic use of operators began in a paper[8] published in 1887, which formally introduced the concept of *resistance operators,* but in fact he had used them much earlier.[9] The resistance operator of a circuit is what modern electrical engineers call the *generalized impedance,* and indeed Heaviside introduced the symbol Z for it, just as it is used today. As he put it in the 1887 paper,

> The resistance-operator Z is a function of the electrical constants of the combination [the circuit components] and of d/dt, the operator of time-differentiation, which will in the following be denoted by p simply.

Then, a typical Heaviside aside followed (from the date, which is years before his problems with the Royal Society, it is clearly a dig at electrical engineers and not at mathematicians):

> ... resistance-operators combine in the same way as if they represented mere resistances. It is this fact that makes them of so much importance, especially to practical men, by whom they will be very much employed in the future [as, in fact, they are]. I do not refer to practical men in the very limited sense of anti- or extratheoretical [like Preece], but to theoretical men who desire to make theory practically workable by the simplification and

systemisation of methods which the employment of resistance-operators and their derivatives allows, and the substitution of simple for more complex ideas.

Here's a simple example of how resistance-operators were used by Heaviside. If presented with a circuit (Fig. 10.1) consisting of discrete, lumped elements, such as a resistor in series with an inductor (with values R and L, respectively), subjected to an externally applied voltage v, then, as shown in Tech Note 1 at the end of this chapter, he wrote the resulting current, i, as

$$i = \frac{v}{Z(p)}, \qquad Z(p) = R + Lp.$$

In his papers Heaviside concentrated on treating constant voltage signals that are suddenly applied at time $t = 0$, e.g., $v = 0$ for $t < 0$ and $v = 1$ for $t > 0$. This, of course, is his famous *step function* (although *he* never called it that). Heaviside wrote it as **1**, and his followers often used $H(t)$ in his honor. To "solve" the operational equation for i Heaviside then expanded the right-hand side in a power series; i.e., treating the *operator p* just as if it were an algebraic quantity having a magnitude, he wrote[10]

$$i = \frac{1}{R+Lp}\mathbf{1} = \frac{1}{Lp(1+R/Lp)}\mathbf{1} = \frac{1}{R}\left[\frac{R}{L}\cdot\frac{1}{p} - \left(\frac{R}{L}\right)^2 \cdot \frac{1}{p^2} + \left(\frac{R}{L}\right)^3 \cdot \frac{1}{p^3} - \cdots\right]\mathbf{1}.$$

This process he called "to algebrize" the problem.[11] To write the explicit time behavior of i, he next individually interpreted each term in the series, which required the decipherment of $(1/p^n)\,\mathbf{1}$, for a positive integer. Heaviside did this in a natural, very compelling way, but a way which unfortunately is not generally true! This flaw (discussed in Tech Note 2 at the end of this chapter) is at the root of many of his troubles with mathematicians. What he did was to argue that since p means to differentiate, and as $p \cdot 1/p = 1$ (which means "do nothing"), then $1/p$ must be the *inverse operator,* i.e., the integration operator. Heaviside thus wrote

$$\frac{1}{p} = \int_0^t du.$$

In particular,

$$\frac{1}{p} \cdot \mathbf{1} = \int_0^t \mathbf{1}\, dt = t, \qquad t \geq 0$$

$$= 0, \qquad t \leq 0.$$

And $1/p^2$ would mean a double integral, and $1/p^n$ would mean an n-fold integral or, in

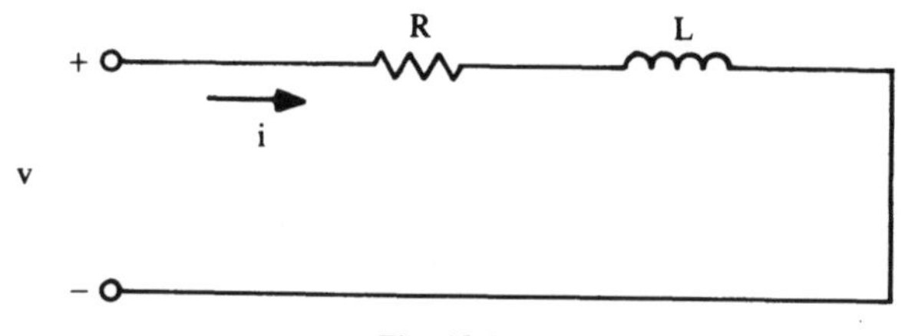

Fig. 10.1.

general,

$$\frac{1}{p^n} \cdot \mathbf{1} = \frac{t^n}{n!}, \qquad t \geq 0$$
$$= 0, \qquad t \leq 0.$$

This result allowed Heaviside to interpret the power series expansion for i as

$$i = \frac{1}{R}\left[\frac{R}{L}t - \left(\frac{R}{L}\right)^2 \frac{t^2}{2!} + \left(\frac{R}{L}\right)^3 \frac{t^3}{3!} - \cdots\right]$$

which he immediately recognized as being almost the power series expansion for an exponential, i.e., he recognized this to be

$$i = \frac{1}{R}\,[1 - e^{-(R/L)t}], \qquad t \geq 0$$

which is, in fact, the correct result.

When used on discrete component systems as in this example, the operator expansion will always give a series in integer powers of $1/p$. When applied to certain continuous systems[12] (distributed parameter systems, such as a telegraph circuit of infinite length), however, Heaviside immediately ran into *fractional* powers of p, e.g., $p^{1/2}$; see, for example, Tech Note 1 at the end of this chapter.

Now this mutilated operator, $p^{1/2} = (d/dt)^{1/2}$, absolutely perplexed those without Heaviside's daring. One can hardly blame the skeptics for their reaction, as it is not really obvious just what (if anything) such a peculiar thing *could* mean! Heaviside, himself, said $p^{1/2}$ is "unintelligible by ordinary notions of differentiation."[13] It is, however, possible to use Euler's generalization of the factorial function (via the gamma function) to attach a useful interpretation to it and, as shown in Tech Note 3 at the end of this chapter,

$$p^{1/2} \cdot \mathbf{1} = 1/\sqrt{\pi t}.$$

This curious formula was actually known to mathematicians long before Heaviside, certainly as early as 1819, but there seems no reason to doubt he discovered it for himself.[14]

The Expansion Theorem

When Heaviside attacked other continuous problems (such as *finite*-length telegraph cables), he found it necessary to employ a more general and sophisticated technique, one beyond the simple power series expansion, to "algebrize" the mathematics. This was his famous *expansion theorem,* which in very simple terms computes a partial fraction expansion of $Z(p)$.[15] Heaviside attached great importance to the expansion theorem, and in his hands it did seem almost like a magic wand that could open the locked doors of incredibly complex problems. To his contemporaries, however, it was a tricky thing. Heaviside's powerful physical intuition led him around the analytical potholes the expansion theorem seemed to dig for those less skilled. When they fell in they blamed the shovel, rather than themselves, or so goes the myth that has spread over the decades.

In fact, the story is more involved than that. By the end of 1892, after years of experimenting with his operators, he had completed the writing of a treatise of major length that presented all he had done. It was perhaps as close as he ever came to writing, from start to finish, an integrated, coherent manuscript on a single topic. He began to present it to the world in early 1893, and his own words warned readers of the strange nature of what they were about to read[16]:

> It [has proven] itself to be a powerful (if somewhat uncertain) kind of mathematical machinery. I have ... convinced myself that the subject is one that deserves to be thoroughly examined and elaborated by mathematicians, so that the method may be brought into general use in mathematical physics, not to supplant ordinary methods, but to supplement them; in short, to be used when it is found useful. As regards the theory of the subject, it is interesting in an unusual degree, and the interest is heightened by the mystery that envelops certain parts of it.

The Royal Society Affair

Heaviside began Part I of his treatise by providing a little personal history on his development of operators. He wrote,

> The sum total of the whole information contained in my mathematical library on the subject of generalized differentiation is contained in the remark made on p. 197 of the second part of Thomson and Tait's *Natural Philosophy* ... . I was somewhat struck ... when I first read it, in trying to plough my way through the fertile though rather heavy field of Thomson and Tait, but as the subject was no sooner mentioned than it was dropped, it passed out of mind. Nor did the absence of any references to the subject in other mathematical works, and in papers concerning mathematical physics generally, tend to preserve my recollection of the remark. Only when the subject was forced upon my attention in the above manner [fractional differentiation, in the form of $p^{1/2}$ and its powers, as they appeared in his series expansions] did I begin to investigate it, and not having access to the authorities quoted [the mathematicians Liouville, Gregory, Kelland], I was compelled to work it out myself. I cannot say that my results are quite the same, though there must, I think, be a general likeness. I can, however, say that it is a very interesting subject, and deserves to be treated in works on Integral Calculus, not merely as a matter concerning differentiation, but because it casts light upon mathematical theory generally, even upon the elements thereof.

And then came the following astonishing denouement to this introduction on how Heaviside "got into" operators:

> And as regards the following brief sketch, however imperfect it may be, it has at least the recommendation of having been worked out in a mind uncontaminated by the prejudices engendered by prior knowledge acquired at second hand. I do not say it is the better for that, however.

Many must have been the readers of the *Proceedings* who ground their teeth after reading *that*!

Part I then presents Heaviside's approach to fractional differentiation via the factorial and gamma functions. While there might be aspects to this that made mathematicians squirm, it wasn't until four months later (with the appearance of Part II[17]) that Heaviside finally went too far. It was here that he introduced his unique interpretation and use of *divergent* series expansions (see Tech Note 4 at the end of this chapter), a treatment nearly devoid of mathematical caution and appreciation for the sensibilities of mathematicians. At one point he declared,

> Mathematics is an experimental science, and definitions do not come first, but later on. They

> make themselves when the nature of the subject has developed itself. It would be absurd to lay down the law beforehand.

Part II was the last straw for the mathematicians, and there would be no Part III. Heaviside did submit it to the Royal Society, but it was squashed by the referee. Parts I and II, by tradition, had not been refereed and, for years *after* Part III, papers submitted by Fellows were not refereed, so it must have taken very special action, indeed, to have imposed a review on Part III. J. L. B. Cooper, in a lecture to the London Mathematical Society, was the first to tell[18] the story behind Part III. E. T. Whittaker's account[19] (as he, in turn, heard it from one involved in the affair) was as follows:

> There was a sort of tradition that a Fellow of the Royal Society could print almost anything he liked in the *Proceedings* (of the Royal Society) untroubled by referees: but when Heaviside had published two papers on his symbolic methods, we felt that the line had to be drawn somewhere, so we put a stop to it.

Heaviside, himself, apparently thought of this action as retribution for his earlier attacks on quaternions, but, in fact, Cooper found (he was allowed to examine the referee's report) that the rejection was based solely on the cavalier use of divergent series. To quote the referee, the author (Heaviside) was "... ignorant of the modern developments of the theory of linear differential equations ..." and was trying to find "a royal road to results which have already been established by exact reasoning." The referee[20] concluded with, "Detailed criticism of results obtained in this way seems out of place. They may or may not be true, but the way in which they are arrived at makes them absolutely valueless."

Lord Rayleigh, then Secretary of the Royal Society, had the unhappy task of telling Heaviside the bad news. On July 26, 1894, he wrote (in fact, it was merely a form rejection, filled in with the date and paper title),

> I am desired to return you the thanks of the Royal Society for your paper "On Operators in Physical Mathematics, Part III" and to inform you that the Committee of Papers, not thinking it expedient to publish it at present, have directed your manuscript to be deposited in the Archives of the Society.

To understand the feelings that must have run through Heaviside when he read this, the words he used in a letter to FitzGerald (February 26, 1894), during the writing of Part III, may help. Then he had called "the task of writing Parts III and IV" a "nightmare of neglected duty." How fast things changed! What had seemed so urgent in February was declared in July to be of no value. On August 5 Heaviside complained to FitzGerald that

> As regards my paper, I think I had better drop it. I don't care to write for any medium where I am not welcome. The way the R.S. behaves is extraordinary. They have lots of money to pay the cost of publication, and they deliberately refuse to use their opportunities. It is such an impractical Society.

Two weeks later (August 20) FitzGerald replied with some good advice:

> I do not think it would be at all dignified or wise to take any extreme step because a referee differed from one as to the value of one's own papers. The best attitude is one of pity and regret that others should be so blind.

Heaviside knew when to throw in the towel. He withdrew Part III, the Royal Society sent the manuscript back on November 5, and Part III saw print only as a (in Oliver's own phrase) "boiled down" presentation some five years later in the second volume of *Electromagnetic Theory.*[21]

William Burnside (1852–1927)

An outstanding mathematician (2nd Wrangler in the 1875 *Tripos*), it was Burnside who served as the Royal Society referee of the rejected operator paper. His obituary notice in the Royal Society *Proceedings* included this comment (with which Oliver would surely have agreed) about his refereeing standards: "He could not be called lenient."

The Aftermath of the Rejection

Heaviside was, of course, bitterly disappointed. It is clear he understood, at an intellectual level, that the major reason for the rejection of Part III was his lack of rigor. As he wrote[22] in November 1894,

> I suppose all workers in mathematical physics have noticed how the mathematics seems made for the physics, the latter suggesting the former ... . This is really the case with resistance operators. It is a fact that their use frequently effects great simplifications and the avoidance of complicated evaluations of definite integrals. But then the rigorous logic of the matter is not plain! Well, what of that? *Shall I refuse my dinner because I do not fully understand the process of digestion* [my emphasis]? No, not if I am satisfied with the result.

But just because he understood his writing was quite clearly lacking in completeness, that didn't stop him from criticizing his critics. He struck back with[23]:

> Of course, I do not write for rigourists (although their attention would be delightful) but for a

> wider circle of readers who have fewer prejudices, although their mathematical knowledge may be to that of the rigourists as a straw to a haystack. It is possible to carry wagon-loads of mathematics under your hat, and yet know nothing whatever about the operational solution of physical differential equations.

and with[24]:

> The discovery of practical methods of manipulating operators is a matter of importance to the future of physical analysis. Objections founded upon want of rigour seem to be narrow-minded, and are not important, *unless passive indifference should be replaced by active obstructiveness* [my emphasis]. In making them rigorists make confession of ignorance. It would be more useful for them to try to extend the matter, and remove the want of rigour, if they want to.

Oliver could even feel sorry for himself in print, and a month after getting Part III back from the Royal Society he wrote,[25]

> We must take the good with the bad, in this as in other matters ... . As regards their want of sympathy with less conventional men [for example, Oliver!], it is not sympathy that is particularly wanted—perhaps it would be unreasonable to expect any at all. What one has a right to expect, however, is a fair field ... so as not to lead to unnecessary obstruction. *For even men who are not Cambridge mathematicians deserve justice, which I very much fear they do not always get, especially the meek and lowly, and those who long suffer under slight* [my emphasis].

For the rest of his life Heaviside would harbor ill feelings for the Royal Society. Ten years after the fact he was still angry enough to pen a letter to *Nature* about it. In reply to an earlier letter extolling the enlightened administration of the French Academy of Sciences, Heaviside wrote,[26]

> What Mr. J. Y. Buchanan says about the French Academy is to me much more wonderful than the revelations of radium. It appears that there is a happy land close by where a scientific man of recognized standing can indulge in the luxury of original research, and then send in an account of his work, *not* to have it rejected by the opinion of, say, a couple of fellow-men, but actually to have it published as a right! This seems impossible. It is the encouragement of original research. Perhaps it is hopeless to expect such freedom in this stick-in-the-mud country, which is so much in love with tradition and antiquated forms. Without any desire to be "contumelious," I would say that our Royal Society reminds me of the House of Lords in many respects.

Heaviside's sad affair with the Royal Society does enjoy today the macabre distinction (from which I believe he would have derived not a little perverse pleasure) of being one of the two most famous 19th century examples of the failure of British physical science at the Royal Society. The first was the rejection of J. J. Waterston's paper on the kinetic theory of gases (in 1845!), with one of the referees declaring[27] it to be "nothing but nonsense, unfit even for reading before the Society." Like Heaviside, Waterston had nothing good to say about the Royal Society after this. In the words[28] of a nephew who knew him well, "I distinctly remember the Royal Society was characterized in very strong terms useless now to repeat."

What makes this incident particularly ironic is its link to Heaviside's similar experience. Waterston's dusty manuscript was finally exhumed from its archival tomb forty years later, because of the efforts of Lord Rayleigh—who would just a few years later preside over the official termination of Heaviside's work in the Royal Society's *Proceedings*!

There can be little doubt that Heaviside came to agree with the opinion of James Swinburne (a noted private consulting engineer, professional "expert witness" in court, and 1902 President of the Institution of Electrical Engineers), expressed in a letter (dated June 21, 1894) received just before the rejection of Part III:

> Why do you secrete important papers at the Royal Society? You have hinted before that you had difficulty in finding a good outlet; the Royal Society is not an outlet, it is an inlet or sink. If you send your papers to the Physical Society there will be a divergence of information.

Interesting words, as Swinburne was a Fellow of the Royal Society, himself!

Heaviside was smart enough at this point in his life to know not to push the controversy too far in public. When *The Electrician* editorial page carried the announcement of the start of a new series of Heaviside articles (which would eventually appear in book form as EMT 2), the editor felt compelled to take advantage of the moment to throw oil on the waters. The editorial, entitled "Mathematics and mathematics," declared[29]

> Not only is Mr. Heaviside the most learned of Maxwell's followers, and the man who wields the most powerful mathematical weapon, but, curiously enough, he is almost the only one who has attempted to make the subject known to practical electricians. A glance through the transactions of learned societies will show how this subject [electromagnetic theory] may be treated by men who are only a little better than mathematicians. We say better, because even a little knowledge of nature *must improve the pure mathematician* [!, my emphasis] and only the very smallest knowledge of nature suffices to differentiate the pure mathematician at Cambridge from the mathematical physicist. Let us look up an elaborate paper on this subject by a mathematical physicist. Well, how much of such a paper can any of us understand? There are three or four people in the whole world at the present time (certainly not more than five) who have practical notions of things, and who are also able to thoroughly understand such a paper.

Then, after complaining about how authors of such papers have no interest in practical matters, and how their papers "bristle with mathematical expressions" (as if Heaviside's didn't), the editorial contrasts this sad state of affairs with Heaviside's new articles:

> With Mr. Heaviside we are in the company of a very clever, good-natured elder brother [Preece's reaction on reading this would have been interesting to see!], who has exactly the same likings as ourselves for all sorts of experimental work, and who tries to explain deep things to us in language that we can understand.

These mushy words, without question, were prompted by Heaviside's comments in the first article[30] of the series:

> It is obvious, I think, that complaints of the want of perfection of the ways and manners of work of explorers on the part of men who are accustomed to more rigorous methods have a considerable element of the ludicrous in them. However harmless in intention, they may operate unfairly in effect, if they lead, as sometimes happens (of which a case was quite lately brought to my notice), to the rejection of honest work which failed to be appreciated by the judges, who have no doubt different ways of working and thinking, and different experiences. When this result arises, it has the effect of putting a learned society in the unfortunate position of *appearing* to exist not merely for the encouragement of research along established lines, but also for the active discouragement of work of a less conventional nature.

However, seeing the editorial must have caused some concern in Heaviside's mind that things were perhaps getting out of control, because in the next article he responded with his famous words,[31]

> Let us above all things try to be just. Even Cambridge mathematicians deserve justice. I cannot join in any general attack upon them. It is to Cambridge mathematicians that we are indebted for most of the mathematico-physical work done in this country. Do not most mathematical physicists hail from Cambridge? Are not Thomson and Tait, Maxwell and Rayleigh Cambridge mathematicians ...?

With this said, the new series of articles commenced an elaborate, extended discussion of the usefulness of operators (beginning with an analysis of one of the more energetically debated issues of the day, the age of the Earth, the topic of the next chapter) in electrical matters. Once finished, the development of the operational calculus came to a temporary halt and wouldn't again spark interest for nearly two decades. Even as late as 1905 he had to explain[32] that his mathematical approach to a problem on the propagation of waves along cables was presented in such a way as

> ... to elucidate the treatment of operators in the solution of differential equations, rational or fractional. This matter has a great future, not merely in electromagnetics, but in all mechanical applications, as academical mathematicians will find out in time, if they live long enough.

And then, in 1906, he apparently washed his hands of the whole business:

> This is enough for the present about the operational treatment of definite integrals, which might go on forever. The above may help others on the way. But perhaps, like the fishes who were preached it by the saint, much edified were they, but preferred the old way. Very well, then there let them stay.

Then, in 1916, there came a fresh interest in the operational calculus, from a most surprising place.

A New Friend at Cambridge

Thomas John I'Anson Bromwich (1875–1929) was, perhaps, an unlikely person to lend support and active, enthusiastic comfort to Heaviside and his mathematics. Arriving at Cambridge in 1892, he was Senior Wrangler in 1895. He was named a permanent lecturer in mathematics at St. John's College in 1907, the year after his election to the Royal Society, of which he was Vice-President in 1919 and 1920.

A closer look at Bromwich's interests shows, however, that he did share common links to Oliver's world. He was, for example, a friend of Searle's, and there are references to Bromwich in Searle's letters to Heaviside from as early as 1909. And in 1908 Bromwich authored a very well known book, still referenced today, on infinite series; Tech Note 4 at the end of this chapter shows how that interest of Bromwich intersected with one of Heaviside's.

Bromwich was not *just* a mathematician, but a man who was also concerned with physical applications. A passage in one of Searle's letters to Heaviside (dated August 21, 1913) gives a little of the flavor of the man: "Bromwich has got a good many optical instruments now and shows some quite good methods of making measurements on compound lenses in his lectures to mathematics students." From an experimental perfectionist like Searle, this was high praise, indeed.

Searle was not the only one at Cambridge to see this side of Bromwich. It is ironic that his colleague, the ultrapure mathematician G. H. Hardy (best known, today, for his friendship with the strange Indian genius Ramanujan), had the insight to write in his obituary notice[33] on Bromwich that he was "The best pure mathematician among the applied mathematicians of

Thomas John I'Anson Bromwich

Cambridge, and the best applied mathematician among the pure mathematicians ... ."

Sometime in 1914 Searle wrote[34] to Heaviside:

> Bromwich seems increasingly interested in your E.M.T. and has been working at it a good deal lately. I wish you knew him. He is really quite a wonderful person. His wife is one of the brightest and happiest women in Cambridge. He is one of my co-examiners in the Math Tripos. There are some very solemn people in Cambridge but he is not one of them.

This might seem to indicate the two men had not yet connected, but Searle may just have been trying to entice Heaviside out of his cloister. In fact, Bromwich and Heaviside had been corresponding since at least 1913.[35] In a letter of January of that year Bromwich wrote to Oliver about the current state of affairs concerning divergent series, and from the tone it seems Bromwich was answering specific questions in response to an earlier query from Heaviside.

On April 17, 1915 Bromwich wrote asking for some of the historical development[36] of the operational calculus. In particular, Bromwich wrote he had been "gradually building up" a different, yet equivalent method of treating Heaviside's operational problems and that

> ... this method of mine is leading me to an independent way of interpreting your symbolic

formulae such as

$$\sqrt{\frac{\partial}{\partial t}} \cdot 1 = 1/\sqrt{\pi t}.$$

Bromwich concluded with words that must have pleased Heaviside:

> I hope this [his new method] will be written up some time soon, so that you can at least see one of the orthodox (or conservatory) type of mathematician doing his best to convert others to heterodox methods: but I am only able to find a little time for writing. I did succeed in teaching one of our young men some of the work last term: and he was most enthusiastic. But it is easier to teach the young than the old, I fear.

Unquestionably, when he wrote of "writing up" his method, Bromwich was referring to a paper[37] he'd already presented to the London Mathematical Society but which didn't appear in print until 1916. It was in this paper that the first step toward replacing Heaviside's mysterious, often seemingly arbitrary symbolic methods, with the contour integral in the complex plane approach, was made. Bromwich's 1916 paper was extremely influential, and his ideas culminated in the now common use of the Laplace transform.[38] Curiously, Bromwich himself seems not to have had even an inkling that his work was the death sentence for that of Heaviside. As he wrote (somewhat contradictorily) in his letter of April 25, 1915:

> My method starts with complex integrals: I fear it sounds rather formidable; but it is really quite simple ... . I am afraid that no physical people will ever try to make out my method: but I am hoping that it may give them confidence to use your methods more, if they know that there is an independent way of getting to the formulae.

And then again, years later on April 5, 1919 after returning to Cambridge from war duty at the Admiralty,

> After coming back to these questions after 2½ years of war-work, I found myself able to work more readily with operators than with complex-integration. Realising that I had probably a better knowledge of these complex-integrals than the average *Philosophical Magazine* reader, I at once saw that I must make the operator-method take the leading place: and complex-integrals have accordingly been pushed into footnotes. I still regard the complex-integral as a useful method for convincing the purest of pure mathematicians that the *p*-method rests on sound foundations: but I am sure that the *p*-method is the working-way of doing these things ... . To give you some idea of how far I have moved toward your own point of view; I had occasion the other day to find one of Fourier's series (for the mean-temperature of a sphere cooling from constant temperature, with its surface kept at zero). I found myself able to get what I wanted *mentally* using the *p*-method; but I would never have dreamed of trying to do it by Fourier-series without pencil or paper (or looking the result up in a book).

These gracious words (with which Bromwich may have intentionally overstated his real position in an attempt to be kind to an old man whom he greatly admired) brought forth what seems to be the only surviving record of a Heaviside letter[39] to Bromwich. It shows Heaviside at his most curmudgeonly:

> Yours 5:4:19, Caesar and Pompey: especially Pompey [Note 40] ... . What a time it takes! I rejoice to know that you have seen the simplicity and advantages of my way ... . Now let the wooden headed rigorists go hang, and stick to differential operators and leave out the rigorous footnotes. It is easy enough if you don't stop to worry. As I said and Lord R.

> [Rayleigh] repeated [Note 41], logic is the very last thing. *I never could stomach your complex integral method* [my emphasis] … .

On this somewhat harsh and final assessment of what Bromwich had created, the two men had little left to say to each other and the correspondence soon died. Others at Cambridge took up Bromwich's interest in Heaviside's mathematics, however, most notably the mathematical physicist Harold Jeffreys. Two years after Heaviside's death Jeffreys authored a book[42] on the subject, published by none other than the Cambridge University Press—an occurrence which would, no doubt, have reduced Oliver to tears of laughter. Jeffreys obviously had a sense of the irony in this, himself, as Heaviside's own words are printed on the frontispiece, "Even Cambridge mathematicians deserve justice."

Shortly after this Jeffreys wrote,[43]

> His methods may appear slipshod, but his precautions were such that they practically always gave the right answer … . He left mathematicians with the problem of explaining why the answers were right, which was duly solved, in time, by Bromwich, and it was then found that many of Heaviside's solutions could be obtained easily by workers without his amazing skill in manipulation, but using the theory of the complex variable.

Later in this book we will pick up the thread of the operational calculus as Bromwich and Jeffreys left it at the end of the 1920s and follow it to modern times, but Bromwich himself will play no further role. This man who was so fascinated by Heaviside's genius, but so unlike him as a human being,[44] soon after Heaviside's death simply ran out of time. Hardy wrote[33] of him,

> No one could have seemed more sane; he was the last man whom anyone would have suspected of any mental instability, and when this developed later it was a great surprise as well as a great shock to his many friends.

But even greater must have been the shock when, on August 24, 1929, Bromwich turned away forever from his wife, his sons, his work, and took his own life.

All through Bromwich's letters to Heaviside, and in his other writings as well, are references to what sparked his original interest in Heaviside's operators. This was the age-of-the-Earth debate of 1895 in which Heaviside's operational mathematics played a role, and as the next chapter will show, the debate had plenty enough to it to challenge the world's best mathematicians.

Tech Note 1: Heaviside's Resistance Operators

In the opening paragraphs of the 1887 paper[8] Heaviside gave a prose definition of a resistance operator: "If we regard for a moment Ohm's law merely from a mathematical point of view, we see that the quantity R, which expresses the resistance, in the equation $V = RC$ [C was the 19th century symbol for current], when the current is steady, is the operator that turns the current C into the voltage V. It seems, therefore, appropriate that the operator which takes the place of R when the current varies should be termed the resistance operator."

So here's what Heaviside did. For the three fundamental, discrete, passive electrical devices (resistors, capacitors, and inductors), we have, respectively, if we use the modern symbol i for the current (and C for capacitance)

$$v = iR, \qquad \text{for a resistor of value } R$$

$$i = C\frac{dv}{dt}, \qquad \text{for a capacitor of value } C$$

$$v = L\frac{di}{dt}, \qquad \text{for an inductor of value } L$$

or, using Heaviside's $p = d/dt$ operator,

$$v = iR$$

$$i = Cpv$$

$$v = Lpi$$

or, defining the ratio v/i as the "resistance,"

$$Z = R, \qquad \text{for a resistor}$$

$$Z = \frac{1}{Cp}, \qquad \text{for a capacitor}$$

$$Z = Lp, \qquad \text{for an inductor.}$$

In particular, the resistance operator for the series resistor–inductor circuit of the text is $Z = R + Lp$. A great merit of this approach is that it allows inductors and capacitors to be treated *just like resistors.* A striking example of this was given by Heaviside himself in the 1887 paper. There he analyzed an infinitely long telegraph circuit with continuously distributed parameters, R, L, K, and S, i.e., per unit length resistance, inductance, leakage conductance (the reciprocal of leakage resistance), and shunt permittance (the reciprocal of shunt capacitance). See Fig. 10.2.

To find $Z(p)$ for such a circuit might seem at first glance to be an incredibly difficult task, and indeed there could not have been but a relative handful of men in the world of 1887 who could have done it. But, as Heaviside wrote, the result

> ... may be obtained directly in a way which is very instructive as regards the structure of resistance-operators. Since the circuit is infinitely long, Z cannot be altered by cutting-off from the beginning, or joining on, any length.

This ingenious observation just says that since the circuit is *already* infinitely long, adding a little more length to it won't make any difference! This clever trick is now routinely taught to electrical engineering and physics students in making transmission line and wave guide analyses, but it is the rare student who learns of its historical origin. In any case, that's what Heaviside did—he stuck on a little bit more of the circuit to the beginning of the circuit.

If the stuck-on portion is of the very short length ℓ, then to a good approximation the circuit in Fig. 10.3 results. Now, since the $K\ell$, $S\ell$, and series $R\ell$, $L\ell$, and Z are in parallel, their respective admittances (reciprocal impedances) are added together. Thus, as Heaviside wrote,

$$\frac{1}{Z} = K\ell + S\ell p + \frac{1}{R\ell + L\ell p + Z}.$$

This is easily manipulated to give

$$Z^2 + Z(R\ell + L\ell p) = \frac{R + Lp}{K + Sp}.$$

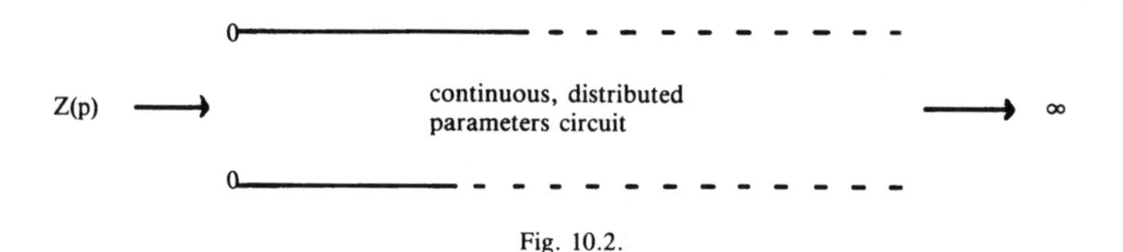

Fig. 10.2.

Then, Heaviside effectively said "let's now suppose we *don't* add a stuck-on portion"! Then $\ell = 0$, the approximation becomes exact, and we immediately have Heaviside's result:

$$Z(p) = \sqrt{\frac{R + Lp}{K + Sp}}.$$

Suppose now that $L/R = S/K$. Then $Z(p) = \sqrt{R/K}$, which is *independent* of p. That is, $Z(p)$ is a constant, time-independent operator, and "treats" all sinusoidal signals the same, irrespective of frequency. This is, in fact, Heaviside's condition for distortionless transmission. This beautiful analysis, worthy of a paper all its own, was embedded in the 1887 paper as two tersely written paragraphs. Heaviside's derivation yields its important result almost without effort, and the action is over nearly as soon as it begins. This may explain why so many failed to appreciate its significance.

TECH NOTE 2: THE PROBLEM WITH THE p AND $1/p$ OPERATORS

Given the simple differential equation $dx/dt = f(t)$, Heaviside would write $px(t) = f(t)$ and then solve for $x(t)$ as

$$x(t) = \frac{1}{p} f(t) = \int_0^t f(u)\, du.$$

That is, he associated the $1/p$ operator with the definite integral over the interval zero to t. This implies, however, that $x(0) = 0$, which may not always be the case. In those situations where this is *not* true, the p and $1/p$ operators are *not* inverse operators; the result of this has often been the calculation of erroneous results by the unwary. Essentially the same problem can occur in far more subtle forms than the blatant example used here; Heaviside often used his intuition and physical insight to guide him. As he wrote[45] concerning the use of operators in an electromagnetic problem,

> ... to avoid error, it is desirable to be guided by the conditions of the physical problem concerned. That will serve to counteract the ambiguity of the purely mathematical machinery.

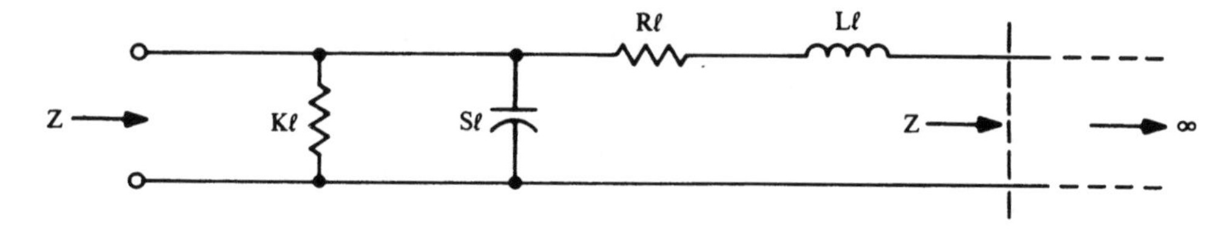

Fig. 10.3.

This bit of wisdom, however practical and useful, certainly won Heaviside few converts among the mathematicians. The noncommutative nature of Heaviside's operators has been discussed in detail by Harold Jeffreys and D. P. Dalzell. In their paper[46] there is a passage (by Jeffreys) that shows how at least one Cambridge *physicist* viewed Heaviside's mathematics:

> ... in actual applications Heaviside got the right answers, ... difficulties ... arising only in extensions to pure mathematics, and in these his object was frankly experimental, to discover by actual trial what methods gave the right answers. Personally I have considerable sympathy with his point of view; the history of mathematics would have been very different if no result had ever been published before a rigorous proof was available, and even now the Gegenbeispiel is an important mathematical method.

In his review of EMT 2 Professor George Minchin wrote,[47]

> The reader is struck with the power and succinctness of the *operational* method of integrating the various differential equations and Mr. Heaviside's complete mastery over the process. It will, however, take the ordinary student of Physics a very considerable time to grasp the logic of this method and to satisfy himself of its validity.

and then, a little later

> Mr. Heaviside is ... always at home with this method, while beginners will have to mind their steps ... .

Minchin was so impressed with the operational calculus, in fact, that he thought knowledge of it would "compare favorably with the wonderful machine in the Academy of Lagado." He was referring to the gadget in *Gulliver's Travels* that, as Jonathan Swift wrote, enabled "the most ignorant person" to write books "in philosophy, poetry, politics, law, mathematics, and theology, without the least assistance from genius or study."

The book was also reviewed[48] by the well-known German physicist Gustav Mie who, after first describing it as having "unusual richness ... in new ideas and important results ..." went on to write,

> Heaviside treats mathematics in highly original fashion as an experimental science. He does not deduce, like the "Rigorists", his methods from general principles and definitions; rather he demonstrates their correctness, after he reaches them by analogy as a kind of hypothesis, inductively through testing on simple examples, which then also show their area of validity. Although the method of presentation, which is very lively and often accompanied by clever and witty remarks, gives extremely fascinating insights into the mental workshop of this famous mathematician, it is also on the other hand often quite difficult to understand, especially if one is not yet accustomed to this experimental treatment of mathematics. Often one finds formulas or complete equations introduced without further explanation; these are then comprehensible only in the later course of the analysis. It is most difficult for him who is looking for a universal point of view, from which to gain an overview of and to master all individual methods, so that they may also be used for other researchers besides Heaviside. By superficial consideration one has the impression that different definitions inductively derived contradict and diverge from one another. It may be assumed, however, that on closer inspection these disharmonies are solved and that the pure mathematician who is interested in the methods of mathematical physics will discover a rich and fruitful field of work.

For a somewhat different view on Heaviside's supposed infallibility, see the paper[49] by F. W. Carter in which he claimed Heaviside's operational solution[50] for the step response of an infinitely long transmission line is incorrect. N. W. McLachlan later published yet another analysis[51] of this problem, using the contour integration technique developed by Heaviside's

Cambridge mathematician friend, T. J. Bromwich. McLachlan, too, said Heaviside's solution "seems" to be in error. Quite honestly, I can't tell which, if any, of the three (Heaviside, Carter, or McLachlan) is right or wrong. All three solutions are discouragingly complex, and I'll leave it to an ambitious reader to check it all out and send the final word to me!

Tech Note 3: The Meaning of Heaviside's Fractional Operator, and Impulses

To understand what $p^{1/2} \cdot \mathbf{1}$ means, we *could* just follow the path laid by Heaviside in EMT 2 (pp. 287–288). There he developed the result *experimentally* (a technique alien to pure mathematicians), which means he compared the ***known*** solution to a problem (obtained by Fourier methods) that has $p^{1/2}$ in the operator solution. For those readers who might not applaud this method he brusquely wrote, "Those who may prefer a more formal and logically-arranged treatment may seek it elsewhere, and find it if they can; or else go and do it themselves."

There are, fortunately, many other ways to arrive at the meaning of $p^{1/2}$ and, as Heaviside said after presenting the experimental approach, "The above is only one way in a thousand." The derivation presented here is the one Heaviside used in Part I of his Royal Society paper (and EMT 2, pp. 288–290), and it is quite straightforward.

The *gamma function* $\Gamma(n)$ has the recursive property

$$\Gamma(n) = (n-1)\Gamma(n-1)$$

which says the gamma and the factorial functions are closely related for integer, positive values of n, i.e.,

$$\Gamma(n) = (n-1)!, \qquad n = 1, 2, 3, \cdots$$

$$0! = 1.$$

It is not clear what (if anything) the factorial operation means for noninteger and/or negative values of n, but Euler generalized the factorial function to include all real values of n with his definition of the gamma function integral[52]

$$\Gamma(n) = \int_0^\infty e^{-u} u^{n-1}\, du.$$

This integral can be evaluated to show that $\Gamma(\frac{1}{2}) = (\pi)^{1/2}$. Setting $n = \frac{1}{2}$ in $\Gamma(n) = (n-1)!$ we arrive at the curious result that $(-\frac{1}{2})! = (\pi)^{1/2}$.

To see what this has to do with $p^{1/2}$, notice that for nonnegative integers m and n ($m < n$) we have what might be called the fundamental relation:

$$\frac{d^m}{dt^m}\, t^n = \frac{n!}{(n-m)!}\, t^{n-m}.$$

Suppose we now see what formally comes from this when m is not an integer; say, $m = \frac{1}{2}$ (i.e., we have the fractional operator $p^{1/2}$). Then,

$$p^{1/2} \cdot t^n = \frac{n!}{\left(n - \dfrac{1}{2}\right)!}\, t^{n-1/2}.$$

Furthermore, if $n = 0$ we have $t^n = 1$ and then

$$p^{1/2} \cdot \mathbf{1} = \frac{0!}{\left(-\frac{1}{2}\right)!} t^{-1/2} = 1/\sqrt{\pi t}.$$

Higher positive half-powers of p are easily found by the process of differentiation, e.g.,

$$p^{3/2} \cdot \mathbf{1} = p(p^{1/2} \cdot \mathbf{1}) = \frac{d}{dt}(\pi t)^{-1/2} = -\frac{t^{-3/2}}{2\sqrt{\pi}}.$$

Negative half-powers of p are easily calculated in a similar fashion. For example, letting $m = -\frac{1}{2}$ gives

$$p^{-1/2} \cdot t^n = \frac{n!}{\left(n+\frac{1}{2}\right)!} t^{n+1/2}$$

and if $n = 0$ then

$$p^{-1/2} \cdot \mathbf{1} = \frac{1}{\left(\frac{1}{2}\right)!} t^{1/2}.$$

Then, to get $p^{-3/2}$ you merely integrate this, i.e.,

$$p^{-3/2} \cdot \mathbf{1} = \frac{1}{p} p^{-1/2} \cdot \mathbf{1}$$

which will give a result varying as $t^{3/2}$. As Heaviside put it,[53] "There is a universe of mathematics lying in between the complex differentiations and integrations." And in a letter[54] to Ludwik Silberstein he wrote, "You will understand that in general we are not usually concerned with the fractional differential operators, only with integral. Nevertheless the fractional ones push themselves forward sometimes, and are just as real as the others."

Heaviside also considered the effect of applying integer powers of p to $\mathbf{1}$. As he wrote,[55] "... $p \cdot \mathbf{1}$ means a function of t which is wholly concentrated at the moment $t = 0$, of total amount 1. It is an impulsive [Note 56] function, so to speak ... . Unlike the function $p^{1/2} \cdot \mathbf{1}$, the function $p \cdot \mathbf{1}$ ... involves only ordinary ideas of differentiation and integration pushed to their limit." Higher integer powers of p were considered, too (e.g., EMT 2, p. 65): $p^2 \cdot \mathbf{1}$ "we see is the time-rate of increase [of $p \cdot \mathbf{1}$], and is therefore a double impulse [modern terminology calls this the *doublet*], first positive and then negative ... ." Heaviside could create some confusion at times, however, by setting $p \cdot \mathbf{1} = 0$ (although in EMT 2, p. 289 he slipped in one quick cautionary note: "... $p \cdot \mathbf{1} = 0$, provided t is positive. It is really an impulse at the moment $t = 0$."). To know when to use which, however, could require the genius of a Heaviside. John Carson, Heaviside's great admirer at AT&T, had a strong negative reaction to this sort of hanky-panky. As he wrote,[57] "His procedure in this respect is quite unsatisfactory and in particular his discarding an entire series without explanation is intellectually repugnant."

Heaviside wasn't oblivious to this problem, however. For example, after getting one problem into the form of an infinite sum of integer powers of p operating on $\mathbf{1}$ (each of which could then presumably be set equal to zero for $t > 0$), he observed that the answer was wrong. He offered the following words[58] (not very helpful ones, in my opinion): "We get a constant term, plus an infinite series of zeros. Now there are zeros *and* zeros. An absolute zero is like the point in geometry, which you cannot see even when you use a magnifying glass, as the schoolboy said. But some zeros can be magnified, and an infinite number of them might make finiteness."

TECH NOTE 4: HEAVISIDE AND DIVERGENT SERIES

In Part III of "Operators in physical mathematics" it was Heaviside's use of divergent series that proved to be his downfall. According to Cooper,[18] the referee's objections turned entirely on the (improper) use of divergent series. In 1826 a mathematician (Abel) wrote,[59] "Divergent series are the invention of the devil, and it is a shame to base on them any demonstration whatsoever." Heaviside's work shows he didn't share this opinion.

Scattered all through EMT 2 are numerous examples of how Heaviside ran into and handled divergent series. Almost every such example had something in it akin to driving slivers under the fingernails of mathematicians. One typical development can be found on pp. 487–488. There he imagined a step input applied through a resistance R_0 into an infinitely long telegraph circuit that is both inductance and leakage free (i.e., in Tech Note 1 in this chapter set $L = K = 0$ and get $Z = \sqrt{R/Sp}$); see Fig. 10.4. Heaviside then wrote the voltage *at the input* to the circuit as

$$v = \frac{e}{1 + R_0/Z}.$$

Next he wrote $e = \mathbf{1}$ and $a = R_0\sqrt{S/R}$, and thus arrived at the *operational* solution

$$v = \frac{1}{1 + ap^{1/2}}\,\mathbf{1}.$$

A simple algebraic twist gives the alternative form

$$v = \frac{a^{-1}p^{-1/2}}{1 + a^{-1}p^{-1/2}}\,\mathbf{1}.$$

Next, using his favorite series expansion from "ordinary" mathematics,

$$\frac{1}{1+x} = 1 - x + x^2 - x^3 + \cdots$$

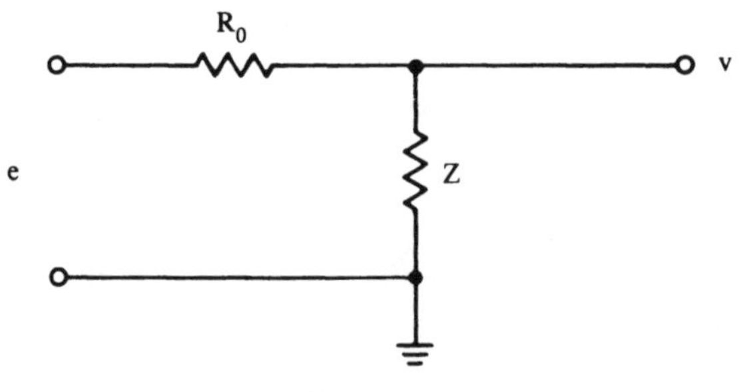

Fig. 10.4.

he expanded both operational solutions, using $x = ap^{1/2}$ for the first one, and $x = a^{-1}p^{-1/2}$ for the second. This gives

$$v=[1-ap^{1/2}+a^2p-a^3p^{3/2}+a^4p^2-\cdots]\mathbf{1}$$

and

$$v=\{[a^{-1}p^{-1/2}+a^{-3}p^{-3/2}+\cdots]-[a^{-2}p^{-1}+a^{-4}p^{-2}+\cdots]\}\mathbf{1}.$$

From the results of Tech Note 3 in this chapter on Heaviside's interpretation of positive and negative half-powers of p, we see the first series involves terms with t in the denominators (which individually blow up to infinity at $t = 0$, a numerical bomb that Heaviside dismissed with "it is not evident what it [the series] means initially") and the second series involves terms with t in the numerators (which individually blow up at $t = \infty$, another numerical puzzle that Heaviside airily waved away with "it is not immediately evident [Note 60] what v rises to finally"). Neither of these concerns stopped Heaviside, and he used such power series to calculate numerical results. Mathematicians, of course, looked on this sort of business with jaundiced eyes.

No matter what the mathematicians might think, however, Heaviside was convinced his ideas were sound. As he wrote[61] in 1906,

> Mathematicians in general are, I find, exceedingly conservative and prejudiced. Nevertheless, I am confident of a great future for the practical use of divergent series, as well as for the generalized analysis which connects them with the convergent ones, because both of these matters are concerned in the operational treatment of physical differential equations.

A half century earlier, however, Cambridge mathematicians would most likely not have been at all concerned about Heaviside's cavalier use of divergent series. William Whewell (1794–1866), for example, associated with Trinity College from 1812 until his death, was a principal architect of the Cambridge mathematics honors program (which culminated in the ordeal of the *Mathematical Tripos*), and was equally "skillful" at twisting such series about to suit his needs. He was also a believer in the proposition of "what is true *up to* the limit is true *at* the limit"[62]; Heaviside felt right at home with this intuitively appealing (but wrong) assertion, too, although he put it in somewhat more poetic (but not very convincing) terms when he wrote,[63] during a limiting operation in a discussion on Fourier series, "I know no reason why a failure should occur just as perfection is reached." A typical example of his casual style in these matters is given by his statement[64] that Euler's constant is

$$\gamma=1+\frac{1}{2}+\frac{1}{3}+\cdots+\frac{1}{\infty}-\log\infty$$

which, as it stands, is meaningless. The proper way to express γ (and it takes no more time and energy to do it right than to do it wrong) is

$$\gamma=\lim_{n\to\infty}\left[\sum_{K=1}^{n}\frac{1}{K}-\log n\right]=0.5772.$$

It was this casual style that prompted one writer to declare,[39] perhaps more from exasperation than from real belief, "the time is ripe to remember *what* the illustrious Heaviside did, but to forget *how* he did it." Another writer has offered[3] at least a partial explanation for some of this sloppiness or, as he put it, for why Heaviside's work "shows signs of hurry":

> Few realise the conditions of publication of Heaviside's *Electromagnetic Theory*. These were most extraordinary for an original work of permanent value. The matter was published week after week and year after year in the *Electrician*. No mathematical author was ever granted so much freedom to publish what, when, and how, he liked, and this freedom was granted by a technical newspaper; *but not a line could be changed after first publication* [my emphasis], since the type was kept set up till the corresponding section of the book was printed... . Heaviside [was] under great time pressure.

Notes and References

1. In March 1922, Heaviside corresponded with John S. Highfield (then President of the Institution of Electrical Engineers) and reminisced about his life. These letters reveal a great deal of anger, bitterness, and feelings of persecution, and the mathematicians of the Royal Society were not the only ones to be called names (including some who had been very public in their praise of Heaviside and of the value of his work).
2. L. Cohen, *Heaviside's Electrical Circuit Theory,* New York, NY: McGraw-Hill, 1928, p. v.
3. W. E. Sumpner, "Heaviside's fractional differentiator," *Proceedings of the Physical Society of London,* vol. 41, pp. 404–425, June 15, 1929.
4. EMT 2, pp. 4–5. Later, on p. 122, he continued this theme with "A man would never get anything done if he had to worry over all the niceties of logical mathematics under severe restrictions; say, for instance, that you are bound to go through a gate, but must on no account jump over it or get through the hedge, although that action would at once bring you to the goal." To a pure mathematician, to whom the rules are, if not *everything,* at least way out in front of whatever is in second place, these words must have seemed (and no doubt still do) those of a dangerous heretic.
5. EP 1, p. 415.
6. E. Koppelman, "The calculus of operations and the rise of abstract algebra," *Archive for History of Exact Sciences,* vol. 8, pp. 155–242, 1971/72.
7. George Boole's use of operators was presented in his 1859 book on differential equations, which Heaviside owned. He was, therefore, *not* totally ignorant of the prehistory of operational methods, a condition which he admitted.
8. EP 2, pp. 355–374.
9. EP 1, p. 125, dated 1881.
10. Recall that $1/(1 + x) = 1 - x + x^2 - x^3 + \cdots$, $|x| < 1$. Heaviside got the power series given in the text by treating R/Lp as x, while ignoring the question of what does it *mean* to do this, i.e., in what sense is $|R/Lp| < 1$, when p is an *operator,* not a magnitude? This devil-may-care approach to operational methods is the hallmark of Heaviside's style. As one writer put it, Heaviside "... always uses p in algebraic formulae just as if it were a numerical quantity. He assumes that he can use such formulae as active operators after reducing them by any kind of analysis to some convenient form. This assumption is so violent that no one would have a good word to say for it were it not for the astonishing success attending its use."—see W. E. Sumpner, "Index operators," *Philosophical Magazine,* Series 7, vol. 12, pp. 201–224, August 1931.
11. As a skilled coiner of new words, Heaviside could also tell when he'd fallen a bit short. Of this rather inelegant word he self-consciously wrote (EMT 2, p. 41), "Perhaps some people will say (as usual) that they do not like 'algebrize'; that it is un-English. People are always saying something." In a weak attempt to explain the merit of his word he then tried to convince the reader that finding a logarithm might well be called to "logarize" (and the reverse process to "delogarize"!).
12. Soon after Heaviside's death several books appeared devoted to demonstrating his operational techniques. Two of them are particularly faithful to his approach, taking it at face value with no criticism. They are, however, quite readable, and reproduce the Master's approach to continuous systems—mostly transmission lines with a wide variety of boundary (termination) conditions. See Cohen (Note 2), and E. J. Berg, *Heaviside's Operational Calculus,* New York, NY: McGraw-Hill, 1929.
13. EMT 2, p. 286.
14. B. Ross, "The development of fractional calculus, 1695–1900," *Historia Mathematica,* vol. 4, pp. 75–89, February 1977. This particular result played an important role in Heaviside's 1894 age-of-the-Earth analyses, which are discussed in the next chapter.
15. Heaviside stated and proved the expansion theorem in EP 2, pp. 371–374. I must admit I find his presentation a bit thick and I highly recommend to all who wish to see the details of the theorem clearly explained that they read the excellent tutorial by J. Lützen, "Heaviside's operational calculus and the attempts to rigorise it," *Archive for History of Exact Sciences,* vol. 21, pp. 161–200, 1979.
16. "On operators in physical mathematics. Part I," *Proceedings of the Royal Society,* vol. 52, pp. 504–529, February 1893.

17. "On operators in physical mathematics. Part II," *Proceedings of the Royal Society,* vol. 54, pp. 105–143, June 1893.
18. J. L. B. Cooper, "Heaviside and the operational calculus," *Mathematical Gazette,* vol. 36, pp. 5–19, 1952.
19. E. T. Whittaker, "Oliver Heaviside," *The Bulletin of the Calcutta Mathematical Society,* 1928; reprinted in the book by D. H. Moore, *Heaviside Operational Calculus,* New York, NY: American-Elsevier, 1971, p. 216.
20. Cooper did not reveal the referee's name, but he was William Burnside (1852–1927) of the Royal Navy College at Greenwich. I am indebted to Professor Bruce J. Hunt, University of Texas (Austin) for this information. In 1875 Burnside was 2nd Wrangler and 1st Smith's Prizeman at Cambridge, and he had no sympathy for Heaviside's rough-and-ready approach to mathematics. A very interesting discussion of mathematical refereeing in general, with some speculations on what might be Heaviside's fate at the hands of modern mathematicians, is in D. V. Lindley, "Refereeing," *The Mathematical Intelligencer,* vol. 6, no. 2, pp. 56–60, 1984.
21. EMT 2, pp. 457–492.
22. EMT 2, p. 9.
23. EMT 2, p. 33.
24. EMT 2, pp. 220–221.
25. EMT 2, p. 11.
26. *Nature,* vol. 69, p. 317, February 4, 1904.
27. M. Jammer, *The Conceptual Development of Quantum Mechanics,* New York, NY: McGraw-Hill, 1966, pp. 12–13.
28. From a letter by George Waterston to Lord Rayleigh, which also mentioned J. J. Waterston's mysterious disappearance (and most likely his death); reprinted in R. J. Strutt, *John William Strutt, Third Baron Rayleigh,* London: Longmans, Green and Co., 1924, pp. 170–171.
29. *The Electrician,* vol. 34, pp. 100–101, November 23, 1894.
30. Ibid, p. 91 (EMT 2, pp. 3–4).
31. EMT 2, pp. 10–11. The IEE in London has a letter (dated January 13, 1928) written by *The Electrician*'s editor, A. P. Trotter (sent to Rollo Appleyard) admitting that he and his assistant editor at the journal (W. G. Bond, who became editor upon Trotter's retirement) had authored the editorial. Trotter wrote, "I don't see how Heaviside, touchy as he was, could have been offended at it."
32. The first quote appeared in print, for the first time, in EMT 3 (Heaviside needed extra material, beyond his journal articles, to fill out the book) on p. 207, and the second quote is from p. 291. The saint referred to is apparently Antony of Padua (1195–1231) who opened his sermon with "Hear the word of God, oh ye fish of the sea and the river, for the infidel heretics despise it!"
33. *The Journal of the London Mathematical Society,* vol. 5, pp. 209–220, July 1930. Hardy is also famous for his odd little book, *A Mathematician's Apology,* in which he declared only useless mathematics to be pure and beautiful. Taking pride in his claim to "have never done anything 'useful,'" Hardy would seem hardly one to appreciate someone like Heaviside. And yet, in his last book just before his death (*Divergent Series,* Oxford: Oxford University Press, 1949), he wrote (p. 36), "Heaviside, whatever his merits as a mathematician, was a man of much talent and originality, and what he says (if often irritating to a mathematician) is always interesting."
34. This letter is undated, but at the top is a note by Searle that he wrote it in 1914. The note itself is dated January 11, 1950, just before the IEE Heaviside Centenary; Searle had evidently been rereading his old letters in preparation for that event.
35. Perhaps Searle was incorrect on the date of his letter (see previous note)—after a lapse of over thirty-five years this would not be a surprise.
36. With a single exception (see Note 39) all of Heaviside's letters to Bromwich appear to be lost. A letter from Bromwich (dated April 25) indicated Heaviside had quickly responded, as he had already received "a very long letter which answers my questions very clearly" for which he offered his thanks. What interesting reading this particular Heaviside letter must have made!
37. "Normal coordinates in dynamical systems," *Proceedings of the London Mathematical Society,* vol. 15, pp. 401–448, 1916.
38. As commonly taught today in sophomore electrical engineering classes the Laplace transform is treated as "merely" an ingenious integral transform with just the right properties to turn the differential equations of linear systems into algebraic equations. The transform variable (usually written as s, but occasionally as Heaviside's p) is related to but *not* thought of as a differentiation operator. Students are usually given no reason to treat the transform differently from any of the other integrals they saw the previous year in freshman calculus. Later, of course, they learn the transform variable is complex, the integral is really a contour integral in the complex plane, and they may even take an undergraduate math class and learn the mechanics of Cauchy's residue theorem (which Bromwich tried to teach to Heaviside, without much luck!). But for another viewpoint on all this, see the paper by C. L. Bohn and R. W. Flynn, "Real variable inversion of Laplace transforms: An application in plasma physics," *American Journal of Physics,* vol. 46, pp. 1250–1254, December 1978. They wrote, "As it is generally taught, the inverse [Laplace] transform is accomplished via the Bromwich integral, and this has led to the misapprehension widely held by physicists (and many mathematicians) that inversion necessarily involves complex variables. This is not so." They went on to describe the 1930 work of Emil L. Post (best remembered

today by electrical engineers and computer scientists for the ideas that paralleled those of Alan Turing's in automata theory), later refined by D. V. Widder, for real inversions.

39. Dated April 7, 1919, this letter is quoted from N. W. McLachlan, "Historical note on Heaviside's operational method," *Mathematical Gazette,* vol. 22, pp. 255–260, June 1938.
40. I can only guess at what Heaviside was alluding to, but perhaps he had in mind Pompey's words in Shakespeare's *Antony and Cleopatra* (Act II, Scene I): "If the great gods be just, they shall assist the deeds of justest men."
41. In a Note appended to a paper by H. S. Carlslaw ("Operational methods in mathematical physics," *Mathematical Gazette,* vol. 14, pp. 216–228, 1928–29), Bromwich mentioned a letter he received from Rayleigh: "Lord Rayleigh expressed considerable interest in this application [evaluating integrals by operational means] of Heaviside's methods; and he took the opportunity to state that (in his opinion) these methods had not received adequate recognition; he also regretted that he had not been able himself to do more to support Heaviside's claims, in view of the fact that the operational method had been severely criticised from the side of Pure Mathematics."
42. *Operational Methods in Mathematical Physics,* Tract No. 23, Cambridge: Cambridge University Press, 1927.
43. In a Note appended to the paper by Carlslaw (Note 41).
44. Hardy (Note 33) wrote "Bromwich was very popular. He had a number of interests outside his work, being an active lawn tennis player, fond of music, and an accomplished dancer. He was sensible and kindly, and always willing to take trouble to oblige his friends or pupils; and whatever he did was done in the most business-like way imaginable." To get a flavor of Bromwich's personality, see his technical, yet very funny, essay "Easy mathematics and lawn tennis," reprinted in *The World of Mathematics,* vol. 4, J. R. Newman, Ed., New York, NY: Simon and Schuster, 1956, pp. 2450–2454. Heaviside, psychologically, could never have written such an essay (his humor was at its best when directed against someone who had displeased him, and was not lighthearted for its own sake).
45. EMT 3, p. 86.
46. "On the Heaviside operational calculus," *Proceedings of the Cambridge Philosophical Society,* vol. 36, pp. 267–282, July 1940.
47. *Philosophical Magazine,* Series 5, vol. 48, pp. 309–312, September 1899.
48. *Beiblätter zu den Annalen der Physik,* vol. 25, pp. 823–846, 1901.
49. "Note on surges of voltage and current in transmission lines," *Proceedings of the Royal Society,* Series A, vol. 156, pp. 1–5, August 1936.
50. EMT 2, pp. 312–315.
51. "Submarine cable problems solved by contour integration," *Mathematical Gazette,* vol. 22, pp. 37–41, February 1938.
52. Heaviside had a fascination for this integral. As he wrote (EMT 3, p. 237), "Of all definite integrals I admire it the most, because of its leading to so many others so easily; not only ordinary integrals, but those of the Fourier and Bessel theorems, and elliptic functions, and all sorts of things ... ."
53. EMT 2, p. 459.
54. Dated October 5, 1909. I thank the Department of Rare Books and Special Collections at the Rush Rhees Library of the University of Rochester for permission to quote from this letter.
55. EMT 2, p. 55.
56. An interesting discussion of this aspect of Heaviside's mathematics is given by W. E. Sumpner, "Impulse functions," *Philosophical Magazine,* Series 7, vol. 11, pp. 345–368, February 1931.
57. *Electric Circuit Theory and the Operational Calculus,* New York, NY: McGraw-Hill, 1926, p. 59.
58. EMT 2, p. 296.
59. Quoted from a longer excerpt in M. Kline, *Mathematics: The Loss of Certainty,* Oxford: Oxford University Press, 1980, p. 170.
60. In an entirely different context McLachlan (Note 39) wrote of $1/(1 + ap^{1/2})$, "The contour integral method yields $1 - e^{t/a^2}(1 - \mathrm{erf}\sqrt{t/a^2})$, but the series for this would not be easily recognized." Tech Note 4 shows, I believe, how true that statement is. It is curious to note that while Heaviside arrived at this operator by studying an infinitely long telegraph circuit, McLachlan found the same mathematical operator also describes the "elasto-viscosity of flour-dough"!
61. EMT 3, p. 287.
62. H. B. Becher, "William Whewell and Cambridge mathematics," *Historical Studies in the Physical Sciences,* vol. 11, part 1, pp. 1–48, 1980. This essay discusses the nature of the *Mathematical Tripos,* and how it evolved through the 19th century.
63. EMT 2, p. 119.
64. EP 2, p. 445. For more on how Oliver used (and abused) power series, see the very nice paper by S. S. Petrova, "Heaviside and the development of the symbolic calculus," *Archive for History of Exact Sciences,* vol. 37 (Part 1), pp. 1–23, 1987.

11
The Age-of-the-Earth Controversy

I am greatly troubled at the short duration of the world according to Sir W. Thomson, for I require for my theoretical views a very long period before *the Cambrian formation.*

— Charles Darwin, 1869

Dwell for one moment on the sublime spectacle of a tide 648 feet high, and see what an agent it would be for the performance of geological work!

— Robert Ball, Royal Astronomer of Ireland, writing in *Nature*, 1881

Assuming the present rate of radiation... the age of the sun does not exceed 18,000,000 years. The earth, of course, is less aged.

— Simon Newcomb, Professor of Mathematics, The Johns Hopkins University, 1887

The age assigned to the sun by Helmholtz and Kelvin communicated a shock from which geologists have never recovered.

— Clarence King, writing in *The American Journal of Science,* 1893

A Royal Theological and Geological Society would probably tend to inanity, if not to profanity....

— *The Electrician,* 1896

The age of the earth as an abode fitted for life is certainly a subject which largely interests mankind in general. For geology it is of vital and fundamental importance—as important as the date of the battle of Hastings is for English history.

— Lord Kelvin, during his 1897 address to the Victoria Institute

Historical Origin of the Debate

Where did the universe come from? In earlier times the answer seemed absolutely clear-cut. God made everything. And for all anyone knows this may be correct in some deep, not yet perfectly understood sense. I will not be so bold, presumptuous, or foolish as to tackle this question, one that will no doubt have scholarly theologians debating a century from now.

One can, however, examine with profit the somewhat less majestic but related issue of the age of the Earth, a question full of geological importance and one, history has shown, containing as much human drama as one could possibly expect from a subject not involving either high treason or passion. Human curiosity over this question has burned for centuries. Surely, for example, the first living creature did not arrive on the scene until after that singular event, the formation of the Earth. Curiously enough, one of the contributors to the 19th century debate over this issue (when it was a particularly hot topic) was Oliver Heaviside. Let me now set the stage for this debate and Heaviside's role in it.

Early Christian theologians, reading the Bible as an historical document to be interpreted as the literal truth, as opposed to a literary masterpiece teaching lessons of moral behavior in the form of allegory, exhibited varying degrees of precision in their determinations of the age of the Earth. Martin Luther argued for 4000 BC as roughly the date of Creation (of everything). Johannes Kepler adjusted this to 4004 BC after stating he'd discovered four years previously overlooked in the chronology of the Christian era. This was fine-tuned in 1650 and 1654 by the Calvinist James Ussher, Archbishop of Armagh and Primate of All Ireland, who declared[1] after years of monumental study of the Hebrew, Greek, Syriac, Ethiopic, and Aramaic versions of the Scriptures "... that from the evening ushering in the first day of the world, to that midnight which began that first day of the Christian era, there were 4003 years, seventy days, and six temporarie howers." He further asserted that Man was created on the sixth day, which was Friday, October 28.

Despite this impressive specificity by the Archbishop, there was never any consensus on the date of Creation. Earlier in 1642, for example, Dr. John Lightfoot, Master of St. Catherine's College and Vice-Chancellor of the University of Cambridge, had narrowed[2] the instant of the Beginning down to the very hours: "... Man was created by the *Trinity* about the third hour of the day, or nine of the clocke in the morning on 23 October 4004 BC."

The proliferation of biblical analyses became so embarrassing that as early as 1646 Sir Thomas Browne, author and physician, publicly despaired and declared[3] the calculation of the Date to be "... beyond the Arithmetick of any but God himself." By the early 19th century more than 120 such analyses were available, offering dates ranging over the depressingly wide interval of 3616 to 6984 BC. Their one point of agreement seemed to be that the Earth had been created essentially as we see it today, only a few thousand years in the past.[4]

The belief that the date of Creation could be calculated from biblical sources, if only you were clever enough to discover the way, was obviously a powerful one, a belief seductive enough to fascinate the imaginations of more than a few brilliant scholars. We may smile with amused indulgence today[5] at such an energetic waste of intellectual talent, but these men were not buffoons. They were searching for knowledge, then as now an admirable goal. Archbishop Ussher, for example, as Knox[6] pointed out, "had a great sense of the sweep of history," and "He believed ... it was necessary to have a reliable chronological structure; this was vital not only for seeing the sequence in the events of Jewish and Christian history, but also for grasping their relation to other peoples and cultures." Hence his years-long quest to calculate the Date. *Genesis*, considered a divinely inspired document written by the hand of Moses (but guided by God, Himself), also enjoyed the support of laws promising a variety of unpleasant earthly punishments for anyone so foolish as to deny the inherent truth of the Scriptures.

The Problem of the Fossils

As Burchfield[7] has pointed out in his admirable book, these biblically based speculations began in the West with the birth of Christianity and continued until nearly the second half of the

19th century. Indeed, only a little more than a century after Christ we find Teophilus of Antioch dating Creation at 5529 BC, a value not much altered, as we've seen, over the centuries that followed. Nearly two thousand years later, however, it was becoming increasingly difficult to ignore the evidence of the fossils of *extinct* animals and their mute testimony that the Earth must have required considerably more than just a few thousand years for its history. Darwin's ideas, as expressed in his 1859 publication of *Origin of Species*, seemed to demand *hundreds of millions* of years, a change in scale so stupendous as to stupify the biblical scholars. Darwin himself was awed by such vast time durations, and several years after the publication of *Origin* temporarily wavered by stating[8] "I now first begin to see what a million means, and I feel quite ashamed of myself at the silly way in which I have spoken of millions of years." The biblical scholars were, of course, scandalized at the obvious implication that *Genesis* must be in error, and that the calculations of Ussher (and those of his number-crunching compatriots) weren't worth the paper they were written on.

The curiosity aroused by fossils has a long history and can be traced back to Xenophanes of Colophon, five hundred years before Christ (as a rejector of anthropomorphic gods, Xenophanes would have denied that "God made man in His own image"; his sin against *Genesis* would thus have been twofold). Until medieval times it was generally believed that the resemblance of these strange artifacts to animal and plant remains was merely a superficial one, and that they were created by one or more mysterious, possibly supernatural, forces. Some mockingly dismissed them as "sports of nature." Imaginative speculation was not unheard of, either. To explain the bones of such massive land animals as the dinosaurs, while still preserving the short history of *Genesis*, it was occasionally suggested, for example, that they might be the remains of Hannibal's elephants. That these artifacts often appeared to be those of sea creatures, and yet were found on mountaintops and deserts, had earlier either been ignored, or explained away in terms of the disruption caused by such supposed worldwide cataclysmic events as Noah's Flood.

But this slavish devotion to *Genesis* was rapidly crumbling by the end of the 18th century. As Cannon[9] wrote, "When, in his *Principles of Geology* of 1830-1833, Charles Lyell revived and extended the geological ideas of William Hutton [a Scottish gentleman-farmer and physician whose 1795 book *Theory of the Earth*, based on years of field work, had put forth such modern ideas as continental elevation, underground heat, the denial of such events as a universal deluge, and the existence of laws of geology as opposed to mysterious forces] he produced almost at once a division of English geologists into 'Uniformitarians' and 'Catastrophists'...."

Lyell declared his belief in the uniformity of the nature of geological forces (rain, rivers, winds, and so on), not only in kind but in intensity, over the entire history of the Earth. Darwin owed a great debt to Lyell's Uniformitarianism because it provided the enormous time durations that his own theory of biological evolution required. And yet, as pointed out by Cannon, Uniformitarianism is, itself, an *anti*-evolutionary theory of geology, since it postulates a never-ending cycle of continent-raising and continent-eroding, with no *cumulative change* over time.

The Catastrophists, on the other hand, believed the Earth had evolved through the action of a series of enormously violent events, separated by periods of relative tranquility. They delighted in the Uniformitarianist embarrassment over the undeniable youth and rugged majesty of the Andes and the Alps which proved, they claimed, that clearly something more than the slow chisel of lazy time had been at work. They even attempted to use the *existence* of the fossils to support their belief in a superviolent past, and eagerly pointed to the discontinuous transistions between strata as proof that nonuniform forces had formed the planet.

Early Catastrophists required stupendous spasms (e.g., giant tsunami able to traverse entire continents) in order to cram the obviously complex topographical development of the Earth into the six thousand years or so allotted by the Bible, but by the 1850s geology could be said to finally be free of Moses. According to Davies,[10] in 1851 Sir Henry De La Beche (Director of the Geological Survey of Great Britain) dismissed Noah's Flood as "that funny story."

The evidence (fossils and immensely thick layers of sedimentary rock) for an almost incomprehensibly old Earth had become so overwhelming that Lyell's Uniformitarianism emerged as the accepted geological theory. It was at this seemingly happy time, with the development of a scientifically based theory of the Earth's history, that Professor William Thomson tossed a pie in the faces of Victorian geologists. Asserting[11] Uniformitarianism was nothing less than believing the Earth to be a perpetual motion machine ("It is quite certain that a great mistake has been made—that British popular geology at the present time is in direct opposition to the principles of Natural Philosophy."), Thomson launched a devastating mathematical attack.

Kelvin's Theory

In 1862 Thomson wrote[12] his famous paper "On the secular cooling of the Earth." In this far-reaching analysis he imagined the Earth, in its Beginning, to have been a uniformly heated ball at a temperature of 7000°F (3900°C), i.e., that of molten rock. Then, assuming a constant value of conductivity based on an experimental study of rock, garden sand, and sandstone from an Edinburgh quarry, he mathematically allowed this blazing globe to cool according to the physical laws of thermodynamics embedded in Fourier's famous partial differential equation for heat flow.[13] From this Thomson was then easily able to calculate the time lapse from the initial instant of cooling until the now observed temperature gradient at the Earth's surface (1°F/50 feet). This value (98 million years) would put an upper bound on the age of the Earth's crust and thus, indirectly, on the first instant at which life could possibly have appeared. The elusiveness of the actual date of the appearance of surface solidity continually tantalized Thomson, as he indicated decades later in an 1894 letter to his former assistant John Perry when he wrote, "I would rather know the date of the *Consistentior Status* [a term originated a century and a half earlier by Gottfried Leibnitz] than of the Norman Conquest."

An age of nearly a hundred million years is a long time, to be sure, but not nearly enough for Darwin. So impeccable was Thomson's analysis, however, that Darwin despaired and called[14] Thomson an "odious spectre" and, as Eiseley wrote,[15] he surely must have thought of Thomson's mathematics as "Satanic."

Thomson's paper was actually the second of a two-punch attack on Uniformitarianism, as earlier in the same year he had written[16] "On the age of the sun's heat." In that paper Thomson speculated on the source of the Sun's energy. Postulating the Sun had come into existence via the gravitational amalgamation of a vast number of colliding meteors (which thus became an intensely hot, molten mass), he calculated the original supply of solar energy. Then, knowing the present rate of energy loss, he was able to determine, assuming a cooling Sun, the age of the Sun. The result, probably 100 million years and certainly no more than 500 million years, obviously put an upper bound on the antiquity of life on Earth. This range of time was also out of joint with the enormous spans required by the Uniformitarians and Darwin. Disconcerting, too, to those who worry about the future in an abstract way, was the conclusion that a cooling Sun, powered by gravitational contraction, could continue to shine for only a few million years more. So dismal were this and other similar Thomson conclusions that Henry Adams wrote, "this young man... thus tossed the universe into the ash-heap."[17]

Both papers were based, of course, on faulty premises.[18] We know today of radioactive decay within the Earth that maintains the thermal gradient without a continual cooling of the planet,[19] and of the thermonuclear reactions deep within the core of the Sun.[20] But all of this was decades in the future, and Thomson's seemingly invulnerable mathematics (as well as his formidable reputation as one of the foremost scientific thinkers of the day) had a chilling effect on debate. It seemed as though there could be no possible rebuttal, and that geology was destined to pass through yet another revisionist phase, and that Darwin's theory must indeed be absurd.[21]

It is very important to realize that Thomson did *not* attack the prevailing theory of geology (and Darwin's theory) merely because he supported a strict biblical interpretation of human history. Kelvin was, indeed, a religious man, and he *did* believe evolution was not sufficiently cognizant of the influence of a "continually guiding and controlling intelligence."[22] But the real origin of his position was, as Sharlin[23] has pointed out,

> He had at first disapproved of the theory because it was contrary to his belief of continuity [Note 24] in the universe. But his main argument against evolution through natural selection was that the theory contradicted well-founded theories of physics.

Perry's Rebuttal of Kelvin's Theory

The confused state between Thomson's mathematics and what geologists could see with their eyes remained in effect for the next thirty-three years, until the publication[25] of a remarkable article in *Nature*. Written by John Perry, once Kelvin's assistant at Glasgow University and now a professor (of mechanical engineering) himself at Finsbury Technical College, it was a cautious yet forceful reexamination of Kelvin's cooling-Earth analysis of 1862. As Perry wrote at the start,

> I have *usually* [my emphasis] said that it is hopeless to expect Lord Kelvin should have made an error in calculation.

Then, just a few sentences later,

> But the best authorities in geology and palaeontology are satisfied with evidences in their sciences of a much greater age than the one hundred million years stated by Lord Kelvin; and if they are right, there must be something wrong in Lord Kelvin's *conditions* [my emphasis].

It is important to note that Perry was setting the stage for directing critical remarks not at Kelvin's Fourier modeling of a cooling Earth, but only at the parameters in that model. Perry even agreed with Kelvin's rejection of Uniformitarianism, writing,[25] "Lord Kelvin completely destroyed the Uniformitarian geologists, and not one now exists. *It was an excellent thing to do* [my emphasis]. They are as extinct as the dodo or the great auk." Still, despite this, and even though Kelvin himself admitted that knowledge of the Earth's internal structure was skimpy at best, Perry quickly found himself under counterattack from Professor Peter Guthrie Tait, Kelvin's close friend and professional colleague.

Tait was, in fact, an old hand at the task of refuting all who questioned Kelvin's age-of-the-Earth analysis. Tait felt himself to be something like Kelvin's intellectual bodyguard, and his relationship with him was perhaps best expressed in an 1871 letter[26] written by the biologist T. H. Huxley:

> Tait worships him with the fidelity of a large dog—which noble beast he much resembles in other ways.

John Perry (1850–1920)

Perry was totally unafraid to speak his mind, and didn't give a hoot about what the possible consequences might be. In March 1902, for example, he wrote one of the first "conservation" letters to *Nature* on "The misuse of coal," warning of the possibility of running out of the fuel. He blamed this on the inefficiency of the steam engine. As a solution he advocated research into the *direct* conversion of coal to electricity. Perry used this letter to demonstrate that his worshipful feelings toward Kelvin had not been changed by the age-of-the-Earth debate. He suggested Lord Kelvin be "entrusted with the expenditure of a million a year for two or three years" so that such an invention would surely be achieved! The response was either to ignore Perry or to pooh-pooh him as an alarmist (as did *The Electrician*).

This was a particularly ironic metaphor for Huxley to have used, as he himself has become best known as "Darwin's Bulldog" for his energetic defense of evolution! Huxley had reason for his jab at Tait, as in 1869 Tait had directly challenged[27] Huxley's critical remarks concerning Kelvin's (William Thomson, then) theory, remarks delivered[28] during Huxley's Anniversary Address as President of the Geological Society of London. Tait's essay was written so powerfully that even Darwin wrote[29] it was "admirably done."

Recalling Tait's previous battles with Heaviside, it is not likely that he was favorably impressed when Perry started his 1895 article with the words:

> Some long mathematical notes... to prove the legitimacy of my approximate method of calculation, are now omitted, as Mr. Heaviside has given exact solutions, and has found that there is practically no difference between mine and the exact numerical answers. That Mr.

A portrait of Faraday, Huxley, Wheatstone (third from left), Brewster, and Tyndall. There is some doubt this is a true group photo, but rather a clever composite (see E. C. Watson, *American Journal of Physics*, vol. 23, p. 157, March 1955).

> Heaviside should have been able, in his letters to me during eleven days, to work out so many problems, all seemingly beyond the highest mathematical analysis, is surely a triumph for his new methods of working [the operational calculus].

High praise for Heaviside, indeed, but it all must have started Tait to salivating afresh for printer's combat.

HEAVISIDE'S ANALYSIS OF KELVIN'S PROBLEM

So now we come, at last, to just what it was that Heaviside did. Kelvin's original problem was that of imagining the Earth as a semi-infinite mass filling all space for $x \geq 0$. The Earth is taken as being at a uniform temperature V_0 at the instant of its creation, and then beginning to cool by reason of its contact with the semi-infinite space $x < 0$ (the cold vacuum of outer space) assumed to be at temperature zero. Kelvin calculated how long the cooling process would require for the temperature gradient at $x = 0$ to match the observed value (and this, of course, was his value for the age of the Earth or, more precisely, the time since the *Consistentior Status*—the "state of greater consistency," i.e., the formation of the Earth's surface crust).

Heaviside began his analysis by redoing Kelvin's calculations, but he did this by the novel approach of solving the inverse problem, and using (of course) his unconventional mathematics. He imagined the semi-infinite Earth as initially at zero temperature, and then suddenly experiencing an elevation of the surface temperature to V_0. The temperature gradient at the surface is initially infinite, but as the Earth heats up, this gradient tends toward zero. Heaviside found the time interval required for the surface gradient to reach the observed value (with reversed algebraic sign, of course). He then argued that if the Earth were allowed to cool ("subside," in his word) from the final state back toward zero (Kelvin's problem), it would take the same time to reach the observed surface gradient (with opposite algebraic sign). He showed this by starting with the one-dimensional heat equation which, it is interesting to note, is exactly the same equation Kelvin had studied forty years before, in 1855, for the electrical behavior of a perfectly insulated, very long induction-free submarine cable:

$$\frac{\partial V}{\partial t} = D \frac{\partial^2 V}{\partial x^2}.$$

$V(x, t)$ is the temperature at distance x inside the Earth, at time $t \geq 0$, while D is the *diffusivity* of the Earth, assumed to be constant throughout the planet. Heaviside next introduced his famous time differentiation operator p, and wrote

$$\frac{\partial^2 V}{\partial x^2} = \frac{c}{k} pV$$

where c is the heat capacity and k is the thermal conductivity of the Earth.[30]

Kelvin had used this as his starting point, and in Tech Note 1 in this chapter the details of the analysis are presented in Heaviside's notation. Call the observed surface gradient g. Then the age of the Earth T is given by

$$T = \frac{V_0^2 c}{\pi k g^2}.$$

Using Kelvin's values for the thermal parameters in CGS units ($V_0 = 3900°\text{C}$, $g = 1°\text{F}$ per 50 ft $= 1°\text{C}$ per 2743 cm, and $D = k/c = 0.01178\ \text{cm}^2/\text{s}$) gives his result of $T = 98$ million years.

PERRY'S THEORY OF DISCONTINUOUS DIFFUSIVITY

Suppose now, as did Perry, that Kelvin's simplifying condition of a constant diffusivity is removed. Perry, in his simplest extension, took the Earth as having two diffusivities: one for the majority of its interior and another (smaller value) in a thin skin layer at the surface. He did this for the original infinite Earth model, and then Heaviside showed him how to do it for the more realistic case of a finite *spherical* Earth. As Heaviside wrote,[31] "Fortunately, in this case, I find that my operational methods lead straight to the solution by a simple process." However, just before presenting his solution to these difficult problems of sphericity and variable diffusivity, Heaviside teased (in his usual biting style) the mathematicians who often looked with disdain at his operators:

> But these are too complicated... and would frighten timid readers, and perhaps some Cambridge mathematicians [Note 32] as well. At the same time I may remark that the solutions can be got through the operators in the form of Fourier series with much less work than by Fourier's way.

These last words, unfortunately, are not strictly true—at least they are not unless one enjoys solving the heat equation in spherical coordinates. As was his style, Heaviside arrived at the correct answer in a hop, skip, and a jump, leaping over vast wastelands of dreary, intermediate calculations. It is, in my opinion, *very* tough going. Fortunately, we can still see how Kelvin's value for the age of the Earth can be greatly increased as a result of the "thin skin effect" even if we stick to Kelvin's simple, infinite Earth model. Tech Note 2 in this chapter shows how Heaviside's operational methods do the job. The end result is that if T is the age of the Earth, and the thermal parameters of the skin are c and k_1 (heat capacity and thermal conductivity, respectively), and those of the interior of the Earth are c and k, then

$$T=\frac{V_0^2ck}{\pi g^2k_1^2}.$$

Lord Kelvin, as part of a reply[33] to Perry published in *Nature*, provided some thermal data on a variety of rocks. Table 11.1 gives some examples in CGS units (the values for heat capacity are mine, calculated as the product of Kelvin's specific heat and density values).

The heat capacity is roughly the same for marble and quartz, but quartz has a conductivity 2.92 times greater. So, suppose the crust is marble (k_1 = 0.0054), the interior of the Earth is quartz (k = 0.01576), and c = 0.506 (the average capacity) everywhere. The age of the Earth is then T = 315 million years, more than triple the age obtained from Kelvin's calculation. Using still other, quite reasonable values for the thermal parameters, one can get values for T in the *billions* of years.

Kelvin's Defense and Perry's Reply

In his note, Kelvin attempted to refute the *details* of Perry's assumed thermal values. Even more interesting is Kelvin's quoting of words favorable to his position, words taken from an

TABLE 11.1

	Marble	Quartz
Conductivity	0.0054	0.01576
Specific heat	0.20279	0.1754
Density	2.7036	2.638
Capacity	0.548	0.463

article[34] by the American geologist Clarence King. Here we have the unusual case of Lord Kelvin attempting to defend himself by quoting "authority"! Perry, unimpressed (he dismissed King's paper as "somewhat inconclusive") replied,[35]

> I showed that, if we assume greater conductivity in the interior than at the surface, we increase this limit of age. I took a number of examples, which could be worked mathematically. I did not pretend that any one of these represented the actual state of the earth. They merely proved that there were *possible* internal conditions which might give enormously greater ages than physicists had been inclined to allow. Of my various results, I did not give one as more correct than another, although some may have seemed more probable than others. It was not my object to obtain a correct estimate. Indeed I tried to show that it was impossible for a physicist to obtain such an estimate, as there were all kinds of possible assumptions which led to many different answers.

The End of the Debate

It is clear that Perry was pleased to be able to quote Heaviside as an authority who agreed with his theoretical arguments neutralizing Kelvin's conclusions concerning the age of the Earth. What Perry had done, as we have seen, was to demonstrate that if one retained Kelvin's cooling Earth model, but replaced the assumption of constant diffusivity with one that increased with temperature (depth), then one could very easily arrive at an age many times greater than Kelvin's. Perry was bold enough, in fact, to actually mention[25] a value of nearly 29 *billion* years(!), but quickly tempered that with the words: "I do not know that this speculation is worth much, except to illustrate in another way the augmented answer when we have higher conductivity inside."

Perry's act of independence of thought was an audacious one. He felt uncomfortable doing it and possibly without Heaviside's mathematical support he would have been much more reluctant to pursue the matter. Indeed, when Tait attacked him, not by rebutting his logic or mathematics but by questioning what right Perry had to modify Kelvin's assumption of constant diffusivity, Perry felt obliged to reply[25] with the astonishing words:

> Some of my friends have blamed me severely for not publishing... sooner. I was Lord Kelvin's pupil, and am still his affectionate pupil. He has been uniformly kind to me, and there have been times he must have found this difficult. One thing has not yet happened: I have not yet received the thirty pieces of silver.

Perry clearly felt that Tait looked upon him as a traitor to his old mentor, and that troubled Perry deeply.

For his own part, Kelvin let Tait do most of the talking for him, and limited himself to characterizing the entire debate as really just a quibble over minor mathematical details. Perry was always frustrated by this, feeling Kelvin did not take him seriously. Perry would surely have welcomed a more vigorous exchange with Kelvin and could have done without Tait's complaint[25] about his "...entire failure to catch the *object* of your paper" and "...I seem to have entirely missed your point." This so irritated Perry that he bluntly replied,[25] "What troubles me is that I cannot see one bit that you have reason on your side...."

Although Tait may have missed the point (something I find rather hard to believe), others who followed the debate on the pages of *Nature* did not. Perry's work had planted the seeds of doubt, and the thirty-three–year grip of Kelvin's analysis on the throats of geology and evolution had been loosened. Soon it would be cast aside as no longer relevant[36] to the business of either geology or Darwin and given back to the pure mathematicians who could admire technical beauty for its own sake.

The principal actors in this episode soon moved on to other concerns, and Heaviside dropped the matter after incorporating his analyses for Perry on the age of the Earth as the lead article in volume 2 of *Electromagnetic Theory*. A discussion on the age of the Earth in a book otherwise devoted to Fourier integrals, electric wave propagation along cables, and travel at speeds faster than that of light may seem odd, and indeed it struck many readers in just that manner. As one book reviewer wrote,[37]

> As before, Mr. Heaviside has plenty of spice for those who do not treat his theories and discoveries with the respect to which they are entitled.... The volume begins with... "Mathematics and the Age of the Earth." This title does not appear exactly germane to the subject of the book, but its introduction is justified on the ground that the solution of the problem of radiation of heat from the earth leads to the solution of similar problems in electrical theory.... A better title... taking the work as a whole, would be "What I Know about Mathematics, and what I think about Mathematical Purists...."

Another remark in the same review tells us quite a lot about Heaviside's image in those days: "...we do not think the lustre of his discoveries would be at all dimmed were he less proficient in the art of scolding."

An Assessment of the Debate

I do not believe that Heaviside himself attached much value to the issue of the Earth's age, *except as a vehicle to display the ability of his operator method to solve problems of astonishing complexity*. The fact that others, especially a "name" like Lord Kelvin, *were* interested in the issue and would, therefore, attract public attention to his calculations, was all that Heaviside cared about. I do not believe, in fact, that Heaviside really had very much of his heart and soul tied to even the *results* of his calculations. As he himself wrote,[38] concerning the meaning of operational methods when applied to this problem:

> ...to illustrate Lord Kelvin's theory of the age of the earth and its recent extension by Professor Perry, the practical import... remains to be discovered, as very uncertain and speculative data are involved.

Would any careful, cautious scientist have written any differently? Perhaps not, knowing his words were for public study in a scholarly manuscript. But what of Heaviside's private thoughts—there, it would seem, we could learn what he *really* believed. From one of his letters I believe it is clear he accepted the theory of evolution (and thus the extreme antiquity of the Earth). His remarks appear as humorous ones, almost throwaway, but that may perhaps enhance the probability that they reflect his true feelings. The letter (dated July 12, 1897) was written to his dear friend, Professor George Francis FitzGerald at Trinity College, Dublin, Ireland. At that time (age forty-seven), Heaviside had been living in Newton Abbot for several months, the first time in his adult life he had lived alone. Never a gregarious man, he now was almost a hermit and had extreme difficulty forming normal relationships with most people. But animals were different:

> A little bird has made friends with me. Not a sparrow nor a robin, but something like a robin. Follows me about, and comes indoors and wanders about; especially present at meal time, when he comes in at the window and watches with interest what is going on, and picks up the crumbs that fall from the rich man's table. But he won't come nearer than a yard. *He* knows what horrid creatures men are. But I wish he was more regular and less frequent in some of his operations. He makes his mark. It is dreadful to think that our ancestors perhaps

> went about doing the same. Evolution is most shocking in its early stages. Wonderful how things have worked out! If it wasn't true, no one could believe it....

There is a more serious piece of evidence that provides some insight into Heaviside's interest in the business of the age of the Earth. Before appearing in EMT 2 in 1899, his analyses had, of course, first been published in the pages of *The Electrician*. Apparently the anti-biblical implications of the mathematical results had caused some concern at the journal because on December 4, 1894 the editor (A. P. Trotter) wrote to Oliver:

> The slight delay in sending out your proof is due to my having heard from my assistant, who received your MS and passed it on to the printers, that there were references in it to religion, and to the origin of life, and other matters outside the field of *The Electrician*. I have read your proof with great interest. I could not allow any other contributor to discuss these matters, important though they are, in *The Electrician*, but I am willing to make an exception in your case.... Will you compromise matters by altering the word "religion" into "system of ethics"? ...I do not think much of a religion which is nothing but a system of ethics, and I do not suggest that the ideas are identical. But for the purpose of your article, I think you will admit that "system of ethics" will do, and we shall thus avoid giving offence to those people who are so jealous of what they call religion and who for the most part do not even know what a system of ethics is.

In addition, Trotter expressed serious reservation about Heaviside's final paragraph, in which Oliver had written:

> As for the origin of life upon this planet, the only reasonable view seems to me to be Topsy's [Note 39] theory. She was a true philosopher, and "she spekt she growed". Any other theory is of the elephant and tortoise kind, a kind of evasion, which explains nothing.... I firmly believe (subject to correction)... of life in so-called dead matter under the influence of the forces of nature.

Trotter's concerns weren't Heaviside's concerns, and he wasn't ready to admit any changes at all "would do." Topsy and "religion" would stay in, while "system of ethics" would stay out. The very next day Heaviside wrote back to say, somewhat undiplomatically,

> I will not *condescend* to notice any pious theories of the earth's age.... I am a little surprised at your being afraid of the parsons. They don't read *The Electrician*.

Trotter had tried to explain the practical consequences: "On one occasion a religious matter was alluded to very indirectly in a Note, and it drew enough correspondence to fill six or seven volumes. You see how difficult it would be for me to refuse to publish... clerical response...." Heaviside, of course, wouldn't see it at all.

In a review of EMT 2 Searle couldn't keep *his* strong religious views from showing. He wrote,[40]

> We are brought by [Heaviside's] mathematical analysis to the irresistible conclusion that, within a finite time, there has been in operation some Agency of which a purely dynamical theory is unable to offer an account.

And then he went on to quote from a recent Kelvin paper,[41]

> Mathematics and dynamics fail us when we contemplate the earth fitted for life, but lifeless, and try to imagine the commencement of life upon it. This certainly did not take place by any action of chemistry, or electricity, or crystalline grouping of molecules under the influence

> of force, or by any possible kind of fortuitous concourse of atoms. We must pause, face to face with the mystery and miracle of the creation of living creatures.

I believe this to be a serious misinterpretation of Heaviside's position; to quote his own words (written in 1912),[42]

> There may be and probably is no ultimate distinction between the living and the dead.

A Final Word

A little later in his letter to Trotter Heaviside wrote,

> As regards my sending you this article at all, it arises from the circumstances of the moment; Perry's work, which he won't publish [but he did, a month later], and my work, which I can't get published (at least by the R. Soc.). So it is good both ways, as Perry took to my operators [Note 43] at once, and I want to show that they have practical value, and not be snuffed out.

It was, apparently, not a secret that Perry was an expert in the manipulation of operators. Many of those who were first introduced to operators by Heaviside's presentation in *The Electrician* were so confused they privately consulted Perry for help! Heaviside's approach in *The Electrician* was the experimental development of $p^{1/2}$, whereas Perry, who of course had read Heaviside's 1893 Royal Society papers, preferred the gamma function approach that Heaviside had used there. Eventually Perry felt compelled to write a letter[44] to *The Electrician* explaining this idea, but he did so fully aware of Heaviside's potentially prickly reaction. Entitled "Mr. Heaviside's operators," it began on a cautious note:

> I know that Mr. Heaviside will probably be a little annoyed at my interference; but I am only going to do publicly what I have done several times privately, and what I believe necessary. Anyhow, I have been greatly thanked for my private services.

Then, after quickly and clearly showing the gamma function approach, he continued

> Of course there are many (to me) obscure points about this subject of operators not touched upon in Mr. Heaviside's Papers read before the Royal Society. I leave all such matters for the mathematicians to speculate about. I rest satisfied with this, that it is quite easy to use these simpler kinds of Heaviside's operators, and I can always test whether my answer is right or not.

And then he finished with

> Mr. Heaviside would, of course, say that all this, where it is not misleading, it is unnecessary, either for men who have read his Royal Society Papers or his previous Papers in *The Electrician*. Well, I know as a fact that some men who are fairly good mathematicians have been glad to get the above explanations. I also know that a number of men have made up their minds to read Mr. Heaviside's Papers more diligently in the future than they have in the past. If he is angry at my writing this letter I shall be very sorry, but I know that the letter ought to be published.

Perry's fears were well-founded. Heaviside's testy reply[45] appeared the very next week in a letter entitled "To Prof. Perry, and Others":

> I must ask Prof. Perry to kindly exercise his forbearance for a time. I have reason to believe

> that I have other readers who have not so much experience with operators as Prof. Perry. For their sake I do not think it desirable to complicate the subject at an early stage by introducing such advanced mathematical notions as the gamma function and generalized differentiations.

Considering the very advanced mathematical level of all the rest of Heaviside's writings in *The Electrician* this was, of course, a ridiculous objection. But Perry did not reply and there the matter dropped.

As a final comment on the Kelvin–Perry debate, Harold Jeffreys (the Cambridge mathematical physicist who often wrote sympathetically about Heaviside's operational methods) many years later published a very good book[46] summarizing the then latest ideas on the age of the Earth. Nowhere does Heaviside's name appear, with a single ironic exception. When discussing the mathematical modeling of the radioactive decay of uranium, Jeffreys arrived at a rather formidable set of differential equations. In a footnote he told his readers: "The solution, and approximations to it, are most easily obtained by Heaviside's method."

No doubt this passing comment, in a book that otherwise ignored his "age-of-the-Earth" work, would have amused Heaviside, a man whose often cynical outlook on life would thus have been justified, at least to him, once again.

Tech Note 1: Heaviside's Operator Solution of Kelvin's Original One-Dimensional Problem

Beginning with the diffusion equation

$$\frac{\partial V}{\partial t}=D\frac{\partial^2 V}{\partial x^2}, \qquad D=\frac{k}{c}$$

Heaviside introduced his time differentiation operator p and wrote

$$\frac{\partial^2 V}{\partial x^2}=\frac{c}{k}pV.$$

Recall that $V(x, t)$ is the temperature at time t at distance x inside the Earth ($x = 0$ is the surface). Heaviside then immediately wrote the general solution, as is easily verified by direct substitution, as

$$V=Fe^{(cp/k)^{1/2}x}+Ge^{-(cp/k)^{1/2}x}$$

where F and G are determined by the boundary conditions. Because V must be bounded for all $x \geq 0$, it must be true that $F = 0$ and $G = V_0$ (using the boundary condition at the surface), and so,

$$V=V_0e^{-(cp/k)^{1/2}x}.$$

The temperature gradient at the surface is just

$$\left(-\frac{dV}{dx}\right)_{x=0}=V_0\left(\frac{cp}{k}\right)^{1/2}=V_0p^{1/2}\left(\frac{c}{k}\right)^{1/2}.$$

Recalling Heaviside's result for the *fractional* operator $p^{1/2}$ from the last chapter, this becomes

$$\left(-\frac{dV}{dx}\right)_{x=0}=V_0\left(\frac{\pi kt}{c}\right)^{-1/2}.$$

By setting this equal to the observed surface gradient (g) and letting $t = T$ (the age of the Earth) the equation given in the text results:

$$T=\frac{V_0^2 c}{\pi k g^2}.$$

Tech Note 2: Heaviside's Operator Solution of Perry's Problem of Discontinuous Diffusivity

Following Perry's lead, Heaviside began by thinking of a thin skin (or crust) of depth ℓ forming the surface of the Earth. He denoted the thermal parameters of this skin by c and k_1 (heat capacity and thermal conductivity, respectively), and those of the interior of the Earth by c and k. Note that the heat capacity is everywhere the same, but since the conductivity can be different in the skin and the interior we can have a discontinuous diffusivity. Heaviside called the temperature at the outer surface of the skin V_0, and that at the inner surface V_1. Imagining an Earth initially at temperature zero, with a sudden elevation of the surface temperature to V_0, he set the heat flux crossing the outer surface equal to the flux crossing the inner surface, on its way into the interior of the Earth:

$$k_1=\frac{V_0-V_1}{\ell}=-k\frac{dV_1}{dx}=kqV_1$$

where $q = -d/dx$, i.e., q is the negative of the *space* differentiation operator. Heaviside was able to express q in terms of p by writing the fundamental heat equation as

$$q^2V=\frac{c}{k}pV$$

and then

$$q=\left(\frac{c}{k}\right)^{1/2}p^{1/2}.$$

Heaviside then defined $R = \ell/k_1$ (he called it the "resistivity" of the skin), substituted it, along with the last expression for the q operator, into the heat flux equation, and arrived at

$$V_i=\frac{V_0}{1+R(ck)^{1/2}p^{1/2}}.$$

This, no doubt, was even more mystifying to most mortals than was $p^{1/2}$, but not to Heaviside—he just went straight ahead and expanded the right-hand side using the power series expansion

$$\frac{1}{1+x}=1-x+x^2-x^3+\cdots, \qquad |x|<1.$$

To do this, he had to treat $p^{1/2}$ as a number, but that was a sin already long since committed. So,

$$V_1=[1-R(ck)^{1/2}p^{1/2}+R^2(ck)p-R^3(ck)^{3/2}p^{3/2}+\cdots]V_0.$$

Since $p^{3/2}\cdot \mathbf{1} = pp^{1/2}\cdot \mathbf{1} = d/dt\,[(\pi t)^{-1/2}] = -(1/2)\,(\pi)^{-1/2}\,t^{-3/2}$, then

$$V_1=[1-R(ck)^{1/2}(\pi t)^{-1/2}+2R^3(ck)^{3/2}(\pi t)^{-3/2}-\cdots]V_0$$

or

$$V_1 - V_0 = \left[-\frac{\ell}{k_1}\left(\frac{ck}{\pi t}\right)^{1/2} + 2\left(\frac{\ell}{k_1}\right)^3\left(\frac{ck}{\pi t}\right)^{3/2} - \cdots \right] V_0.$$

Since the surface gradient is the temperature drop across the skin ($V_0 - V_1$) divided by the skin depth, we get (from the "thin skin" assumption of $\ell \cong 0$, and the fact that we are interested in a value of $t \gg 0$, then only the first term is significant):

$$g = \frac{V_0}{k_1}\left(\frac{ck}{\pi t}\right)^{1/2}.$$

If g is the observed value, then we can solve for T, the age of the Earth, and find

$$T = \frac{V_0^2 ck}{\pi g^2 k_1^2}.$$

Even before Heaviside had solved this and related problems with absolute mathematical precision, Perry had worked out approximate solutions by himself. Heaviside found that Perry had done quite well, in fact, coming very close to the exact results. So astonished was he at this that Heaviside wrote in one of his notebooks,[47] "Is he a witch, or is this one of the most remarkable coincidences in ancient or modern history?"

Notes and References

1. G. Daniel, *A Short History of Archaeology,* New York, NY: Thames and Hudson, 1983, p. 34.
2. Ibid, p. 34. See also W. R. Brice, "Bishop Ussher, John Lightfoot and the age of creation," *Journal of Geological Education*, vol. 30, pp. 18–24, January 1982.
3. G. L. Davies, *The Earth in Decay: A History of British Geomorphology*, New York, NY: American-Elsevier, 1969, p. 13. See also D. R. Dean, "The age of the Earth controversy: Beginnings to Hutton," *Annals of Science*, vol. 38, pp. 435–456, July 1981.
4. While a few thousand years would soon prove to be a mere drop in the bucket of time required for the emerging modern theories of geology and evolution, it was already far too long for the philosophers; hence La Bruyere's lament, "All is said, we have come too late; for more than seven thousand years there have been men, thinking."
5. Today's equivalent "biblical chronology obsession" appears to be that of determining the exact date of Christ's death.
6. R. B. Knox, *James Ussher, Archbishop of Armagh*, Cardiff: University of Wales Press, 1967, p. 105.
7. J. D. Burchfield, *Lord Kelvin and the Age of the Earth*, New York, NY: Science History, 1975, p. ix.
8. In an 1868 letter reprinted in *More Letters of Charles Darwin*, vol. 2, New York, NY: D. Appleton and Co., 1903, p. 211.
9. W. F. Cannon, "The uniformitarian-catastrophist debate," *Isis*, vol. 51, pp. 38–55, March 1960.
10. Davies (Note 3), p. 208. A wonderfully romantic, nonmathematical Victorian description of the possibility of giant tides in ancient times is given in the essay by Professor Robert Ball (then Royal Astronomer of Ireland), "A glimpse through the corridors of time," *Nature*, vol. 25, pp. 79–82, November 24, 1881, and pp. 103–107, December 1, 1881. As he enthusiastically wrote, "...the mathematicians have discovered the new and stupendous tidal grinding-machine. With this powerful aid the geologists can get through their work in a reasonable period of time, and the geologists and the mathematicians may be reconciled."
11. "On geological time," in *Popular Lectures and Addresses*, vol. 2, London: Macmillan, 1894, pp. 10–72, and in particular, p. 44.
12. *Mathematical and Physical Papers,* vol. 3, Cambridge: Cambridge University Press, 1890, pp. 295–311.
13. Thomson, however, was not the originator of the idea of the Earth cooling from an initial state of elevated temperature. A century before, Georges Louis Leclerc, Comte de Buffon, a French naturalist well known to students of probability theory for his famous demonstration of how to determine *experimentally* the value of pi by tossing a needle on a regularly lined surface, had experimented with iron spheres cooling from incandescence. Using his hands as a "thermometer" to estimate cooling rates, and boldly extrapolating from spheres only inches in diameter to a globe the size of the Earth, Buffon arrived at an age of 75,000 years.
14. L. Eiseley, *Darwin's Century*, Garden City, NY: Anchor, 1961, p. 235. Even at home Darwin couldn't escape

the gloom of Thomson's mathematics. His son George, who would become a distinguished scientist in his own right (he originated the tidal theory mentioned in Ball's essay of Note 10), tried to convince him of the merits of Professor Thomson's views, to judge from an 1870 letter J. D. Hooker wrote to Darwin: "Tell George not to sit upon you with his mathematics... . Because a scarecrow of $x + y$ has been raised... you boo-boo. Take another dose of Huxley's penultimate G. S. Address [see Note 28] and send George back to college."—quoted from *More Letters of Charles Darwin* (Note 8), vol. 2, pp. 6–7. Going back to college apparently didn't change George's mind, as years later he used Thomson's theory to put forth an explanation for mountain building—see "The formation of mountains and the secular cooling of the Earth," *Nature*, vol. 19, p. 313, February 6, 1879.

15. L. Eiseley, *Darwin and the Mysterious Mr. X,* San Diego, CA: Harcourt, Brace, Jovanovich, 1979, p. 215.
16. *Mathematical and Physical Papers*, vol. 5, Cambridge: Cambridge University Press, 1911, pp. 141–144.
17. The contraction theory was originated by the German Hermann von Helmholtz, who advanced it in 1854. Helmholtz's work was translated into English and appeared as "On the interaction of natural forces," *Philosophical Magazine*, Series 4, vol. 11, pp. 489–518, 1856. Adams' remark is in an essay in *The Degradation of the Democratic Dogma*, London: Macmillan, 1919, p. 142.
18. As one writer put it, "The application of Fourier was beautiful, but the result was disastrous,...." Quoted from B. A. Behrend, "The work of Oliver Heaviside," *The Electric Journal*, vol. 28, pp. 26–31 (January) and pp. 71–77 (February), 1928; reprinted in E. J. Berg, *Heaviside's Operational Calculus*, New York, NY: McGraw-Hill, 1936, p. 203.
19. A nice summary of Kelvin's problem, and of radioactivity's role, can be found in E. Rutherford's "Radium—The cause of the Earth's heat," *Harper's Monthly Magazine,* vol. 110, pp. 390–396, February 1905. Kelvin never believed in radioactive decay and regarded gamma rays, for example, as merely a *vapor*!
20. To see how the question of the energy source of stars bedeviled astronomers, see S. Newcomb, "The new problems of the universe," *Harper's Monthly Magazine*, vol. 107, pp. 872–876, November 1903. The author of the 1878 best-seller *Popular Astronomy*, Newcombe was the Carl Sagan of his time. The perplexing state of affairs was nicely summed up by Stokes in a July 1900 letter (*Memoir and Scientific Correspondence by the Late Sir George Gabriel Stokes*, vol. 1, Cambridge: Cambridge University Press, 1907, p. 81): "The sun is continually giving out an enormous amount of energy, which can be measured like the energy given out from a steam-engine. But where is the supply?"
21. A good summary, pro and con, of the short and long theories of the Earth's age, is in J. Prestwich, "The position of geology," *The Nineteenth Century,* vol. 34, pp. 551–559, October 1893. His opening words are gloomy: "The position of geology... is anomalous and possibly without precedent. On one side its advance is barred by the doctrine of Uniformity, and on the other side by the teaching of Physicists."
22. D. B. Wilson, "Kelvin's scientific realism: The theological context," *Philosophical Journal*, vol. 11, pp. 41–60, 1974. This interesting paper shows how strongly Kelvin believed God plays a role in humankind's destiny; for example, Kelvin declared God would exempt humans from the "heat death" of the universe that is apparently predicted by the second law of thermodynamics (the entropy of the universe, as a whole, is inexorably increasing). Heaviside took a more pragmatic, certainly far less sanguine, view on this particular issue. As he wrote (EP 1, p. 488) in 1885, "The perpetual running down of the available energy of the universe is a matter that must be cleared up. It is incredible that it can always have been going on, and dismal in its final result if uninterrupted. It is therefore the duty of every thermodynamician to look for a way of escape."
23. H. I. Sharlin, "On being scientific: A critique of evolutionary geology and biology in the nineteenth century," *Annals of Science*, vol. 29, pp. 271–285, 1972.
24. Sharlin argued that Kelvin believed that for life to have had a *beginning* (as evolution teaches) would mean a step change had occurred (life from non-life), and this is a violation of *continuity* in the universe. This argument, of course, forced Kelvin to offer a counterexplanation as to how life *did* get established on Earth, since he was arguing the Earth *wasn't* eternal (or even very old). Kelvin's response was that the universe (and life) had existed from "time immemorial" (i.e., there was no beginning) and living matter traveled to Earth by meteors from some other world. On this point he clashed with T. H. Huxley, who wrote in 1870 (*Biogenesis and Abiogenesis*), "If it were given to me to look beyond the abyss of geologically recorded time... I should expect to be a witness of the evolution of living protoplasm from not-living matter."—Quoted from I. S. Shklovskii and C. Sagan, *Intelligent Life in the Universe,* New York, NY: Holden-Day, 1966, p. 214.
25. "On the age of the Earth," *Nature*, vol. 51, pp. 224–227, January 3, 1895. In preliminary form, this material had been circulated privately among some of Perry's friends (including G. F. FitzGerald and Oliver Heaviside) in October and November of 1894. As Heaviside wrote in NB 10:273, "Professor Perry has sent me a paper (privately printed) giving good reasons why Lord K's estimate of the age of the earth is an underestimate. His [Perry's] method is a little obscure, so I have worked out some exact solutions."
26. L. Huxley, *Life and Letters of Sir Joseph Dalton Hooker,* vol. 2, London: Murray, 1918, pp. 165–166.
27. *The North British Review*, vol. 50, pp. 215–233, July 1869. This essay is unsigned, but from the style and tone there seems little doubt that Tait was the author.
28. *The Quarterly Journal of the Geological Society of London*, vol. 25, pp. xxxviii–liii, 1869.
29. In a letter to J. D. Hooker (dated July 24, 1869), in *More Letters of Charles Darwin* (Note 8), vol. 1, p. 313.
30. The diffusivity D is the *thermal conductivity* k divided by the product of the specific heat (the energy required to raise one gram by 1 °C) and the density. This product has CGS units of calories/cm^3 · °C and is called the *heat capacity* c. The diffusivity, therefore, is k/c with CGS units of cm^2/s.

31. EMT 2, p. 25.
32. There is an amusing sequel to this "challenge" from Heaviside. About ten years later a question concerning a cooling globe was inserted on the *Mathematical Tripos* examination at Cambridge. Unfortunately, the examiners incorrectly solved their own question! It wasn't until many years later that Bromwich and H. S. Carslaw each published the correct solution. See their consecutive papers in the *Transactions of the Cambridge Philosophical Society*, vol. 20, 1921 (Carslaw's "The cooling of a solid sphere with a concentric core of a different material," pp. 399–410, and Bromwich's "Symbolical methods in the theory of conduction of heat," pp. 411–427).
33. "The age of the Earth," *Nature*, vol. 51, pp. 438–440, March 7, 1895.
34. "The age of the Earth," *The American Journal of Science*, vol. 45, pp. 1–20, January 1893.
35. "The age of the Earth," *Nature*, vol. 51, pp. 582–585, April 18, 1895.
36. And yet thirteen years later we find G. F. Becker of the U.S. Geological Survey refusing to give up the battle, using Fourier to conclude the Earth cannot be more than 65 million years old—he explicitly rejects the importance of radioactive decay, and there is not a single mention of either Perry or Heaviside! This is all the more odd since for at least five years serious scientific speculation on the role of radioactivity had been appearing in the literature. Becker's article, "Age of a cooling globe in which the initial temperature increases directly as the distance from the surface," appeared in *Science*, vol. 27, p. 227–233, February 7, 1908.
37. *Electrical World and Engineer*, vol. 34, p. 427, September 16, 1899.
38. EMT 2, p. 28.
39. Topsy is a character in Harriet Beecher Stowe's *Uncle Tom's Cabin*. Heaviside held firm on what he had written, and these words appeared unaltered in *The Electrician*, vol. 34, p. 185, December 14, 1894 (EMT 2, p. 20).
40. *The Electrician*, vol. 43, pp. 668–669, September 1, 1899.
41. "The age of the Earth as an abode fitted for life," *Philosophical Magazine*, Series 5, vol. 47, pp. 60–90, January 1899. This was reprinted in two parts by *Science*, vol. 9, pp. 665–674, May 12, 1899, and pp. 704–711, May 19, 1899.
42. EMT 3, p. 519.
43. See, for example, Perry's book *The Calculus for Engineers*, London: Edward Arnold, 1897, pp. 230–242, which discusses operational techniques, including fractional differentiation. Perry realized, however, that operator mathematics wasn't clear to most of those following the exchanges between himself and Kelvin. In a letter to *Nature* (vol. 51, pp. 341–342, February 7, 1895), for example, in which he showed how Kelvin's value could easily be too small by a factor of 121, he opened with: "Three weeks ago I sent him [Kelvin] the solution of the problem of a cooling sphere whose conductivity k and volumetric capacity c are any functions whatsoever of the temperature v, but which are always proportional to one another. As Mr. Heaviside had been writing to me, and had shown me that under certain circumstances the differential equation became linear, and as I had used his *operators* much as he himself uses them, I cannot say to what extent I can claim credit for the work. I now venture to send you the more general case." Perry then proceeded to present his analysis *without* operators, using only "conventional" differential equations.
44. *The Electrician*, vol. 34, pp. 375–376, January 25, 1895.
45. *The Electrician*, vol. 34, p. 407, February 1, 1895.
46. H. Jeffreys, *The Earth: Its Origin, History and Physical Constitution*, New York, NY: Macmillan, 1929, p. 67.
47. NB 10:318.

12
The Final Years of the Hermit

He recalled the days when he used to read The Electrician *and its articles by Mr. Oliver Heaviside. At that time he did not fully understand their manner, but he realised that more of that sort of thing was wanted. He has since come to the conclusion that Mr. Heaviside's work formed the basis of all the practical work which had come after him.*

— From *The Electrician*'s report of Professor Ernest Rutherford's address to the Manchester Section of the IEE (March 7, 1913)

He was a queer old fish, but there was something very attractive about him.

— G. F. C. Searle, in a letter to the President of the IEE soon after Oliver's death

I make so many blunders in the figures that I can't be sure of anything.

&

I have become as stupid as an owl.

&

What a dreadful thing it is to have no memory for my own work.

&

Right at last. A lot of Warworm mistakes, but it goes perfectly at last. Very slow work, too. But I must try hard to recover my power of execution.

&

It goes so stiffly that I think I must have gout in my brain.

— Excerpts from the research notebooks of Oliver Heaviside, after age 60

Oliver Heaviside, W.O.R.M., Inventor of the Devilization and Balancing of Transmission Lines.

— Oliver's signature on a 1922 letter to the President of the IEE

I believe in Ghosts, Everybody does, though my Ghosts are not Thy Ghosts....

— From a 1924 letter from Oliver to Professor Ernst Berg at Union College

A "Gentleman" with a Pension

A potentially significant change in Heaviside's technical life occurred in 1895, with the resignation of A. P. Trotter as editor of *The Electrician*. On March 30 the journal sent out an announcement of Trotter's departure, giving as the reason his desire to return to full-time consulting. The real reason, however, was more like Biggs' situation (although Oliver appears not to have been involved this time), as Trotter revealed in a letter to Heaviside:

March 27, 1895

> Dear Mr. Heaviside,
>
> I regret to inform you that owing to serious interference with my editorial work by the secretary and the publishers, I am compelled to retire from *The Electrician*. The Proprietors are too busily occupied with their own business to give any attention to my affairs; they are quite out of sympathy with my views, and I have no course but to resign.
>
> It is a great wrench to wipe out of my life my connection with *The Electrician*, and no longer to have before me that ideal journal towards which I have been working for nearly five years. I have been writing too much and editing too little. My relations with *The Electrician* Printing and Publishing Company are too strained to allow me to think of contributing to the paper, but perhaps I shall forget the unpleasantness before long [Note 1].
>
> I am succeeded by W. G. Bond who was assistant to W. H. Snell. We have agreed together on all matters....

So Oliver lost another editor. The new editor, however, would fortunately prove to be as much a Heaviside supporter as Biggs and Trotter had been. But the following year *did* bring big changes in his life.

With the death of his father in late 1896, Heaviside found himself on his own for the first time in his life. Financial concerns, never very far from the surface, had been helped somewhat by the intervention in the spring of 1896 by FitzGerald and Perry, who obtained a Civil List pension[2] of 120 pounds a year for him. This time they had better success than they had had with the Royal Society Relief Fund episode, in overcoming Heaviside's fear of charity, and Oliver accepted the money with a good deal of pleasure. It was with the announcement[3] of the pension in *The Electrician* that Oliver's increasingly isolated life was first reflected in the title of "Hermit of Paignton." The lead editorial[4] that week was devoted to the pension, and it shows clearly that Heaviside's horror of being on the dole was well known:

> We are informed that certain very distinguished and very enthusiastic men of science threatened to visit the philosopher in his Devonshire retreat—or rather, threatened to picnic for weeks around his retreat, so that he might be compelled to say he would accept of the pension in case it were offered. His knowledge of the ordinary affairs of life must be as limited as his power of applying mathematics to physical problems is unlimited, if he thinks that this honor—an honor that the Duke of Wellington felt to be an honor after he had been acclaimed as the Saviour of Europe—tends in the slightest degree to tie his hands, to blunt his pen, or to affect his independence.

The editor also couldn't resist one little stab at the pure mathematicians who liked to sneer at Oliver's operators:

> Mr. Heaviside dared to write about things that even rigorous mathematicians themselves did not understand.... It was easy to say that he had invented no new mathematical weapons—every one of his weapons was invented before. That is true, but how many of them were used, and not only used, but used with learned skill in attacking problems that had never been attacked before?

With Oliver now a *government*-recognized authority, *The Electrician* did not hesitate to ask for his opinion in all electrical disagreements. Sometimes, however, even as the "solicited expert," Heaviside still could (or would) not contain his odd brand of humor, and perhaps hurt himself a bit by not taking things more seriously. For example, all through 1896 a very hot debate ("a great whetting of swords and sharpening of spears," as *The Electrician* so nicely put it) raged over the discovery that the current-voltage characteristic of the electric arc displayed a "negative resistance," i.e., raising the voltage drop across the arc gap, at a certain voltage, could actually cause the arc current to *decrease*. Asked for his technical opinion on this then startling observation, Oliver started off his reply[5] reasonably enough, but he then gave it up with a flippant closing:

> I am asked my opinion about negative resistance. ...The effects produced by the negativity of *R* (and other quantities) have occupied my attention in certain papers, and are interesting and instructive. But I have no faith whatever in the permanent existence of a body with negative resistance on account of the general instability. At the same time I am not prepared to deny that a substance might temporarily, under suitable circumstances, behave as a negative resistance.... Whether the arc may be conveniently regarded in this light is not for me to say. I do not know enough about the arc. I prefer gas for personal use.

In his reply to FitzGerald's letter of congratulations concerning the pension, Heaviside revealed his past feelings of not being accorded, by his family and local townspeople, the respect he felt his due, and of his joyful anticipation of how things would now be different:

> March 3, 1896
>
> My dear Heaviside,
>
> I was immensely pleased when I got your and Perry's letters this morning announcing that you had accepted the recognition of your services by your country. I am only sorry that it is not 1200 [pounds] a year as it would be more in accordance with your [just] deserts. But I suppose if such large sums were at the disposal of H.M. ministers no deserving candidate would ever get anything at all. It is a very great relief to me personally that you have been thus recognized. I guessed what you had to put up with though I did not know before that your relations had not recognized your position in the scientific world. Perhaps they did, but did not recognize that the scientific world had any position itself.
>
> As to who had any hand in it, Perry has done all the work. A few of the rest of us have occasionally urged him on when he was despondent of success. It is easy to do that when the other person has to set to again. Lord Rayleigh was great help and Lord Kelvin took an interest in it. But Perry is the person who has really stuck to it and worked for it and I am very glad to see has succeeded in persuading H.M. ministers that the country ought to show some recognition of your very valuable services. I sincerely trust that it will enable you to avoid in future the disagreeablenesses of the past and show your neighbors that you are really somebody whom your country appreciates.

> March 4, 1896
>
> Dear FitzGerald,
>
> When the servant gal puts on her Missis's bonnet and smirks before the glass, she is as

satisfied for the time as if it were her own. I am in the position of the gal when I read your most flattering remarks. I only hope that the grateful country will not... *want to know* what it has to be grateful about!—I expected from Perry's last that he was the worker, because he did not include himself as such. It is remarkable that people should be so kind and friendly.

As regards my last, I am afraid you must have thought it rather ill-conditioned. But if I can only get a decent sleep (I have usually to be satisfied with 4 to 5 hours; I should say dissatisfied) there is plenty of Mark Tapley [Note 6] in me, and I can see the comic side of the inconveniences of being a fish out of water. Here are one or two comic incidents:—I am sitting in my room engaged in writing immortal works (for a time), when in comes a man in his shirtsleeves with a pail. He does not knock at the door. He leaves it open. He says nothing, but goes to the window and throws it open. "What are you up to?" I say. "Going to clean the window" all in one word, is his reply. "No you are not", I say. "You must not intrude in this way". "Miss Eviside [Note 7] says I was to clean the window". "Now you hear what I say. Leave the room". He does not go, but stares and begins again "Miss Eviside says I was to clean the window". I then repeat my order and get him to go at last, still going on about Miss Eviside says....

Here is another one. I take something to the shop to be repaired. Only a little job, half an hour or so. Done by tomorrow for certain, if I will call. Call. Not done. Call again. Not done, more excuses, more assurances. At end of a week call again, and see myself the machine in corner, nothing done. So I desperately venture to remind of his last promise, (to say nothing of former ones). "Oh! did I?" he says, "Yes, I believe I did". Then I bring the matter to a point and enquire if he had any *special reason* for not doing it. (A little irony there but he didn't see it.) He considered for a moment, and then drawled out. "Oh, yes, I remember now. A little while after you left, a gentleman came, who wanted something done at once, so of course I had to...." Was *that* clever irony at my expense? Not a bit of it. Only a thickhead. I told him it was something to put in *Punch*. He didn't see that. The real meaning of it was the tradesman versus gentry business.

Three months later Heaviside wrote to announce he was at last a man of means, and that he believed with money would come the respect he wanted:

June 6, 1896

Dear FitzGerald,

I have received a document entitling me to 90 pounds, three quarters *back pay*. This is quite a windfall, and puts me straight. I am sure that you will be glad to hear of the liberality of this grateful country, stimulated by yourself. My conversion to a man of wealth will I feel sure have a very beneficial effect on my relations *with* my relations. In fact it has effectuated in that way already. Palm oil is the best lubricant for the wheels of common life. Use no other.

FitzGerald, with obvious relief, replied on June 8: "I am very glad to hear that the provision of the country for its great man is relieving your anxieties and greasing the wheels of life for you."

During this period between the deaths of his parents Heaviside was not technically inactive, by any means, and he had a fair amount of correspondence with Bond at *The Electrician*. On February 1, 1896, for example, Bond wrote to ask for "a luminous mathematical demonstration of [the] probability, improbability or impossibility as the case may be" of Jaumann's claim that the new Roentgen rays are longitudinal vibrations[8] of the ether. We don't have Heaviside's reply, but he of course must have eventually accepted the request because of his subsequent writing on the topic. Bond wrote again two days later and quoted Oliver's words that Heaviside's contribution will (when done) be a "smashing, pulverizing and destroying" of Jaumann, but Bond then lamented that it was one thing to get opinions in letters and quite another to get, in Bond's phrase, "a brief appreciation" of Jaumann. It's clear Bond did not want the normal Heaviside treatment of heavy mathematics, but rather what he called (in a letter of March 31) an "impressionist sketch." Bond tried to assure Oliver that "there is no sin in being 'a la mode'." In the end, however, Bond had to settle for more.

And in the fall of 1896, another nasty business occurred with Preece, just before Heaviside's father died. At the B.A. meeting that year at Liverpool, Preece presented[9] a new type of long-distance transoceanic cable. This cable was a metallic circuit (no earth return) formed by making each wire in the pair with a semicircular cross section and then bringing their flat sides as close together as possible, separated only by a thin piece of paper. In his paper Preece said "electrostatic induction" (his term for capacity) is the problem of such cables and "electromagnetic induction [i.e., inductance] may have a very beneficial influence on telephone working"! This was a *very big* turnabout for Preece, and there is no mention of his earlier pronouncements of induction as "evil" and as a "bug-a-boo." But, of course, his cable *design* achieves just the opposite effect as it *increases* capacity and *decreases* induction.

Heaviside was kept informed of these developments by FitzGerald who wrote on September 28:

> On the [next to the last day of the B.A. meeting] Preece surprised us all by saying he had taken up an Italian adventurer [Marconi, and we will return to this later] who had done no more than Lodge and others had done in observing Hertzian vibrations at a distance. Many of us were very indignant at this overlooking of British work for an Italian manufacture. Science "made in Germany" we are accustomed to but "made in Italy" by an unknown firm was too bad. This was capped on the last day by that extraordinary paper which is in last week's *Electrician*. It is awful to contemplate that the head of our telegraph department should make such extraordinary statements as that the capacity of a pair of concentric cylinders is independent of the difference of their radii [it actually varies inversely with the logarithm of their ratio] and should spend public money in trying to verify it by experiment. Why it makes us the laughing stock of all European Science.... Lord Kelvin came in just at the end and contented himself with expressing a mild dissent.

Bond asked Heaviside for his opinion on what Preece was proposing, and he replied in a short letter[10] to the journal:

> ...it is a long way to America, and one may never get there; but it is as well to go in the right direction. Those who are acquainted with my theory will recognize that the direction taken [by Preece] is the opposite one to that which I have so often advocated....

This prompted Bond to write an editorial[11] entitled "A conflict of authorities" which really put Preece on the spot. Quoting Preece's own words at the B.A. meeting, where he declared he *wanted* and *hoped for* a vigorous discussion of his new cable, Bond then went on to point out that "... Lord Kelvin, with his matchless authority in matters of this kind, publicly stated that the theory propounded led to proposals which he thought would not have the effects desired." And then Bond threw the mongoose in with the cobra when he next wrote,

> ...and Mr. Oliver Heaviside, than whom there is no one in this country, or, indeed, in the world, more competent to express an opinion on the theory of submarine telegraphy and telephony, stated [Note 10] in a letter [that Preece was wrong]. ... As far as we can understand his meaning, Mr. Preece is anxious to reduce capacity and increase inductance, and thus would seem to have become a convert to the "distortionless circuit" theory. But... Mr. Preece has reduced inductance to a minimum and increased capacity very nearly to a maximum.

This opened the floodgate, and Preece at last got his discussion. First in was Preece, himself, who answered[12] Bond's editorial by flatly denying his design increased capacitance (a statement that is obviously wrong, one that could be shown wrong by any Victorian physics *student*), and stating furthermore that he had had a 300-foot test cable made and had its capacitance *measured*. He was (he wrongly said), right.

Rollo Appleyard wrote[13] to point out that Preece may have measured his cable's capacity, but he had failed to say *how*. Then FitzGerald got into it by writing[14] in support and amplification of Appleyard's comments:

> ...if Mr. Preece really desires to discuss this question, he must describe fully the dimensions, arrangement,... of his new cables, and how his tests have been made.

There was, in fact, little love lost between Preece and FitzGerald. On November 23, for example, FitzGerald wrote to Oliver about his plan to *trap* Preece into revealing his ignorance; his words reveal that Oliver was helping by feeding FitzGerald inside information that he was getting from Arthur (almost certainly without his brother being aware of it!):

> But I want him to say that this [what FitzGerald thinks is Preece's misconception about how to measure the capacitance of his cable] is what he means just because he is crafty. I must have him carefully pinned down for fear of his wriggling. I am very glad to be coached up about it.... Would you think it well for me not to quote you as an authority for fear of its reacting on your brother? I will do just as you like. I am glad to hear the great man is waxy because it shows that he is in doubt but I hope he is getting sufficiently entangled not to be able to escape without crawling down.

This plot to entrap Preece apparently came to nothing, or at least I have been unable to follow its thread beyond this single letter.

Two weeks later, however, Bond published[15] a very sarcastic note (which may have helped cost him his job less than a year later):

> Again, the Post Office and the Professors seem to be at issue as how to measure the capacity of a wire forming part of a metallic loop. No doubt the Professors are hopelessly wrong, but until one knows exactly what the Post Office engineers mean when they talk of... capacity... and exactly how they measure it, we hesitate to condemn... these presumptuous Professors.... A scientific discussion cannot be satisfactorily conducted when the side which courts criticism does not publish its data.

In America the reaction was, if possible, even more severe. In an editorial,[16] "Official science," *Electrical World* had in fact beaten Bond to the punch in ripping Preece apart. First reminding readers of the "*KR*-law" and all its faults, it went on to say,

> Mr. Preece... is a doughty champion for official science, particularly when such science originates with himself.... Commenting upon this latest official theory, Mr. Heaviside flatly states that, according to both theory and practice [induction should be increased]. In a reply to some criticisms of... *The Electrician*, Mr. Preece does not design to directly meet the issue raised by Heaviside—perhaps the greatest authority in the world on the subject—but merely confines himself to denying [that capacity is increased].... This latest outbreak of official science gives us cause once more to congratulate ourselves upon the fact that on this side of the Atlantic the *personnel* of government departments do not consider scientific authority as one of their official prerogatives.

Preece could not have been a happy man at the start of 1897, but Oliver no doubt was—he was about to become, at the advanced age of forty-six, an *independent* man of leisure (or so he thought).

Life in the Country

One hundred twenty pounds per year was just one-fifth of Arthur's GPO salary, and only slightly more than a tenth of Preece's, but for Oliver the pension was enough to make him think of himself as "a man of wealth," worthy of more than life in a room over a music store. In 1897, therefore, he moved to Newton Abbot, a few miles from Paignton, and took up residence (with an elderly lady as a servant) in a house called (the British gentry are fond of giving their houses names) Bradley View.[17] His first letter to FitzGerald after the move shows a happy and excited Heaviside, but gradually he became taxed by the unfamiliar pressures of taking care of himself and a house, as well as by increasing tension with his neighbors. In addition, perhaps as diversions from these everyday problems, he developed an interest in a strange but noisy creature (the "old croaker") in his garden and worried about going insane. He also comes across as a bit of a prejudiced snob.

May 23, 1897

Dear FitzGerald,

...After a weary search, I found a suitable place at 35 pounds. Behold a transformation! The man "Ollie" of Paignton, who lives in the garret at the music shop, is transformed into

Oliver and his bicycle, probably in the early 1890s. In September 1899 Searle wrote, in advance of a visit, "I hope to bring my bicycle so that we can have some rides together. You must not scorch for I am not in very good practice just yet." One definition of Victorian *scorching* described it as the "impulse overruling the cyclist's reason, compelling him to overtake any and all moving objects which may be in front of him," with the end result of "slaying miles and pedestrians." In 1950 Searle recalled his friend's daring: "We used to put our feet on the foot-rests on the front forks and then let the cycle run down hill. Oliver put his feet up, folded his arms, and let the thing rip down steep and quite rough lanes, leaving me far behind."

Mr. Heaviside the gentleman who has taken Bradley View, Newton Abbot, (the town is enough for addressing), that disagreeable residence which has been empty so long. Built for a gentleman's house (small) it was occupied for very long by a farmer, and got into a very disreputable condition in consequence. But 30 pounds or so has been spent on it, mostly inside, and now it is not so bad. It is such a change! Very pretty place. Fields gloriously arrayed with buttercups and fringed with proper hedges with plenty of trees (it would shock a Scotch farmer) come right up to the garden, and a wood and hills behind them make up the picture. Birds singing like mad. Outside, the road, with a high bank on the far side, and houses and trees above. One of these trees, if it falls in the right direction, will smash my little house, I fear, but it is not an old tree yet....

The climate is quite different from Torquay, quite bracing in comparison. Being a thin skinned fatless man, very sensitive to cold, I shall have to acclimatize myself. The contrast between day and night temperature is far greater than at Paignton, ditto winter and summer, and I am not sure that I can stand it in the winter; then I must move again. At present, in this fine season, it is lovely, and the change from the music shop is most pleasant. I look at the buttercups instead of bill posters, and hear the chorus of birds instead of the Salvation Army, or the horrid row of hammering in the shop! Let me not borrow evil.

June 3, 1897

Dear FitzGerald,

As you like my unscientific writings, here's another...

There is a mysterious animal in the buttercup field, which has an oak in the middle (a Scotchman would cut it down instantly to get sixpennyworth more grass!). He begins at sunset and goes on for hours—perhaps all night; also Sunday morning, which is irreligious, but there must be a reason for it. I thought at first it was the beating of carpets whack-whack, whack-whack, very regular, by two beaters.... It is really a mixture of a bleat and a croak. Tree about 150 yards away seems to be the place. They say it is a bird, probably a woodpecker, but I'll wait till I see it before I believe it. Then a horrid tramp came in and sneaked round my estate, seeking what he could devour. Moving from the room to watch his little game, a great horrid cat came and killed my canary! Then the tramp was spied by the gal, and forthwith begged for bread which he didn't get. No beggars allowed on my premises for bread or anything else, on principle. Bread indeed! I know better than that what they want....

There is a nice looking vine in my greenhouse, but I can't see any grapes on it yet. Perhaps they have to be grafted on.... It is rather a cold place. I have been here long enough to see that my initial intense satisfaction (I felt at home at once; I never felt at home at P.) was the same thing as the calm pleasure which follows the removal of long continued pain, and not the positive pleasure of great delight. But it is good for all that, and will last in a measure for some time.... As I expected, I feel the change of climate. In fact, I took a dreadful chill... and it flew to the stomach and bowels as usual, and then to the brain, which was blown up over and over again. It is wonderful what the brain will stand; break and mend again. I have often wondered that I am not in a madhouse, incurably imbecile, brains all smashed and mixed up....

With this unpleasant ending, Heaviside's letters now began to take on an increasingly depressing tone.

June 19, 1897

Dear FitzGerald,

There is in the next garden an infant, terrible and prodigious, with a most remarkably vociferous voice. "Dada, there's, that, man, again! Dada! Dada! there's, that, man, again!

> Dada, that, man, is going, into, the greenhouse! Dada! Dada! That, man, etc. etc." So I go and look over the wall to try if that row can't be stopped, and have a few words with the Dada about the weather and how nice his garden is looking, and so on. He is, I find, awfully deaf; conversation is only possible by shouting. That accounts for the extraordinary development of the infant's voice. He is a retired grocer, and is selling off his remainder of cheese on the premises at 7-1/2 pence a pound. Had a pound, to soften *mores*. Perhaps you may have observed in your career that the judicious distribution of small sums of money has an emollient effect on some people. The British workman for instance. His moral character is instantaneously transformed by 6 [pence] especially if he hasn't earned it! However, as regards the grocer and his wife (who is also very deaf) the result was to make them inconveniently friendly. ... She is one of those persons who go about the streets without a bonnet, of course I must draw the line somewhere; though I am decidedly democratic in principle, it is not always pleasant in practice....
>
> The singular animal supposed to reside in the oak tree is still going on, with increased vigour. The blank space between his two croaks is shortened almost down to nothing. He is also heard, to the extent of an occasional few croaks, in the daytime. I think that is done in his sleep, as he begins regular work only at dusk. People going by at night say "What a row!" and make imitative noises. Perhaps, however, it is a nocturnal phenomenon of some kind, not an animal....
>
> There is a bat now flying about in my sitting room. It must have got in at the window upstairs, and then come down. Shall try and catch him; never examined one at close quarters. He goes round and round and may smash something before he has done. Not having any salt at hand, obliged to let him go. Opened front door, and he soon found the way out. But he went out too straight for a bat. Perhaps it was the old croaker [Note 18] out for a holiday....
>
> P.S. I hope you will not get drowned in crossing the Atlantic [Note 19].

In a letter of July 12 Heaviside indicated gardening was quickly losing its initial appeal:

> ...Wild horses were used [by the previous owner] to eat up the grass, so the "lawn" is, all over, holes, deep holes. It must be dug, or else filled up. What dreadful work cutting grass is, with a scissors. 90 feet by 12, say. Hard on the back, very. Lawn mower no good. Shears not good either, when the grass has to be lifted to cut it.

Heaviside's relationships with his neighbors were going downhill now, too. In September 1898 FitzGerald visited Oliver at Bradley View, and upon his return to Dublin wrote a thank-you note for his host's trouble. Heaviside's reply shows feelings of hostility and persecution.

> September 24, 1898
>
> Dear FitzGerald,
>
> Do not talk of trouble. I should have been glad to have had the trouble and pleasure of supporting you for a length of time, it is not a state of luxury exactly, but at any rate in a manner that would ensure you against starving. For example, I could give you something better than a thing *called* a chop, one inch of meat and 8 inches of bone! But no doubt you would find the Hotel more comfortable.
>
> In spite of the secretive manner in which you entered and left this town, you did not escape notice, either on your arrival at the railway bridge in the morning, or your departure up the hill in the evening. "Who did the ... go out with? Was it his father? Was that the old man's bike standing outside?" Certainly the rudest lot of impertinent, prying people that I ever had the misfortune to live near. They talk the language of the sewer, and seem to glory in it. You would be astonished if I were to go into detail about the way they have baited me. But I have not the least doubt of some hanky panky behind it.

97

new form: but it would not work as could n't steer. I am afraid one must raise body above level of wheels.

1 Windsor Terrace
Malahide
14 . VII . 96

Dear Heaviside

I must try and get a copy of Berkeland's paper. I think Swinton's results and Jaumann's can be explained by the action on the invisible return molecules being important. If a molecule be going slowly it is more disturbed by a magnet than when going fast $\frac{mv^2}{r} = Hev$ ∴ $mv \propto r$ or the curvature is great with small velocities. Now in Swinton's experiment the return molecules would be set whirling into a sort of vortex down the middle of which the cathode discharge could pass freely. The ease of discharge would be increased by this separation of the outgoing and return molecules in much the same way as the draught in a chimney is facilitated by separating the downgoing and return currents. Jaumann's result may similarly be due to a crowding of these return molecules away from the negatively charged body driving the cathode discharge towards that body. I am not satisfied however with either of these suppositions.

Yours sincerely Geo. Fras. FitzGerald

P.S. Air cushions are the best kind of spring. What is wanted is a spring handle not to shake ones arms. Likewise, I think, a completely lying down position would be a great advantage as decreasing windage which is the greatest cause of resistance on flat and smooth roads.

FitzGerald was, like Heaviside, a great bicycle enthusiast. He once wrote to Oliver, ''I consider bicycle riding the salvation of the body.'' In this letter of July 14, 1896, FitzGerald sketched his latest idea at the top—the words beneath read: ''new form: but it would not work as could not steer. I am afraid one must raise body above level of wheels.'' The postscript starting at the bottom and running up the left margin says: ''Air cushions are the best kind of spring. What is wanted is a spring handle not to shake ones arms. Likewise, I think, a completely lying down position would be a great advantage as decreasing windage which is the greatest cause of resistance on flat and smooth roads.''

ANOTHER CHANGE AT *THE ELECTRICIAN*

During his short tenure at the journal, Bond kept up a rather chatty correspondence with Heaviside, not always about technical matters, sometimes even bordering on just friendly gossip. On August 17, 1896, for example, Bond brought Oliver up-to-date on Trotter (Bond and Trotter kept in touch for decades, long after both had left *The Electrician* behind):

> You may like to hear that Mr. Trotter is rapidly finding his bearings out at the Cape [Note 20]. He is becoming an ardent Cape politician and is fighting with all his strength against the imposition of cheap and nasty [electrical standards] upon the simple minded inhabitants of Cape Colony.

And on April 28, 1897, in what appears to be a response to some comments by Heaviside on Tesla:

> I am not usually given over to the evil spirit of betting, but feel very much inclined to wager you 10^6 shillings to a bad penny that Tesla's "discovery" will prove to be a myth [there is no mention in the letter of what the "discovery" was]. Personally he is modest and charming, but entirely without scientific training and possessing no extraordinary scientific *instinct* to make good that deficiency. Once again he has got into the words of a lot of "technical journals", who hang upon his lips and extol everything he does and says without stint. I cannot conceive how genuine science and sound accurate work can be expected in such a deleterious atmosphere.

Just six months later, however, in October 1897, Bond was out as editor; on November 10 he wrote to Oliver to explain his resignation. Just as Biggs and Trotter had had serious difficulties with the owners, so had he, and he had learned that "faithful service is not by itself sufficient to save one's soul when... one must fawn and flatter [the owners] and *that* I couldn't. Hence, as in the case of Mr. Trotter I knew my ejection was a mere matter of time...."

Bond was replaced by Edward Tremlett Carter, who had been with the journal since 1893, and who had been serving as assistant editor since the beginning of Bond's tenure. Like Trotter, Carter came with a technical background,[21] and like Snell he was a youthful editor (taking over at age thirty-one) who would die young (in April 1903, not yet thirty-seven), but who would nevertheless have time to make editorial decisions having significant impact on Heaviside.

Carter had barely settled into the editor's chair when Heaviside proposed a third volume of *Electromagnetic Theory* be contracted for, even though the second volume was not yet done. Carter tossed this hot potato to the publisher, George Tucker, who wrote[22] Oliver (March 14, 1898) to remind him of the realities of the publishing world (and of Oliver's own obligation to get the second volume finished before worrying about a third):

> ...my particular anxiety for the moment is to complete the second [volume] and get this on the market.... I am only a brutal publisher with the strong desire to get the book on the market at the earliest possible moment... [and] ...to offer the fruits of the author's hard work to the unsympathetic public at so much per copy.

Even as they tried to light a fire under Heaviside to get volume 2 done (the journal's masthead, each week, now carried the notice "The second volume of this important work is in an advanced stage, and will be ready shortly."), the editorial staff used him as a technical consultant for evaluating new developments. On May 23, for example, Carter sent Oliver a copy of a paper[23] that had just appeared in America which, in Carter's words, "gives expression to [the author's] wonderful doubts of the correctness of the ordinary electromagnetic theory of light." Being asked for his opinion of the paper, as either an article or an editorial, was typical of the positive image Heaviside enjoyed at *The Electrician*.

Nikola Tesla (1856–1943)

This illustration appeared in the New York *Sunday World* of July 22, 1894, "Showing the inventor in the effulgent glory of myriad tongues of electric flame after he has saturated himself with electricity." The "star" of electrical engineers on both sides of the Atlantic in the 1890s, many of Tesla's comments sound crazy today (at best he was a seriously disturbed individual), but in his time they were taken with great solemnity. Some, however, like *The Electrician*'s editor, W. G. Bond, had their doubts even then. A myth has grown around Tesla that he was nearly awarded the 1912 Nobel prize in physics, but wasn't because he let it be known he would turn it down. In fact, he was never in contention. But one year Heaviside *was*—in 1912!

In view of this effective, ongoing relationship between Heaviside and the journal, it is revealing of his touchy nature that he then risked alienating Carter over a rather silly matter. In October 1898 *The Electrician* ran a short editorial note[24] that Heaviside thought was directed at himself, and which he found greatly offensive:

> Though many a dictionary and *thesaurus* may ignore him, the "freelance" is an individual who has long been known to the newspaper world, and he has found his way into the electrical press. Unthinking persons envy him his independence of thought and action, and take his doings seriously.... It is popularly, yet quite illogically, assumed that the individual who is dissociated from, say, the telephone industry or telegraph enterprise, must be more worthy of a hearing than anyone whose attainments and experience have perforce placed him in the position of an *ex parte* [i.e., biased] advocate. The choice, in fact, generally lies between the well-informed individual whose valuable services have been retained, and the badly-informed person whose futile assistance is required by no one....

Heaviside immediately sent[25] a postcard to Carter demanding to know what was the meaning of this note. Carter just as quickly replied (October 17) with a letter marked *Private and Confidential* to deny any reference to Oliver was intended, and that such a thing "never entered my head or into the head of anyone associated with *The Electrician*." That Heaviside could have believed the journal would attack one of its own authors, especially one whose views were regularly *solicited*, is a bizarre indication that his obsession with Preece (and what Oliver believed were his suppressive tactics) was alive as ever as the 19th century drew to a close.

Also unchanged was Preece himself, who was still up to his old tricks. Just two weeks after the "free-lance" business, on November 1, he took over the office of President of the Institution of Civil Engineers (Preece was undeniably a popular man who could win just about any professional elective office he went after). His inaugural address[26] contained yet another attack on theory, and also on *professors*—he almost certainly had Lodge and FitzGerald in mind:

> We are suffering from a lack of competent teachers. A teacher who has had no training in the practical world is worse than useless, for he imparts ideas derived from his inner consciousness or from the false teaching of his own abstract professor, which lead to mischief. *In my experience I have met with very serious inconveniences from this cause* [my emphasis]. The ideal professor of pure abstract science is a very charming personage, but he is a very arrogant and dogmatic individual, and, being a sort of little monarch in his own laboratory and lecture-room, surrounded by his devoted subjects, his word is law, and he regards the world at large, especially the practical world, as being outside his domain and beneath his notice. He is generally behind the age.

In his own way, Preece was feeling just as vulnerable as Heaviside, due in his case to his increasing inability to understand the new theoretical developments, and his words show he believed in a strong offense as the best defense. The lead editorial[27] of *The Electrician*, however, didn't let his comments pass without notice:

> There are many professors, we think, who do not come up to [Preece's] ideal, but who have conferred upon practical engineering many a valuable service, by virtue of their knowledge of the bearing of theory upon practice. Such men are the Lodge and Heaviside... of today, and the Faraday, of yesterday.

From this we see that Oliver, by 1898, had become one of the "great men" of Victorian electrical science. Indeed, Heaviside was now a known and respected quantity on both sides of the Atlantic. In America, for example, his work establishing the true nature of energy transmission by a conductor and its surrounding dielectric was routinely acknowledged in prestigious forums.[28] In August 1895 the publisher of *The Electrical World* in New York City wrote in an attempt to establish a relationship similar to the one Oliver enjoyed with *The Electrician*, inviting him to write (for pay) articles on electrical topics of his choice. In 1899 he was elected a Foreign Honorary Member of the American Academy of Arts and Sciences. And in his own country his transmission line theory was considered worthy of careful experimental study, if not by the Post Office, then by scientists of reputation.[29] To be compared with Faraday, Preece's lifelong hero, must have been sweet for Heaviside (and *very* sour for Preece).

THE PASSING OF THE CENTURY—AND OF A FRIEND AND A FOE

The first part of 1899 finally saw the appearance of the second volume of *Electromagnetic Theory*. In an undated letter, Trotter wrote from Cape Town both to congratulate Heaviside and to criticize the book. He opened his letter with "May the fact that I cannot understand 19/20 of your *Electromagnetic Theory* prevent me from congratulating you on the completion of Vol. II?", and then followed that with a lecture. Trotter believed mathematics to be a "shorthand" which is "unintelligible to those who have not studied it," and that the "*ultimate purpose* of most shorthand writing is in transcription into language" that a person with "ordinary education can understand."

It is safe to assume he had no better luck with this lecture[30] than he'd had with the earlier ones he'd delivered as Heaviside's editor!

FitzGerald wrote, too, on May 7, to thank Oliver for a copy of the book, as well as to say "I am afraid your remarks on the subject of Mr. W.P. will not tend to smooth asperities—but I suppose it is well to have the truth permanently recorded." What he was referring to was the forever smoldering Heaviside/Preece fire. The Preface of the book showed the fire was still nice and hot in Oliver's soul:

> I regret that I have been able to make so little impression upon British official science as expressed by its late leader [Note 31]. It is true that the "*K.R.* law", which set such unnecessary and unwarrantable restrictions upon telephony, is not much heard of now. With advancing practice it became so ridiculously wrong (say 1,000 per cent.) that it was impossible to save appearances by any manipulation of figures. But a dangerous and alarming official error has been pressed forward, even to the extent of experimentation with the public funds. I refer to Mr. Preece's proposal to increase the capacity of telephone cables, with a view to Atlantic telephony, by bringing the twin conductors as close together as possible. It is, indeed, very true that by Mr. Preece's ingenious plan of flattening the wires on one side, and bringing the flat sides closely together, the capacity may be considerably, and even greatly increased. But it is not the working capacity [i.e., symbol transmission speed] that is increased, but the electrostatic capacity! Faraday knew that much.
>
> And this blundering is so unnecessary. For if it be beneath the dignity of one who sat at the feet of Faraday and afterwards rose to be the leading authority on electrical matters (according to *Answers* [Note 32]), to consult the works of an insignificant person, still there are other ways. Why not ask someone else? It may be too late to consult the family doctor; but there are many young gentlemen going about who have been to technical colleges and are quite competent to give information concerning the capacity of condensers.
>
> It is to be hoped and expected that the late important removals in the British Telegraph Department will lead to much improvement in the quality of official science. The above two examples show how much improvement has been needed. Others could be given. This volume may help.

After starting to read his own printed copy of it, Oliver wrote the next day to FitzGerald of his assessment, as well as to give his friend some advice on cycling, health, and diet. All in all, it's a pretty depressing letter.

> May 8, 1899
>
> Dear FitzGerald,
>
> ...As regards my vol. 2 E.M.T., well, I had a go at it Saturday and Sunday, having received a copy, and got a very bad headache. It is a dreadfully dull book, unless you can go into it thoroughly; I was impelled to write it because someone or more than one must develop it. I really think it will in the long run have a considerable influence on the theory and practice of physical mathematics; still I cannot recommend any man of many cares and interests to trouble about it. I am sure it will not sell well. Vol. 1 has sold nearly 600 copies; vol. 2 I expect say 300 in same time. If the Editor will take vol. 3, it will be of a different type; much more varied in content; first chapter on convection current problems and connected matter....

I really think the bike is a noble thing for sedentary people. It can, of course, be made an instrument for athletic uses; you can *toil* at it if you like; but that is not necessary. Something was really wanted to take the place of the violent exercises indulged in in a man's youth; they can't go on indefinitely; walking is too tame and tiring; you don't get any further, so most people settle down to hardly any exercise at all, and so degenerate... and go off too often in middle age—heart, liver, kidney and so on. The bike seems to me exactly the right thing to keep a man going on physically active to old age. I am delighted to see an old gentleman of 75 still at it.... Idiots consider me a madman [Note 33] about the bike; I ride *every day*....

Measles, no measles here [in his letter FitzGerald had mentioned four of his eight children had recently had measles], but I have had my domestic troubles too.... Can't get a suitable housekeeper. So I do my own housework mostly. Simplified it considerably. Find that potatoes is the only food that can be eaten in quantity regularly. Easily cooked too. So when I get home at 1, I put on the potatoes immediately; then 1*st* course;-one glass milk, one slice cake. Read paper. Then [in] 1/2 hour 2*nd* course:--potatoes and butter. The butter is essential. An egg and rasher of bacon for breakfast supply quite enough nitrogenous food for the day. Sometimes I have a treat. A cauliflower, etc. But generally I can't be bothered. But the housekeeping is the worst. I do it as often as I can avoid, as the Irishman said (Irishman, of course) when asked how often he went to mass. In default of a proper working housekeeper I am inquiring for a respectable woman to come for the day, and then take her leave after tea. Even that is hard to find. I can't put up with common persons, wives of millhands and bricklayers and so forth; there are such a lot of vile blackguards going about here. I could tell such tales. But never mind....

There was, in fact, little time left for FitzGerald to hear tales from anyone. Still, he managed, at least one more time, to plant the seed of curiosity in his friend's mind. In his May 7 letter there is the following passage:

Have you worked at the propagation of waves around a sphere? A case of this is troubling speculators as to the possibility of telegraphing by electromagnetic free waves to America. It is evidently a question of diffraction [FitzGerald was wrong on this point] and I think must be soluble. Perhaps the case of propagation around a cylinder would be easier and I think must have been done by Lord Rayleigh in some of his papers on wave motion past obstacles though he has probably only worked at the case of the obstacle small compared with the wave length.

And then again on May 20:

I am afraid that wave propagation round a sphere is rather a complicated piece of mathematics, but somehow it seems to me as if it should be workable.

The question of a third volume for *Electromagnetic Theory* had dragged on for months by now, and on November 12, 1899 Oliver wrote to Carter about it. On November 17 Carter's assistant, F. Charles Raphael, answered that the issue still was not decided—Carter's health was now on the decline and he had gone on a Mediterranean cruise in the hope of a recovery. It wasn't until the first day of 1900 that Carter was able to respond to Oliver, and then it was to ask for a "brief synopsis" of the new book. He also reminded Heaviside that the typical reader of *The Electrician* was "of a more practical turn of mind," and asked if the theoretical treatments Oliver had in mind for the proposed volume 3 might also have some applications to immediate practical concerns. These seemingly reasonable questions from an editor brought forth the following response[34]:

January 3, 1900

Dear Mr. Carter,

...Synopsis? Can't. The Lord will provide. He always does. Besides that, I know I can make a third volume and very likely a fourth as well. I can't say more synoptically than, that, broadly speaking, vol. 3 would relate to electrical waves in general....

You are kind enough to refer to some theoretical anticipations [Carter had mentioned how earlier Heaviside's *theoretical* articles had come in advance of practical implementation]. I am reminded when my first description of surface induction along wires [the skin effect] was followed by Hughes' experiments; my description of nearly distortionless propagation of waves along wires was followed by Lodge and Hertz's ditto; you mention wireless telegraphy—any partial anticipation of its theory was not intentional, I can assure you. I have no idea whatever whether anything I may give in vol. 3, apparently unrelated to practice, may soon or ever become directly related. I may add that I think it is wrong not to make a broad distinction between advanced theory and actual practice, and to restrict ones researches to practice would be to stop them....

I am aware and so is everybody that the practicians who take *The Electrician* only glance at my articles and that the readers thereof are a small minority. It was always so save with a few exceptional articles, and it always will be so.

I am afraid you will think the above very unsatisfactory.... I can't help that, though I am sorry, being much indebted to the Editor of *The Electrician* in the past for the opportunities given me. The best I can do is to suggest that you give me *carte blanche* [!] and I will try to make the best use of it.

On January 8 Carter replied that this would be acceptable. We can only hope Heaviside appreciated his good fortune in having yet another saint for an editor. And so as the new year of 1900 began Oliver had a new book to write, and the articles that would eventually form it began to appear with the issue of February 23. Carter's editorial announcement of the new series shows the effect of Oliver's arguments:

[Readers] will now be able to follow him into remoter recesses of the labyrinth of electromagnetic theory, and perchance some of them will be inspired by the vigour and originality of his treatment to investigate for themselves along new lines of applied theory. For it is not the truth that theory must always follow the lead of practice; on the contrary, fresh fields for practical research are not infrequently revealed by the light of newly-developed theory.

Two weeks after Heaviside got the go-ahead from Carter, David Hughes died on January 22 of a series of strokes, and left a huge fortune of nearly half-a-million pounds as proof of his inventive skills. FitzGerald, himself, had only slightly more than a year left.

There are many more interesting letters between Heaviside and FitzGerald, telling us, for example, of Oliver's complaints about having to make do without proper servants, that at one time or another both had ridden their bicycles over chickens, and of many other personal matters that Heaviside either couldn't or wouldn't tell anyone else. Technical discussions traveled back and forth, too, addressing such matters as FitzGerald's *ad hoc* suggestion of the contraction of moving bodies (the Lorentz–FitzGerald contraction, soon to be *deduced* by Einstein from his special theory of relativity) and the possibility that gravity might well be an effect that propagates with a *finite* speed. And then the turn of the century came, and with it, during the first month of 1901 and exactly one year after Hughes' death, the long reign of Queen Victoria at last ended. The national tragedy of the Queen's death was followed by one far more personal to Oliver just one month later—on February 22, 1901 FitzGerald suddenly died soon after submitting to surgery to find the cause of his chronic and severe indigestion. He was not yet 50.

In a moving obituary[35] Oliver Lodge wrote as "one who loved him as a brother." Heaviside took the loss equally hard, even with some bitterness. In an article footnote,[36] the following month, he wrote:

> George Francis FitzGerald is dead. The premature loss of a man of striking original genius and such wide sympathies will be considered by those who knew him and his work to be a national misfortune. Of course, the "nation" knows nothing about it, or why it should be so.

The passing of his friend was soon followed, on July 4, by the death of Peter Tait, who died a broken man after the loss of a beloved son in the South African Boer War. Tait had not been a friend, but he had been a fellow traveler through common trials, one whom Heaviside had at least respected.

The new century brought other changes to Heaviside's world, some that showed it was the Victorian *age*, itself, that had died, even before Victoria. Others felt it, too. In November, for example, at the Society of Arts, William Preece observed,[37] "Automobiles are the rage of the day. The fashion has taken some time to reach England, but now motors grunt and smell everywhere." And two letters from Searle at Cambridge show how science was becoming ever stranger, too. On November 10, 1901 he wrote of the still fresh idea of the quantized charge of the electron:

> It seems very much as if Maxwell's notion of a sort of continuous fluid kind of electricity of which you could have 1, 2, 2.00034, 2.0000000 parts etc. must be replaced by one in which you can only have 1, 2, 3, 4... parts. It is something like the difference in buying milk and eggs. I hope some day to look more closely at the electron idea.

And two years later, on September 26, 1903, he wrote:

> I had 20 pounds [value, not weight!] worth of radium lent me for an evening, recently. I should not like to write a book on Chemistry just now. What could one say about Radium?

The world was changing, quickly, from the one of 1887 with its horses in the street and flickering yellow gaslights.

But Oliver wasn't.

The World Catches Up to Heaviside—and Leaves Him Behind

Others, besides S. P. Thompson, had been thinking about how to apply Oliver's ideas for improving telephone transmission. In December 1898, for example, Stephen Dudley Field obtained a U.S. patent for continuously loading a cable by winding soft iron wire strands among the copper conductors.[38] Field stated[39] his work was entirely experimental and "attributed" all the claimed good effects to reasons given in a convoluted bit of legal prose that I must admit I can't figure out—no matter, we hear no more of Mr. Field in the history of loaded cables!

Even *before* Thompson, however, there was at least one man who read and understood Heaviside—the American John Stone Stone (1869–1943) who was hired in 1890 straight out of college (The Johns Hopkins University) by the American Bell Company in Boston. Stone was soon in correspondence[40] with Heaviside (before the end of 1891) asking for help on his work for Bell on long-distance telephone circuits. Heaviside's letters to Stone are not known to exist today, but it is clear from subsequent letters from Stone that Oliver not only replied, but provided the requested aid.[41]

Stone's major contribution was as a strong advocate of Heaviside's ideas at an American telephone company, a lone voice in a wilderness of engineers who still accepted Preece's *KR*-law (or variants of it). His advocacy was strengthened in 1897 when George Ashley Campbell (1870–1954) was hired as his assistant. Together, Stone and Campbell attempted to apply Heaviside's theory of energy propagation along wires to the problem of wave reflections at the junctions of high-resistance overhead circuits to lower-resistance underground cables. Such "mismatched" junctions can reflect a significant portion of incident energy back in the direction it comes from, causing seriously degraded telephony. So independent of mind was Stone, in fact, that he would spend his time on the problems that interested *him*, not necessarily those his supervisors might wish to have addressed. In the end, it cost him his job.[42] Stone left American Bell in 1899, and the Harvard and MIT educated Campbell continued with this highly technical work. It was eventually to lead him to the invention of *practical* lumped loading coils.

At least as interesting as the technical development of the loading coil was the resulting legal battle over it between Campbell and Columbia University Professor Michael Idvorsky Pupin (1858–1935), who also claimed to have invented loading. Heaviside was aware of Pupin, as on July 7, 1900 Raphael at *The Electrician* had sent him a copy of the professor's paper[43] on the matter. At this time Oliver had no inkling of Pupin's goal of claiming all the credit for loading, and Heaviside wrote favorably to FitzGerald of the fact that at least the Americans were taking him seriously. In his paper Pupin had, in fact, written,

> Mr. Oliver Heaviside, of England, to whose profound researches most of the existing mathematical theory of electrical wave propagation is due, was the originator and most ardent advocate of wave conductors of high inductance. His counsel did not seem to prevail as much as it deserved, certainly not in his own country.

On August 12, in his reply to Heaviside, FitzGerald wrote:

> I was very glad to see Pupin's work and hope he has got the right sow by the right ear. Anyway he has faith and that is more than half the battle.

FitzGerald need not have worried about the professor—he would prove to be more resourceful than anybody could then imagine. The story of Pupin and the loading coil is a fascinating one, well told by Professor Brittain,[40] but the details aren't of direct interest here because Heaviside played only the most indirect role. Both Campbell and Pupin, however, had their initial patent claims rejected because of Heaviside's prior publications. In the end, though, Pupin carried the day in the patent office. It is clear, today, that the intellectual achievement was Heaviside's and the engineering talent was Campbell's (loaded telephone circuits were in operation, using Campbell's coils, *before*[44] Pupin published his 1900 paper).

Campbell was the victim both of poor legal support from the patent lawyers of American Bell, and of the shrewd Pupin who seems to have been highly skilled in self-promotion. Years later Pupin wrote a rags-to-riches autobiography (it won a 1924 Pulitzer prize), but in it he seemed to have developed some memory lapses. His story made no mention of Campbell (or, for that matter, that there even was a legal dispute over the loading coil). The only mention of Heaviside was the assertion that even in his intellectual developments Oliver had been preceded by another (Aime Vaschy, who, ironically, was with the *French* Post Office) and in any case it was all obvious anyway[45]! Vaschy did indeed independently travel much of the same path as Heaviside, but it was Heaviside who had priority. In his autobiography Pupin forgot his gracious words for Oliver in his 1900 paper, as well as similar praise from one[46] published the year before:

> Most of the mathematical investigations dealing with this subject [wave propagation on wires] are purely symbolic. Mr. Oliver Heaviside has done much to introduce the living language of physics in place of the sign language of mathematical analysis.

Rather than be troubled by fighting Pupin in court, and perhaps negating *all* patent claims to loading, thereby destroying the possibility of *monopolizing* loading, American Bell recognized

Pupin's patents (thus thumbing its nose at its own man, Campbell). It bought out Pupin (as well as the essentially worthless patents held by S. P. Thompson). In the end the professor received a total of $445,000 (remember, there were no income taxes then!). Campbell went to his grave convinced Pupin had received the recognition that was rightfully his own. Oliver, too, received absolutely nothing, and his bitter feelings toward Pupin were with him, as we'll see, until the day he died.

After Pupin's death Vannevar Bush made an attempt to do justice, when he wrote,[47] "The vision was Heaviside's. The design formulae were Campbell's." To which, of course, he might have added "But the glory and money were Pupin's."

Within a few years the fantastic success of Campbell's loading coils (which were always credited to Pupin) caused even Preece to change his position. At the 1907 B.A. meeting at Leicester he delivered a paper[48] that exceeded everything he had ever done before in revising history. Preece was now (at last) an expert on inductive loading, and as such felt it necessary to explain it to everyone else. After giving a brief overview of *his* earlier work (Heaviside appeared nowhere in his account), he then said of the once inviolate "*KR*-law":

> And the calculations [in the earlier work] were based on the *KR* law. It is quite certain that this law is not absolutely true, for it neglects the factor of self-induction.... It is known that the effect of this electromagnetic induction is beneficial in one sense, as I pointed out in 1896....

Not a single word about Heaviside's distortionless circuit of 1887, and it got even worse. Preece continued with

> The factors that are involved in the problem of long-distance speech are so numerous and so diverse that all attempts to solve them by mathematics, except on very general grounds, have proved fruitless and only approximately true. It has been only by practical experiment that success has been attained. The defect in the mathematical investigation of the problem appears to be the neglect of the prime and principal factor, *energy*, and the assumption that there is in Nature something analogous to *negative* energy. Energy, like matter, is always positive and it cannot be annihilated by a negative sign. But mathematical reasoning undoubtedly guided Prof. Pupin in going as far as he has.

After Preece was done with this bizarre[49] presentation, Oliver Lodge spoke up. He reminded those present that it was Oliver Heaviside who had predicted the benefit of self-induction. S. P. Thompson added his voice to Lodge's, admitting he had patented his "ocean telephony" cable only after being inspired by Heaviside's writings. To all this Preece could only give the weak reply that "he did not agree that Heaviside had rendered service to the science of telegraphy." This prompted Lodge to have the last word, so he answered with the rejoinder that "he referred rather to what was disclosed in Oliver Heaviside's writings than to what telegraph engineers had learned from the same."[50]

Oliver Puts His Name on the Atmosphere

At about the same time that Pupin and Campbell were fighting over the loading coil, Heaviside wrote a new paper that came to be recognized as a contribution to the theory of *wireless*. Despite his comment to Carter about not having had much to do with wireless, he did follow its developments with interest. This work has a double irony to it, in that he *is* often remembered for it, and that of all he did it ranks as among the least in importance (he wasn't even the first to do it!).

Until the turn of the century it had been assumed that wireless "rays" would travel in straight lines and thus eventually shoot off into space as the Earth's surface curved away from beneath them. To say otherwise could bring a swift comeuppance. S. P. Thompson, for example, was said by one upset correspondent to *The Electrician* to have asserted "he was confident it would be possible to establish electrical communication between England and America across intervening space." Thompson's critic[51] asserted that such statements are a "jump at conclusions, and do not take into consideration that we are dwelling on a *globe* and not on a 'flat earth'." The critic went on to write that "in the case of even so short a distance from England as St. John's, Newfoundland, it [wireless energy] would have to travel in the direction of the chord joining the two places through at least 1,000 miles of water [unless] the ray should be endowed with a certain prescience of the difficulty to be encountered, and seeking to avoid it... took a trajectorial flight *over* the curved surface of the water instead of going *through* it."

For Thompson's critic this was absurd. *The Electrician* agreed, and a year later an editorial note[52] stated, "[wireless] over very long distances appears to us impossible. That ether waves inherently tend to follow the curvature of the earth, as suggested by some writers, is an absurdity."

These words are doubly interesting because they were written in response to a lecture[53] by Preece a few days earlier. During this lecture Preece asserted that wireless waves *would* follow the curvature of the Earth, and so here I must admit Preece was indeed ahead of most of his contemporaries. The mystery in all the controversy, of course, came from *how* the waves could (if they did) bend with the curve of the planet.

Initially, the only possibility seemed to be diffraction (recall FitzGerald's letter to Heaviside), but subsequent detailed calculations showed this effect could not carry a wave even a quarter of the way round the globe. The mystery became even deeper in 1902 when Marconi reported,[54]

> ...I had the opportunity [in February 1902] of noticing for the first time in my experience, considerable differences in the distances at which it is possible to detect the received oscillations during daylight [as compared with at night].

Lord Rayleigh wrote,[55]

> The remarkable success of Marconi in signaling across the Atlantic suggests a more decided bending or diffraction of the waves round the protuberant earth than had been expected, and it imparts a great interest to the theoretical problem.

Marconi himself speculated that this might be the result of what he called the "diselectrification" of the antenna by daylight which prevented "the electrical oscillations [in the antenna] from acquiring so great an amplitude as they attain during darkness."

Heaviside proposed another possibility.

The actual correspondence is not available, but an account[56] has been given by Eccles[57]:

> In the spring of 1902 I was writing from time to time on wireless telegraphy in the pages of *The Electrician*, and one day Mr. Tremlett Carter, the editor, showed me a letter from Mr. Oliver Heaviside which, while discussing other things, asked if the recent success of Mr. Marconi in telegraphing from Cornwall to Newfoundland might not be due to the presence of a permanently conducting upper layer in the atmosphere. I believe this letter was shown to various friends of the editor, but I think it was not published [indeed, it was not].

Whatever the reason for the rejection of this letter for publication (I have been unable to find any reference to it in the surviving Heaviside editorial correspondence[58] with *The Electrician*), it merely made Oliver look for an alternative outlet. He found this outlet in the form of an invited contribution[59] to the new (10th) edition of the *Encyclopedia Britannica*. Entitled "The theory of electric telegraphy'' its scope is actually a bit broader than that implies. The article did indeed develop the wave propagation theory of telegraphy up through the distortionless circuit and lumped loading. He even attempted to use it as a vehicle to relegate Pupin to the role of a mere *follower* whose work was done at his suggestion:

> The *writer* [my emphasis] invented a way of carrying out the principle [of increasing induction] other than uniformly, and *recommended it for trial* [my emphasis]; viz., by the insertion of inductance coils in the main circuit at regular intervals.... In America, some progress has been made by Dr. Pupin, who has described an experiment....

As we've already seen, this was far too little and much too late.

But then there is the fâmous part of this article. Leaping free of conductors, Heaviside extended the scope of his essay to include ''wireless.'' He began with a very general geometry (two coaxial, conical conductors with a common apex, and axis normal to the horizontal), and then examined the radiation field in the space between the cones. By assigning particular values to the cone half-angles he reduced the geometry to several interesting, special cases. For example, the case of zero and 90 degrees for the inner and outer cone half-angles, respectively, corresponds to a semi-infinite wire (the inner cone) perpendicular to an infinite conducting plane (the outer cone). This is the geometry Heaviside then used for a simple explanation of ''wireless telegraphy.''

Heaviside wrote that signals will be generated by this particular geometry as hemispherical waves centered at the apex of the cones, and that ''They (the signals) are guided by the wire and by the plane representing the surface of the sea.''

Heaviside then continued to elaborate on the ''guiding'' of waves by wires, and pointed out that seawater ''has quite enough conductivity to make it behave as a conductor for Hertzian waves and the same is true in a more imperfect manner of the earth. The irregularities make confusion, no doubt, but the main waves are pulled round by the curvature of the earth, and do not jump off.''

And then comes the *most* famous part of the article. ''There is another consideration. There may possibly be a sufficiently conducting layer in the upper air. If so, the waves will, so to speak, catch on to it more or less. Then the guidance will be by the sea on one side and the upper layer on the other.''

Heaviside was, in fact, not the first to make this suggestion. His article was dated June of 1902 (and it was published in December), but Arthur Kennelly in America had made the same suggestion in an article published[60] in March. But *both* Heaviside and Kennelly had been preceded in their speculations of a conducting layer in the upper atmosphere by Balfour Stewart (1882), and others, who were looking for a mechanism to explain the observed daily fluctuations in the Earth's magnetic field.[61]

It was the names of Heaviside and Kennelly, however, that became attached to this supposed layer (it was alternately called the ''Kennelly-Heaviside'' or just the ''Heaviside'' layer). It was Eccles who first did this[62] in a paper that was the first attempt to quantitatively model the layer as an electrically conductive region, i.e., as a plasma of ions. Not everybody, however, thought such an upper-atmospheric layer was a reasonable conclusion. For example, in 1913 Eccles delivered a paper at that year's B.A. meeting at Birmingham which, when printed[63] in *The Electrician*, stimulated a great deal of correspondence to the journal that lasted well into 1915. The letters appeared so regularly, in fact, that at last Lee de Forest felt compelled to write[64] one of his own:

> Many of us in America who are at all familiar with radio-transmission... have read with amazed amusement the series of letters appearing from time to time in your columns under

> the title "The Heaviside Layer". It is rather difficult to understand just what it is all about.... Every day since its first discovery has added proof of the existence of the reflecting layer, and yet we read... letters which would indicate that many radio speculators are utterly unaware of these facts.

No matter that de Forest was convinced—other people continued to look for alternative explanations. In 1914, for example, J. A. Fleming presented a very interesting *refractive* analysis[65] in which he modeled the varying refractive index of the atmosphere (taken to be a single gas at a uniform temperature, with refractive index decreasing with gas pressure, i.e., with altitude). He then worked out the necessary relationship between the physical parameters of the gas and the size of the Earth to achieve "circular refraction," i.e., to achieve a bending of a tangential ray with the same radius of curvature as the Earth's surface. This ingenious idea led Fleming to the conclusion that a pure atmosphere of krypton could just do the job! He of course appreciated that this didn't agree with certain known and obvious facts, and so he ended his analysis with the following interesting observation: "Having regard to the great variations which exist in planetary atmospheres it is quite possible our earth is unique in this respect, as in many others, in being perhaps the only planet on which long-distance wireless telegraphy is possible."

As late as four years *after* de Forest's letter, *diffraction* was still not dead as a possible long-distance mechanism; it was still being studied in the hope of avoiding the acceptance[66] of the "Heaviside layer." The author of one such analysis,[67] after concluding diffraction just would not do, then accepted the reflecting layer idea, re-did his calculations, and gloomily concluded[68]: "The results of the paper therefore tend to confirm the existence of the reflecting layer. A consequence of its presence is that it places grave obstacles in the way of communications with Mars and Venus, if the desirability of communicating with those planets should ever arise."

So, it is thus an interesting little observation that, with *or* without the Heaviside layer, prospects looked poor for interplanetary radiotelephony in the 1910s! It wasn't until the mid-1920s, in fact, with Joseph Larmor's modifications[69] of Eccles' 1912 theory (a low-density electron gas replaced the ions), and Edward Appleton's subsequent inclusion of the effect of the Earth's magnetic field[61] (as well as his direct detection of a reflecting layer at an altitude of 90 km, with M. A. F. Barnett), that matters began to make sense.

As the years passed, the names of Heaviside and Kennelly gradually fell into disuse. The first edition of Frederick Terman's classic *Radio Engineering*, published in 1932, mentioned the "Kennelly–Heaviside" layer literally dozens of times. By the time of the fourth edition in the 1950s, however, their names had totally vanished.

And so today we have forgotten all the history[70] and merely call it the "ionosphere."

Heaviside's *only* contribution to "wireless," that of suggesting an upper-atmosphere conducting layer, was not a very significant one in the technical sense of putting a new idea on the table. It is an interesting fact, however, that Oliver had at least one chance (perhaps in part due to the visibility he gained by his suggestion) to make other contributions to wireless theory—which he declined. On October 4, 1904 the American radio experimenter Reginald Fessenden wrote a letter to Heaviside after meeting with John Perry in Washington, D.C. Perry hand-carried this letter back to England and posted it to Oliver. It was nothing less than an offer of employment as a technical consultant on "difficult mathematical problems" connected with wireless and enclosed with this letter was an up-front retainer fee of 100 pounds! Fessenden also offered to provide Heaviside with any experimental equipment he might need. The general goals would be for Oliver to study ways of reducing atmospheric disturbances and other forms of signal interference, and methods of improving signal transmission.

Heaviside, via a reply relayed through Perry, said no.

Not one to give up easily, Fessenden then wrote directly to Oliver on January 16, 1905, making the offer anew and adding the information that the U.S. Navy was interested in the work and, to show big developments were in the making, that "We are constructing our transatlantic stations...." Nothing came of it, and it is interesting to note that three years later Fessenden was *still* trying to entice somebody of mathematical skill (and engineering interest) to attack the analytical wireless problems he was continually running up against.[71]

While Heaviside may have been reluctant to claim any credit for the theory of wireless, or to accept offers to participate in its continuing development, William Preece was not so shy. It was Preece, in fact, who acted as the key link in Marconi's success in England and, as FitzGerald's letter to Oliver indicated, this support came as a great surprise and shock to many. Marconi came to Preece in the spring of 1896, an unknown youth with no formal training in anything and possessing essentially no theoretical knowledge at all of electrodynamics, and yet Preece quickly threw himself (and the GPO) behind him. Why did Preece do this, all the while ignoring Lodge who had done as much as Marconi, and sometime earlier? Interesting reasons of a nonpersonal nature are given by Professor Aitken in his excellent book[72] *Syntony and Spark: The Origins of Radio*, but I believe he has overlooked an obvious yet very important one—Preece just was not about to provide support to the very man who had so dramatically challenged him in 1888, and with whom he "had met with very serious inconveniences."

Later, after Marconi had cleverly outmaneuvered Preece by managing to keep his wireless developments out of the control of the GPO and the British government (this story is nicely told by Aitken), Preece had less enthusiastic words for Marconi. Even in 1897, just a year after the start of their relationship, Preece made sure *his* role was never overlooked, as shown in this passage from an interview[73]:

> It will be understood, therefore, that it is due to Mr. Preece that Marconi has received the fullest recognition in England and that engineers from four different departments of the English government are now supervising his work.

Professor Aitken stated[74] that Preece supported Marconi because "William Preece in 1895–1896 found himself at a technological dead end and it says much for his intelligence that he recognized the fact." But in 1900, at the B.A. meeting at Bradford, we find Preece declaring,[75]

> The sensation created in 1897 by Mr. Marconi's application of Hertzian waves distracted attention from the more practical, simpler, and older method [Preece's *induction*-field system].

By 1906, in a letter to *The Times* of London (printed October 12), he was even more assertive:

> Ten years ago the Post Office was actively engaged in developing wireless telegraphy. It had been doing so since 1884. In 1892 there was as much fuss made about my experiments as there is now about those of Mr. Marconi. They seem to be forgotten.

And finally, in 1907, in testimony to the Radio-Telegraph Committee, Preece was reported as declaring,[76]

> ...he had been at work on the subject of wireless telegraphy for 12 years before Mr. Marconi came to England—when Mr. Marconi came here he only came with a new way of doing an old thing.... [He can] not patent wireless telegraphy.

Increasing Trouble with Life

Carter at *The Electrician* died in April 1903. The decline in his health had been so sharp that Raphael had been appointed joint editor the previous October; he became sole editor in October

1903. Despite Raphael's fond remembrance[34] of Oliver years later, it doesn't seem as if he was really a strong supporter of Heaviside. Even before Carter's death it had been decided to discontinue Heaviside's articles, the ones destined for the third volume of *Electromagnetic Theory*. Over the next several years there was a great deal of correspondence between the publisher, George Tucker, asking where the rest of the book was and when it might be done, and Oliver, who always said there was still more. Tucker admitted part of the problem was the long delays in getting proofs back to Heaviside, claiming it was due to the increasing difficulty in getting print compositors who could set the difficult mathematics Heaviside made such extensive use of—to this, on Tucker's letters, Oliver would scribble "rot," "humbug," and "Perhaps losing his head."

After so many years together, Heaviside and *The Electrician* were nearing a parting of the ways. Sometime after Raphael was replaced as editor in 1906, the new editor (W. R. Cooper) tried to write a letter of rapprochement, but Oliver would have none of it.

April 19, 1907

Dear Sir,

I notice that it is a long time since *The Electrician* has contained any contribution from you. Will you kindly let me know if you can see your way to contribute articles occasionally on certain subjects, as you have done in the past. I know that they are appreciated, and although I cannot give a very great deal of space to purely theoretical articles, at the same time I should be very sorry to see such articles neglected.

Yours Faithfully,
W. R. Cooper
(Editor)

April 20, 1907

Dear Sir,

I am sorry I cannot accede to your request. I remark that your letter is written apparently in ignorance of the facts of the case.

(1) It has been entirely exceptional for me to contribute anything of an "occasional" nature. (2) I have written serials for publication in book form. Vol. 1 of E.M.T. took 3 years to put through. Vol. 2 took a little over 4 years. It was meant that Vol. 3 was to go through *The Electrician* in the same way; but I found that under the late Editor it was impossible. It would have taken 12 years. So I gave it up. (3) At present the Publisher is having the rest of the book put in type; but so slowly that it would be unsafe to make any guess as to when it will be finished.

Yours truly,
Oliver Heaviside

This withdrawal from the only public part of his life was mirrored in his private one, as well. Even Biggs and Preece had eventually managed to reach an accommodation,[77] but Heaviside *never* forgot an injury, be it real or imagined. On November 3, 1904, for example, Joseph Larmor wrote as Secretary of the Royal Society to tell Oliver that he had been selected to receive the Society's Hughes Medal for "your contributions to the mathematical theory of electricity." It was hoped he would be able to attend the award ceremony on November 30. For whatever reason, be it his unhappy feelings for Hughes, or lingering bitterness about the rejected operational mathematics paper of nearly ten years in the past, he turned this important honor down! Almost in desperation Larmor wrote again, assuring Heaviside it would *not* be necessary for him to actually appear in person, and that "I write this private remonstrance because I should be extremely sorry that the list of awards of what will be the electrical medal should be without your name." Heaviside once more said no, and the affair was kept private. The medal went to someone else, and was not offered to him again.

There were two reasons for this increasingly common, curmudgeonly behavior. His correspondence with Searle shows that his health was seriously declining, with complaints of malaria "with internal ruptures," jaundice, and gallstones. His health was not helped by his neglect of a proper diet. As he described the situation in a letter[78] to Searle, either in 1900 or 1901: "I made some jam the other day out of some apples the boys had not stolen and some blackberries which I could not eat. But I am not fit for a cook; I forget. Then it all goes to cinder, to be discovered hours later. Or if I boil an egg, I am startled by a loud report; either I did not put any water in or else it has all boiled away."

In addition to health and diet problems, his striking and eccentric nature had attracted the attention of the local Newton Abbot juvenile delinquents and both his letters and notebooks show the toll of their harassment. On the corner of one of his copies of *Nature* from 1905 is the note,

> Boys in field frequently calling [obscene names].... Sent note to Police. Didn't come. Spoke to Policeman. Didn't come.

In a 1906 notebook there are many entries recording the taunts of "insolently rude imbeciles," and of objects being thrown at his windows, doors, and over the walls around Bradley View. One July evening he even heard what he then thought to be the "report of a gun" and soon after discovered that a large window appeared to have been shot out (later he discovered it had been broken by someone who had "merely" torn a hunk of metal from a nearby street lamp and thrown it at this window). Another favorite trick was to plug up the sewage drainpipe out of the house, an act that eventually attracted enough nasal attention to be mentioned in the local newspaper. So desperate for help did he become that he offered 150 pounds of apples to the local police sergeant to provide a constable to watch Bradley View.

The fact that Heaviside claimed the leader of the hoodlums was a local blindman, and that they were careful to make trouble only when nobody but Oliver was about, perhaps caused those he complained to not to take the matter as seriously as they should have. His relationship with his brother Charles became entangled in these pathetic matters, and he even blamed Charles for much of his torment. As Oliver wrote of his daily trials in one of his notebooks[79]:

> Of course all this in detail is trivial. The mischief is the way it spreads, by the imbecility of the people. That amiable fool C. is responsible for a lot of it. Telling people I am "afraid" to go through N.A. That is his own invention, for one thing. Then, he had no right to say it at all, or anything, without consulting me. Then it gets repeated and spread all over, and I am insulted more than ever.... I mentioned to him I had refused medal [the Hughes Medal]. No reason given. He said contemptuously, he thought I was very foolish. Then he goes and tells it at home, no doubt with further expressions of contumely. Then his wife goes and talks to the Man [probably a local tradesperson]. And then the hatchet faced man is impudent to me. "He, He, He... He He He. Arnt yer going to take yer Medal? He he he!" Then I try to make complaint to C., but he is up in arms at the very notion of my having anything to say against the man.... It is a wretched state of things, to be reviled and slandered and insulted here and outside and to have my own brother actually promoting it. "Well, *I* think it's very amusing!" was his nice remark on my telling him of some beastly behaviour of boys outside. And then no doubt he went and made game of it, at home. There is no way out of it, without money, that is the eternal fact which always brings me to a stoppage.

The desire for *money*, and what he believed it could do for him, perhaps explains his decision to turn one award down and yet covet another. The Hughes Medal, for example, was just that—a mere medal. But when Searle wrote on December 1, 1906 and mentioned in passing that J. J. Thomson had just been announced as the new Nobel prize winner in physics, Heaviside's reply evidently expressed the hope that *he*, too, might one day receive the prize (which, of course, includes a substantial amount of money). A letter from Searle (December 17) had a single line standing apart from everything else in the letter, an obvious reply to Oliver's hope: "I should be delighted to see you get a Nobel Prize and I hope it may be awarded you before long." Indeed, in 1912 Oliver *was* a final contender[80] for the physics prize. However, along with such great men as Einstein, Mach, Lorentz, and Planck who joined him on the final list of choices, *all* lost out to Nils Gustaf Dalen (who won for developing automatic regulators for feeding gas fuel to lamps in lighthouses and buoys!).

Heaviside's situation at Bradley View continued to deteriorate, and by Christmas of 1907—just after the death of Lord Kelvin—when Searle and his wife visited him, they found him "in a miserable state" and Searle recalled,[78]

> He was yellow with jaundice or some such complaint and was shaking all over. He had been sitting upstairs watching to see if boys would throw stones and break another window. He had suffered a good deal in that way. Boys would tease him and write things on his gate.

Things became so bad that by July 1908 Charles made other arrangements for his brother. It was agreed that Oliver would leave Bradley View and take up residence in the home of Mary Way, the unmarried sister of Charles' wife. She would have the downstairs of her home (called Homefield, situated high on a hill overlooking the beautiful bay at Torquay[81]), while he would have the upstairs. And that is where he stayed for the seventeen years that remained of his life.

Life at Homefield

Life at Homefield was easier for Oliver than for Mary Way. As remembered by Searle she was a kind, good-natured woman in her middle 60s when Oliver came to her, and she displayed extraordinary patience and tolerance for her sharp-tongued, crotchety housemate. She also provided the human touch, as well as food cooked by someone who knew what she was doing, that he had been without at Newton Abbot. His situation brightened to the point that sometime in 1910 he wrote to Searle and predicted he would live another 25 years. Heaviside generally appreciated what she did for him, but always with a selfish touch to it. He could write to Searle of her as "mulier bestissima"—bad Latin for the very best woman—but he also demanded that she go outside only with his permission, and expressed strong displeasure when her friends visited her (in her own home!). As a man who loved heat he felt all others should be warm, too, and constantly worried that "Miss Way" might not be properly dressed and so would catch cold and die—and *then* what would he do? He once asked Searle's wife if she could check on "Miss Way's" underclothing!

Living with such a man on a daily basis, with his constant demands to come before anything else, put a terrible strain on the poor lady. Because of their regular visits with Oliver during these few years before the First World War, the Searles could see what he was doing to her, and on September 3, 1913, Searle wrote,

> ...just remember this. Not all people have your extraordinary toughness of constitution and a long strain of nervous kind may cause a breakdown. What would you do if M.W. gave way under the strain and broke down in health? It might be a *long* time before she recovered [Note 82], as recovery from that sort of strain is very slow, indeed.... She took you into her house when you were at your wits end at N.A. Surely she is "Good Mary Way" still.

Homefield

Just a few months before, on April 3, Searle had also written of her, in an obvious reference to what a saint she had to be to put up with Oliver, "her name is written in golden letters in the list of the Great Ones." Such remonstrances from Searle received nothing but scribbled Heaviside complaints on the letters themselves, such as "rot," "This is very ill natured," and "Man of nonsense."

The years at Homefield before the war were not yet devoid of all interaction with the outside world (which characterized his very last years), even though his creative time was past and his health remained poor. He kept in touch with many of his correspondents from the "old days," and his letters were full of everything from science to health gossip to grumblings about ancient grievances. For example, in one letter of February 1910 to S. P. Thompson, we learn both that he was suffering from swollen joints and had difficulty in walking, and that he believed the British Government owed him 25,000 pounds for the invention of inductive loading. At the same time he was writing to W. H. Bragg (who shared, with his son, the 1915 Nobel prize in physics) about the nature of x-rays and radiation burns. And it must have been a happy time, at least for a while, when the long-delayed third volume of *Electromagnetic Theory* at last was published in late 1912, and sold at 23 shillings a copy. It was dedicated to his long-dead Irish friend, bearing the inscription[83]

IN MEMORY OF
George Francis FitzGerald, F.R.S.
"We needs must love the highest when we" know him.

It received a good review[84] in *The Electrician* (not surprisingly), which opened with

> This new volume by Heaviside will prove a welcome addition to the libraries of mathematical electricians and physicists. Experience has proved that his writings will repay study, although many of us are often unable to soar with him into the empyrean. It is a matter of history that he was the physicist who laid the foundations of the modern theory of telephonic transmission—a theory which has proved a veritable gold mine for the practical telephonist.

The only other review,[85] which was also favorable, appeared somewhat later; as the reviewer put it,

> ...the present Notice has been delayed over four years. But the consolation is that the reviewed volume—as, in fact, every work of Heaviside—is never too old. It has the freshness and the life of originality, and is, both as regards contents and form, stimulating beyond saying. In a very short Preface [just two brief paragraphs, including his hope to present a bust of Preece to sit beneath that of Faraday, at the IEE] the author explains that he has "excluded parts of the third volume and included parts of the fourth". This confession is regrettable, inasmuch as it seems to imply that the plan of publishing a fourth volume has, for the time at least, been abandoned.

In fact, there would be no fourth volume. Whatever he planned for it, others were now beginning to publish similar work and it was too late. And even after transferring what parts of the fourth volume still remained new into the third, it is obvious that Heaviside nonetheless felt a need to "pad-out" the third volume with several nontechnical, philosophical essays. However interesting these essays may be, they wouldn't have been taking up page space if Oliver had had anything of *solid mathematics* left to say. The "mystery of the fourth volume" has grown into a bit of a legend over the years; Chapter 13 will pursue the issue as it developed after Heaviside's death. But even while he was still alive, his fourth volume created interest. In a January 1917 letter, for example, the American scientist Ludwik Silberstein actually offered to journey to Homefield for a week or two to help get the manuscript ready for publication. In March Heaviside replied[86] to decline, declaring Silberstein's enthusiasm showed the difference "between 40 and 67." Heaviside, in fact, did not think of what he had as a manuscript at all, but rather referred to it as a "Scrapheap"—"It is raw material, and *no one can do the work save myself.*"

Death Takes the Past—and the Present

On November 6, 1913, William Henry Preece died at age seventy-nine, leaving a set of *Electrical Papers* in his library to his heirs, who also no doubt didn't read it. A man of enormous reputation, his funeral was attended by Knights and Fellows of the Royal Society, messages of condolence were received from high government officials (and one Viscount), and directors of the American Institute of Electrical Engineers journeyed across the Atlantic to pay their last respects. Oliver was not so generous in his "praise."

He clipped Preece's obituary[87] from *The Daily Telegraph* and sent it, marked up with marginal comments, to a friend at Oxford, the physicist Dr. Charles Vandeleur Burton[88] (whose reply called the comments "really quite entertaining"). Burton's clipping hasn't survived, but it is easy to imagine what Heaviside might have written next to such lines as "Sir Humphrey Davy's greatest discovery was Michael Faraday, and Faraday discovered William Henry Preece"; "He inspired so much confidence in the official world and among his colleagues that it became a motto, 'Preece is always right'"; and "Sir William Preece has been styled the 'Father of Wireless Telegraphy'." We do, however, know what Heaviside wrote on his copy of *Nature*, when that journal ran S. P. Thompson's obituary[89] of Preece:

S.P.T.: Sir William Preece will probably be best remembered in after time by the pioneering work he carried out for a number of years on the subject of telegraphy without wires...

O.H.: This was not wireless telegraphy at all. It wasn't thought of. These conduction and induction experiments had been done before. No mention is made of experiments *to* 1886 which he had not allowed to be published. He wanted to make out they were his experiments and on them he rested his claim to be the Father of Wireless telegraphy.

S.P.T.: ...even in his severest contentions with opponents he bore no malice.

O.H.: He did!

S.P.T.: ...a deserving subordinate who had some technical improvement to suggest found in him a sympathetic listener.

O.H.: (to steal from if he could)

S.P.T.: His entire inability to appreciate the work of Oliver Heaviside...

O.H.: quite so; O.H. interfered with his stealings!

At the top of one page of this obituary Heaviside summed up his appraisal of "The Father of Wireless": "He was the Father of Wiles... an intensely greedy, grasping man." There was to be *no* forgiveness in Oliver's heart, *ever*.

Preece's death was the real start of the passing of those who had played important roles in Heaviside's life. The "War to end War" began for England in August 1914 and it would be a terrible time for all. Some of Heaviside's friends would not live to see it through. Poynting died a diabetic even before it began in March 1914, within days of George Minchin's death. George Tucker[90] at *The Electrician* died "still in harness" in January 1916, and S. P. Thompson followed him out of this world not long after, in June 1916, just days after appearing to be in perfect health.

Heaviside's health just before the war was none too good, either. Searle's letters to him in 1913 indicate Oliver was suffering from gout, internal bleeding, heart trouble, and at one time he could walk only with the aid of crutches. The situation became so bad that Charles had to hire a nurse to help Miss Way care for Oliver. Heaviside himself referred to this time of crisis, in a letter to Searle, as "going down into the shadow"—he might easily have died.

The war, with its food and fuel rationing, was a difficult time for Heaviside. Some help had come just before it started, on March 24, 1914, with an increase in his Civil List Pension by 100 pounds, to 220 pounds per year. It was granted "in recognition of the importance of his researches in the theory of high-speed telegraphy and long-distance telephony." Other help came from money sent by admirers from all over the world (via Searle), including the very generous aid of several hundred pounds from the French electrical engineer Joseph Bethenod. When Heaviside learned of this he was both grateful to Bethenod and angry with Searle, writing (to the President of the IEE on January 4, 1922), "But for Mr. B. at Paris, I should have gone under from 1914 onward.", but also that it was all a result of Searle's "begging letters" sent out "entirely against my wishes." Caught in the middle, with none of the credit and all the blame, poor Searle couldn't win.

Sometime in 1916 Mary Way had had enough—she departed Homefield and Oliver (who had, in fact, become the legal owner in 1911) was again left alone. Miss Way went to live in the home of Charles' family, where she remained until her death in March 1927. To some of his correspondents Oliver wrote she had gone mad and had to be put away, but according to Searle this was simply not true, and it tells us more about Heaviside's state of mind than of Miss Way's. It was about this time that he really began to live as a hermit, and he chose a suitable alternative name for Homefield—"The Inexhaustible Cavity." Rollo Appleyard speculated[91] that he did this in reference to the cave of Adullam, into which everyone in distress gathered (i.e., those with technical questions could query Oliver about them), but Karl Willy Wagner had, I think, the *real* explanation. As he wrote,[92]

> In a discussion of the third volume of Heaviside's *Electromagnetic Theory*, I called it an "inexhaustible fund of stimulation"; since the expression amused him, he thereafter in his correspondence repeatedly called his hermit's cell "The Inexhaustible Cavity".

The war didn't keep Heaviside from his old habit of tirelessly penning letters. He corresponded with Bromwich on operational mathematics (as discussed in Chapter 10), and with C. V. Burton on the philosophical aspects of the war. The surviving letters from Burton show that Oliver had a cynical view of it all, one that if widely known would not have been popular with most of his countrymen. The following excerpts from two of Burton's replies to Heavisidean pronouncements show how extreme this view must have been:

> January 2, 1915: I can't agree that *war* has a tonic or bracing effect, or that there is a single word to be said for it....
>
> Undated: You have not convinced me that this war is anything but a calamity....

And at the end of a letter to B. A. Behrend in Boston (April 5, 1918) he wrote a note for the eyes of any wartime mail inspector who might have taken a look: "For information of Censor. Kaiser Bill [Wilhelm II, Emperor of Germany] is one of the best friends we have ever had, because he is waking us up."

It was through the efforts of Behrend, in fact, that Oliver was made, in the spring of 1918, just the fifth Foreign Honorary Member of the American Institute of Electrical Engineers (the IEE in London had already made him one of *its* honorary members ten years earlier, apparently forgiving him for his previous nonpayment of dues). These honors were in recognition, a bit late, of his inductive loading ideas. Nothing done with Heaviside was done easily and quickly, of course, and Behrend was forced to *negotiate* with Oliver over the precise nature of how and why he should be "honored." An excerpt from one of these Heaviside letters (dated February 20, 1918) to Behrend shows how delicate a dance the American had to step to:

> I think honors have been so much overdone. The more honor, the less value. It is depreciating the currency.... The value of the work I have done was long ago well recognised, beginning with Rayleigh, FitzGerald, Kelvin, Lodge and other big men. Preece was ultimately goaded into saying that he did not see that Heaviside had anything to do with it [loading]; and his jackal Kempe [Note 93] declared that it was something quite different, that Heaviside's way had been proved to be a failure. I took notice that although no one questioned that the loading coil system was my invention *before* Pupin sold his patent (though it was sometimes laughed at by some practicians who miscalled themselves practical men), yet *after* a great change quickly occurred and I was repudiated in a most emphatic manner. Now, being an innocent in relation to dollars, from want of experience possibly, it may be malignity on my part to suspect that the reason of the change was based on commercial considerations. Prior publication, of a distinct kind, if admitted, might have interfered with the flow of dollars in the proper direction.... What is the practical remedy? To perform, in my idea, a belated act of bare justice. If your great Institute, through its head or friends is prepared to announce... that I am the inventor of the loading-coil system... I don't care to be tacked on to Pupin, either for or aft. Pupin got his dollars, and I do not see what more he is entitled to, save relatively insignificant "improvements" in practice. I want explicit recognition, as well as implied. ...A Committee might settle [the question of nomenclature]. Heavy and loading are closely connected. It was pre-ordained [and then Oliver mentioned the idea of the analogy of heavy masses on a rope helping in propagating waves]. Heavify and Heavification seem to me best [instead of the then common term of *Pupinize*]. The fame of Pupin reached Newton Abbot when I lived there some 12 years ago. The [workhouse boys] called out Poop, Poop, Poopin, and flung mud and stones. Such is fame....

Behrend had some sympathy for Oliver's feelings of being ignored, of course, and in so expressing it got the following amusing story from Heaviside (in a letter dated April 5, 1918):

> You speak of my having been forgotten. Hardly that.... A friend of mine now working at the aircraft factory at Farnborough [Note 94] told me a funny story about his visit to the United States some years ago. He had mentioned me and received this startling information: "Heaviside? Is he still living? I thought he was one of the classics!" Now that is real fame, isn't it? And I may live 20 years more if I can keep out the rheumatism and damp cold which is the plague of England, and have the pleasure of hearing such remarks again....

In the end what Oliver got was the honorary AIEE membership,[95] but of course there was no public humiliation of Pupin. It was naive of Heaviside to have even asked. Aware of Heaviside's continual need of money, Behrend suggested Oliver write a book. The reply (June 24, 1918) shows Oliver still had the idea of a fourth volume for *Electromagnetic Theory*, but was simultaneously discouraged by it and yet thought even so he should be paid for it *in advance*!:

> ...As regards your proposition about a book, I really couldn't do it. I have neither the time nor the inclination; the spirit doesn't move that way. I want to get on with vol. 4 of E.M.T. and may be able to do it bye and bye. It is a question of health and freedom from disturbance and petty worries and too much other work. At present a domestic calamity (the old lady with whom I have lived 9-1/2 years [Miss Way] went mad, and has gone away...). Throws a great amount of work upon one, to get the house, very much neglected, into order and rearranged to suit my convenience and to try to sell in a profitable way portions of the stock to recompense me for my great money loss through her. So even vol. 4 is at a stop now and it is

> slow work for an old man in any case. Now, if you took vol. 4 instead of the book you propose it would do very well... but then my conditions would be far too stringent I think to meet your desires, payment in advance [Note 96], like a retaining fee, and yet to send copy at my own convenience, etc., etc.!

Nothing came of this, but at least, at last, the Great War ended. The following year, on June 30, 1919, Rayleigh died in his bed of a heart attack.

The End of the Hermit

With the war over, Searle and his wife visited Oliver at Christmas 1919, the first time in four years. Searle later recalled[97] that nearly all the crockery was by then broken and unreplaced (Searle had to drink his tea from a slop basin), that a sheet of *The Times* served as a tablecloth, and that Heaviside enlisted him in a slightly crazy hunt for gas leaks with lighted candles! Oliver could never get enough gas to run both his lights *and* fires, and was continually trying to get better performance than was possible out of the ancient piping of Homefield. Once he even removed the main pipe fitting near the meter, and then lit the gas that came roaring out. Frightened, understandably, by the enormous tongue of flame thus produced, he managed to put it out with a wet cloth but burned his hands and face in the process. Soon after the Searles visited again, and were greeted by a blackened Heaviside peering at them, with one eye, through an opening in a blanket-bandage wrapped around his head and held in place by a rope tied around his neck!

John Perry died August 4, 1920.

The last five years of Heaviside's life, with both hearing and sight failing, struggling on 220 pounds a year and a helpful "gift" now and then to meet the mortgage on Homefield (well over 1000 pounds) as well as the cost of food, clothing, and pipe tobacco, and cut off by his own wishes from the outside world, were years of great privation and some mystery. What *did* he do? The only clues we have are from his letters in 1922 to John S. Highfield, then President of the IEE, and the memories[98] of a fortunate visit to Homefield, less than a year before Oliver's death, by E. J. Berg, Professor of Electrical Engineering at Union College.

During these years, while Oliver retained an ability to think clearly with his old wit as sharp as ever, there were disturbing signs, too, that his mental state was taking on some odd aspects. Near the end of the war, for example, he awarded himself the title or degree of W.O.R.M., which he included with his signature on letters. Searle tried to make light of this many years after the fact, saying[99] Oliver's "chief motive was impish fun." He would do this, however, on quite serious letters to people who could not possibly have understood any such motive. I am inclined to agree less with Searle and more with the assessment[100] that this and other peculiar behavior patterns "indicate that his brain was tragically abnormal" in his last years.

Heaviside's 1922 correspondence with Highfield began as the result of the IEE's decision to present Oliver with its newest (and highest) award, the Faraday Medal. It required delicate negotiation, of course, as Oliver could be expected to be touchy about nearly anything. He even found the *name* of the medal to have fault: "If the Faraday medal is to be your highest, then you can't have a Newton Medal, of course." And as usual, *Heaviside* set the conditions on just *how* he was to be honored: "I wish it to be clearly and *explicitly* recognized and stated that I am The Inventor of the loading-in-lumps systems. I presented that to the world." He repeated this demand in a funny follow-up letter in April:

> ...I hereby declare that I will *not* wear the medal, whether gold or bronze, solid or hollow, nor have it pinned upon me by anyone.... I will *accept* the medal, and look at it, and then

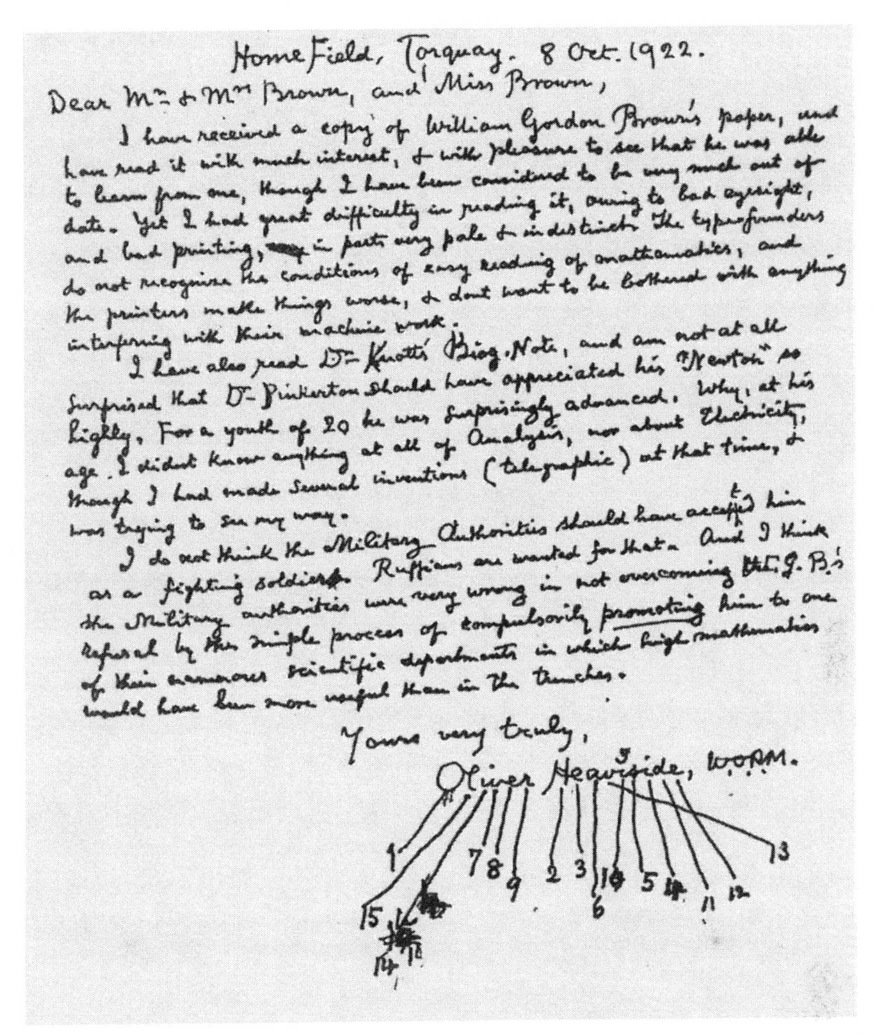

Home Field, Torquay. 8 Oct. 1922.

Dear Mr. & Mrs Brown, and Miss Brown,

I have received a copy of William Gordon Brown's paper, and have read it with much interest, & with pleasure to see that he was able to learn from me, though I have been considered to be very much out of date. Yet I had great difficulty in reading it, owing to bad eyesight, and bad printing, in parts very pale & indistinct. The typefounders do not recognise the conditions of easy reading of mathematics, and the printers make things worse, & dont want to be bothered with anything interfering with their machine work.

I have also read Dr. Knott's Biog. Note, and am not at all surprised that Dr. Pinkerton should have appreciated his "Newton" so highly. For a youth of 20 he was surprisingly advanced. Why, at his age I didnt know anything at all of Analysis, nor about Electricity, though I had made several inventions (telegraphic) at that time, & was trying to see my way.

I do not think the Military Authorities should have accepted him as a fighting soldier. Ruffians are wanted for that. And I think the Military authorities were very wrong in not overcoming W. G. B.'s refusal by the simple process of compulsorily promoting him to one of their numerous scientific departments in which high mathematics would have been more useful than in the trenches.

Yours very truly,

Oliver Heaviside, W.O.R.M.

The letter sent by Oliver to the parents of William Gordon Brown (see Chapter 2), in sympathy of their son's death in the First World War. What *could* they have thought when they read that signature? This pathetic title of W.O.R.M. was his view of how the world thought of him.

> hang it on a nail, for the sake of the prestige the [IEE Council] believe it will communicate by my name being at the top of the list. PROVIDED we can come to a clear understanding about the explicit (not implicit) recognition that I am The Inventor of the Lumped Loading system. . . .

In May Oliver wrote to tell Highfield he was not well; his words were both brave and sad:

> My eyes are very bad. That is nothing new. It is connected with my scarlet fever ears. The disease is in the innermost parts, in the bones close to the brain. Not to be got at. But warmth always does good. One result is inflammation of the eyeballs, and muscles connected, and an almost constant discharge. I have special specs. But I think large goggles would suit me better. . . . But just now my feet are worst. Rheumatic discharges, distorting the bones and swelling the feet. Yet I don't think I have come to the "evening" of my life, as Lodge says.

1963 or thereabout, after going about in a pram, pushed by a impudent boy making faces at me! That is my present estimate.

In this same letter he referred to the Faraday Medal as "the dammedal," and gave his feelings about awards, in general:

> My objection to medals and things of that sort has been a quite deep-seated one all my life, and the late *enormous* multiplication of so called "honors", by the million, removes the honor of distinction entirely. It is now the honor to be left out!... I am afraid it is too late for reverence.... My W.O.R.M. title is the *only* one that I care for. It is quite a solitary distinction.

By late July it had been agreed that Highfield would come to Homefield to present the Medal personally. Highfield originally planned to bring several colleagues with him, but this greatly upset Heaviside:

> I don't know what you mean exactly by heading a deputation. Who are they? And I can't talk to more than one at a time, and that is not easy. It is brainwork mostly and eyework.... And I may not be able to get a room cleaned of the damcoal dust... in time. Then you must *sit* in the hall or lobby. You must *sit*. It is very rude, to look down to a man. That's not "paying respect" to me is it? Nor is staring at me like the Zoo. Hadn't you better come 1 at a time on 4 successive days?

Heaviside's health was no better in July than it had been in May. The pain in his eyes was now so bad he could read only with great difficulty. In addition, he had gout, swollen tonsils, and "violent bumping about of my big [i.e., enlarged] heart." Heaviside refused to see a doctor, but rather treated himself, wrapping himself in a "polar nightcap," i.e., in several layers of blankets from the upper chest to the top of his head.

Highfield showed this letter to one of the other members of the deputation, who replied:

> It [Oliver's letter] would be amusing were it not rather tragic, but at all events I quite agree with you that he would be more bothered than gratified by having a number of healthy and sane individuals coming to see him in his polar head dress and suffering and ill balanced, bodily and mentally, so I leave it to you to be the member of a deputation of one.

When informed that he would not be flooded with visitors, Oliver replied on July 29, evidently in good spirit:

> ...Very good.... Alone, or with a lady to protect you against my notorious violence.... I usually get on very well with ladies, with clear soprano voices that are so distinct and so unlike the throaty voices of gruff men. And they like me too, I think, even too much, though I don't flatter them. But they must have some sense of humor.... No Deputation. A lady for protection allowed.

And so, at last, Highfield traveled to Torquay, sailing his yacht into the beautiful bay that lay beneath Homefield high on a hill. Highfield remembered that time in detail[101]:

> It was my duty as President to present the Medal, and the duty was an interesting and pathetic one. Heaviside lived entirely alone in a pleasant house in Torquay—a house decaying from long neglect. I called first by appointment and found him waiting in the weed-covered drive in an old dressing-gown, armed with a broom, trying rather vainly to sweep up the fallen leaf. He was pleased to see me in a queer shy way and took me through a furniture-laden hall, all covered with dust, to his own room. He had papered the walls with prints and reproductions from many publications, and pointed out to me old Presidents of the Institution of Electrical Engineers and all the recent ones, and asked me all about them. The walls held a

> pictorial record of his life's interests. He was in all ways entirely competent and still preserved his power of impish criticism. Often he was amusing in a caustic way; he was... genuinely pleased that he had been recognised as one of our famous men. His way of life made a pathetic background to his mental activity, but I am sure he did not regard it as pathetic. He talked freely of all recent work and development, but his talk was interfered with by many petty personal complaints of the most homely description. I saw him several times and tried to improve his menage and especially his food supply, which was inadequate, but with little result. He was quite content to live as an old-world hermit, and so indeed he lived.
>
> I saw him again in the August of 1922 and again when I formally presented the Faraday Medal early in September. He seemed greatly improved in health, and there is no doubt that the interest taken and shown in him by his fellow workers of the Institution had really cheered him; in fact, he seemed quite happy. The presentation of the Medal was interesting. He vigorously criticised the wasteful expense of the leather-covered vellum document which accompanies the Medal, but was consoled by the Medal being of bronze and not of gold. He read every word of both document and Medal and was especially pleased to see the name of Alexander Siemens, who signed as the oldest Past President. He talked much about telephony and wireless, all interspersed with homely grumbles at the many defects of his neighbours. He seemed to know all that went on in the town. It is impossible to give any adequate account of one who so despised what most men desire, but when I left him I felt that he was content, that he respected the Institution, and that it had pleased and made happy one of its famous men.

It was during these visits that Highfield learned the intensity of Heaviside's desire for warmth, and of his inability to get it. Oliver was constantly at odds with the local gas company (he called them the "Gas Barbarians") for nonpayment. And little wonder, as at one time he was using gas at the incredible rate of 800,000 cubic feet per year (with a cost of about 200 pounds, essentially his entire pension)! Through the intervention of Mr. R. H. Tree, the Institute's Chief Clerk, Highfield made arrangements for the IEE to pay Oliver's gas bill.

Tree apparently struck a chord in Heaviside, and the two men got along nicely. On October 10, 1922, for example, Heaviside wrote to Highfield that

> Mr. Tree brought me a fine fat sole, which I cooked on Saturday on a special fire of Lipton's thin boxes. To prevent its getting cold, I ate right out of the pan on the fire and having recovered taste, found it delicious.

In his own letter to Highfield, written the next day, Tree also related the sole incident, adding that Oliver had been so pleased with the fish that he had insisted Tree select a picture from his gallery. Tree took one and on the back Oliver wrote

> October 6, 1922
> Mr. R. Tree from
> W.O.R.M.
> Torquay

The W.O.R.M. signature on all his letters wasn't the only hint of an odd side to his mental state. At some time before the start of his correspondence with Highfield, Heaviside became fascinated with the very letters in his name, often writing them in alphabetical order with the number of times a letter appears in "Oliver Heaviside" written beneath. And in his letter of September 23, 1922, he signed off by writing,

> I am, Sir, Yours very truly and anagrammatically,
> O! He is a very Devil, W.O.R.M.

He thought that if "Heavification" wasn't quite right for referring to loading (certainly it was far superior to "Pupinize"!), then perhaps "Devilization" would do.

The correspondence between Highfield and Heaviside eventually died out but, as late as the last day of 1922, Oliver wrote to inform his rich yachting correspondent in London that his poor bedroom stove in Torquay, previously damaged by an explosion, had at last "broke down for good."

His brother Arthur died on September 22, 1923, and just three months later his old and faithful editor at *The Electrician*, C. H. W. Biggs, was gone, too.

A Last Look

The last view we have of Heaviside at Homefield is from Professor Berg's reminiscence[98] of a visit he made not long before Oliver died. A longtime admirer of Oliver's work, Berg was then teaching a course on operational mathematics to electrical engineering students at Union College.[102] In 1923 he met Ludwik Silberstein while giving a paper in Rochester, New York (Silberstein spent a major portion of his life at Kodak) on the operational calculus, and got a letter of introduction from him. He arrived in Torquay on June 7, 1924, and sent the letter to Oliver, along with a note asking to see him the next day. Arriving at Homefield on the morning of June 8, Berg found the front door plastered over with old bills marked "Paid" in red pencil, as well as a small envelope on which was written "To be called for":

> *Prof. E. J. Berg,*
> Being deaf and living upstairs I leave this out in case I do not hear you. My full address is Mr. Oliver Heaviside, Homefield, Torquay, and letters always reach me by the Letter Box. I have been very ill the last winter, and am waiting for some fine dry weather to begin summer convalescence. Things are much changed here and I have nothing to offer you. So you must bring your own grub if you are likely to feel hungry.
> Yours, etc., O. Heaviside

After some amusing adventures (including getting past the doorstep of the initially reluctant Heaviside by "begging" for some pipe tobacco), the two men struck up a friendship. One rule of visiting was made quite clear, however: "He never discussed his writings. He had forgotten all about them."

After Berg returned to America there followed an extensive correspondence. On September 22, 1924 Berg wrote from Union College to tell Heaviside "I am lecturing on your mathematics and find very great enthusiasm on part of the students." He also reported that the D. Van Nostrand Company was selling his EMT series in America at $35 a set. Oliver's reply of October 3 shows that his cynical evaluation of mathematicians hadn't changed in the more than thirty years since his war with the "Cambridge school of thought":

> "Great enthusiasm". But that has its dangers. They must be mathematicians first, or it is no use to them. Or, if they *are* mathematicians they have had a lot of stuff put into them which ought not to be there at all, for the proper appreciation of *my* mathematics. And, even though they may be delighted at the beginning to find how easily I do work of extraordinary difficulty by old fashioned ways, yet when they go further and find what a lot of hard thinking is required, they may become discouraged and throw it up.

And as for EMT, he added in a postscript "Pray do not forget that my *Electrical Papers* are actually my Great Work... out of which my E.M.T. grew."

Another letter to Berg, on November 9, shows Heaviside was also still reading at least some technical literature: "You may remember what I said about academical mathematics. There is a remarkable example of it in the Sept. or Oct. Journal of A.I.E.E. by an Associate Junior near the beginning [Note 103]. But that is *not* the way to do Heaviside's Differential Operators."

On November 16, 1924, Berg received the last letter he would get from Oliver, sent in thanks for a gift of shoes—his feet were then in an absolutely wretched condition:

November 16, 1924

Prof. E. J. Berg

Dear Professor,

I duly received the boots marked "5 dollars" and I think they are as near as possible without orthopedic operations; I cant remember whether I *ackd* receipt before unless it was in Bobby [Note 104] letter. He is just the same, roars at me louder and louder, explains things to my stupidity, etc. as though he were trying to force a quarrel on me. I am very meek, mild, and humble of course, and also very ill again. The arrival of winter has caused Gas Co. trouble again. Hardly ever enough gas and rapid failure of the old, small mains by naphthalene and other chokage.

Worse than that I had to have some repairs done to the windows and roof slates and gutters, to keep out the wind and rain. The single man I got was most willing, but was too confident, and he was not well, though he said he was. I was quite grateful for my practical way of helping the out of works, by giving them work they *can* do, treating them liberally as well. But he didn't understand the construction of the roof (his master does mostly repairs to small workmen's houses). So I had to direct him in many respects, and be out of doors, and I had an accident. In coming down a ladder my coat caught in something, and down I went, a fall of 11 feet on the broad of back. The man did the usual stupid thing by trying to get me on my feet to see if I had hurt myself. I said, let me lie, till the pain in back goes. In about five minutes it was gone, and I was able to be pulled up, and actually walked indoors shakily. The pain had gone through me to the front, to the muscles and bones all the way down from the top of the chest. Well, by warmth and in bed I would cure that in a few days. But I could not get the warmth or proper food, and the window mending had to go on, so the house chilled, and the man had to be looked after. The result was a violent attack of internal derangement....

I have another fight on. I order Milkmaids Pure Cream with no Preservatives. Liptons send me the common heavily sugared and adulterated tins they call "Milks." I refuse them. They try to make me take them. I wrote to Switzerland. They sent my letter to London. The London firm said they were greatly obliged to me. They were astonished to learn that Liptons didn't stock the Pure Cream Milkmaid Brand, and would do their best for me. Liptons here made a special requisition for me and got 24 Milkmaid from their London firm. With lump sugar added according to taste they are splendid, and I ordered here 48 more, with 6 pounds of lump. They sent the lump but no more Milkmaids. Yet I have to proceed cautiously. I have 167 "Milks" and I want them all taken back in exchange for "Milkmaids," and I wont pay their bill unless they will allow for difference in price; so as Xmas is coming on when business is blocked, I want a settlement quickly. I am aiming at Lipton, the largest grocer in the world, permanently stocking Pure Unadulterated goods, so pray do not send me any as I know in your kindness you might want to for they would come too late.

Yrs tr.
Oliver Heaviside.

It wasn't like the battles with Preece or Tait, but at least when Oliver went out he went out fighting. The Searles visited with him on New Year's Day 1925 (Charles had died just four days before), and he was yellow with jaundice. Returning the next day, Searle couldn't get Oliver to open the door, and two days later, on the fourth, he learned that that morning Heaviside had been found unconscious by his policeman friend, "Bobby." Two doctors were quickly called in by relatives, and that evening Searle rode with Oliver in an ambulance to a nursing home. A quarter of the way through the new century, and it was his *first* ride in a motor vehicle!

For a while, with gentle care and good food, he seemed to rally. But decades of substandard living conditions and the total absence of even minimal health care had taken their toll. Heaviside's rugged natural constitution wasn't enough, as it had been in 1913, to bring him

"The End"

through. He died Tuesday, February 3, 1925. His official death certificate listed a variety of chronic ailments, but falling 11 feet off a ladder and landing on his back is more likely what did him in. If he hadn't been so eager to run up and down ladders, perhaps he *would* have made it to 1963, even if it was in a pram.

The funeral took place on Friday, February 6. In addition to family members, the Chief Clerk of the IEE attended, the same Mr. R. H. Tree who had once brought Heaviside "a fine fat sole" for dinner. Several of the obituaries that appeared said these were the only ones present. If one believes the notice[105] published by *The Electrician*, however, then for what it was worth ("Not very dammuch!", I can imagine Oliver grumbling), the AT&T and Western Electric Companies also sent representatives to observe the passing into the next world of a man they had overlooked while he was still in this one.

He lies buried in the grave of his parents, Number 346, at the Paignton Cemetery, near the junction of Colley End and Ailescombe Roads. But if you go looking for it you have to keep a sharp eye out. As the last one in, his name is on the bottom of the marker, and if the grass isn't kept cut the "Oliver Heaviside, F.R.S." can disappear from view. I suspect this wouldn't have come as much of a surprise to him. *Sic transit gloria mundi.*

Postscript: Ten years later, on April 25, 1935, the first round-the-world telephone conversation took place.[106] Traversing a total of 23,300 miles (19,500 by radio), and using 980 vacuum tubes to provide 2000 dB of gain in each direction, two men in New York City enjoyed a nice chat (by way of San Francisco, Java, Amsterdam, and London). The late Dr. Oliver Heaviside, FRS, would have been impressed.

Notes and References

1. Trotter did *not* forget. In May 1899 he wrote to Oliver from Cape Town, South Africa (see Note 20) to say "I regret that I have not forgiven the Pender parasites who drove me from *The Electrician*." It is unlikely that it was with Sir John Pender himself (see Chapter 7) that Trotter had his troubles. On Pender's death in 1896, Bond (Trotter's friend and assistant) ran a note praising the late Proprietor for having left the journal's management to the editors' "unfettered discretion."
2. The "Civil List," dating from 1689, is a list of the annual sums Parliament allocates to pay for the expenses of the English sovereign. Starting with Queen Victoria, pensions could be included in this list, granted on the recommendation of state ministers for those who had achieved distinction in the arts, literature, or science.
3. *The Electrician*, vol. 37, p. 331, July 10, 1896. The pension was officially granted on March 5, 1896.
4. Ibid, pp. 346–347.
5. *The Electrician*, vol. 37, p. 452, July 31, 1896. The evolution of the concept of negative resistance is a fascinating topic, one I believe has not yet been treated by the historians of science. Today, of course, electronics couldn't even exist without the concept of negative resistance, but just 90 years ago it was an alien concept.
6. Mark Tapley is a good-natured, energetic young man in Dickens' novel *Martin Chuzzlewit*.
7. "Miss Eviside" was almost certainly Miss Rachel Elizabeth Way Heaviside, the eldest daughter of Charles Heaviside, who lived with Oliver and her grandparents in Paignton, as both a companion and a helper of the chores.
8. See Notes 9 and 43 in Chapter 6, and Note 52 in Chapter 7.
9. W. H. Preece, "Electrical disturbances in submarine cables," *The Electrician*, vol. 37, pp. 689–691, September 25, 1896.
10. *The Electrician*, vol. 37, p. 741, October 2, 1896.
11. *The Electrician*, vol. 37, pp. 796–797, October 16, 1896.
12. *The Electrician*, vol. 38, p. 24, October 30, 1896.
13. *The Electrician*, vol. 38, p. 94, November 13, 1896.
14. *The Electrician*, vol. 38, p. 128, November 20, 1896. FitzGerald, for *years*, kept after Preece's error on the capacitance of his cable. For example, in response to yet another Preece claim that *science* is inferior to *engineering* (*The Electrician*, vol. 43, pp. 233–234, June 9, 1899), FitzGerald called (*The Electrician*, vol. 43, p. 346, June 30, 1899) Preece a "bit of a wag" and suggested that readers shouldn't pay too much attention

to someone who could so terribly misunderstand the theory of his own design. Or perhaps, concluded FitzGerald, Preece's words had only been meant as a joke.

15. *The Electrician*, vol. 38, p. 172, December 4, 1896.
16. *Electrical World*, vol. 28, pp. 617–618, November 21, 1896.
17. So called because it had a "view" of the woods of nearby Bradley Manor.
18. The "old croaker" eventually disappeared without ever being identified, and Heaviside felt the loss enough to write FitzGerald (July 6): "Old croaker gone away. Sad loss. He left one night when piano was played and blinds up. Couldn't stand the noise."
19. This was Heaviside's way of wishing FitzGerald well on his upcoming trip to the B.A. meeting that year in Toronto.
20. Trotter didn't last long in private practice after leaving *The Electrician*—from January 1896 to mid-1899 he was Government Electrician and Inspector in Cape Town, South Africa. He had gone off to Cape Town full of adventure after a grand farewell party at the Whitehall Club, at which he was presented with a gold watch. It was a job, however, that he described in a May 1899 letter to Heaviside as "Dismally cut off from scientific thought, and most of one's zest for scientific work oozing away through the waxing climate." Trotter returned to London, for good, just two months later, barely missing the opening of the Boer War.
21. Carter had studied under Silvanus P. Thompson at University College, Bristol, and had held a lectureship at Hanover-square School of Submarine Telegraphy and Electrical Engineering in London. He was also the author of a book on gearing for electrical machinery, published by *The Electrician* Printing and Publishing Company.
22. There is an interesting secretarial notation at the top of this typed letter that it was "Dictated to the Phonograph" by Tucker—office technology had taken a big step into the future at *The Electrician*.
23. The paper in question was "A curious inversion in the wave mechanism of the electromagnetic theory of light," *American Journal of Science*, vol. 5 (4th series), pp. 343-348, May 1898, by Carl Barus of Brown University. It dealt with Barus' view that the field equations can be interpreted in such a way as to "make the wave run backward." Heaviside didn't write a reply to the Barus paper, but he generally had little patience for those who dared to question Maxwell. Nobody was too big for him, not even Rayleigh, on this matter. When Rayleigh published a paper in the June 1898 *Philosophical Magazine* raising such a concern, Oliver quickly replied to point out the great man's error ("Note on an alleged 'Apparent failure of the usual electromagnetic equations'," *The Electrician*, vol. 41, p. 255, June 17, 1898). Heaviside and Rayleigh had had their disagreements over Maxwell before. Oliver's hurt feelings at being misunderstood by "Lord R.," as far back as 1888, are documented in J. Z. Buchwald, "Oliver Heaviside, Maxwell's apostle and Maxwellian apostate," *Centaurus*, vol. 28, pp. 288-330, 1985.
24. *The Electrician*, vol. 41, p. 802, October 14, 1898.
25. So did FitzGerald, who got the same response from Carter. FitzGerald then wrote (October 26) to Oliver to say that without such a reply he had been prepared to cancel his subscription to *The Electrician*!
26. *The Electrician*, vol. 42, p. 49, November 4, 1898.
27. "Electricity in civil engineering," *The Electrician*, vol. 42, pp. 44–45, November 4, 1898.
28. E. J. Houston, Jr. and A. E. Kennelly, "The insulating medium surrounding a conductor, the real path of its current," *Proceedings of the American Philosophical Society*, vol. 36, pp. 140–170, May 1897.
29. E. H. Barton, "Attenuation of electric waves along a line of negligible leakage," *Philosophical Magazine*, vol. 46 (5th series), pp. 296–305, September 1898, and W. B. Morton, "On the propagation of damped electrical oscillations along parallel wires," *Philosophical Magazine*, vol. 47 (5th series), pp. 296–302, March 1899. When Barton reported, at the June 10, 1898 meeting of the Physical Society, on some discrepancies between his experimental results and Heaviside's theoretical predictions, Oliver sent a note (he did *not* show up in person, of course!) suggesting some possible reasons (*The Electrician*, vol. 41, p. 286, June 24, 1898). He concluded this note with an obvious slam at Preece's ability to always get *his* numbers to come out right—"if even more correction is wanted, try the *KR*-law"!
30. Trotter expanded on his "shorthand" analogy in a paper he wrote four years later ("Useful mathematics from the engineer's point of view," *Engineering*, vol. 76, pp. 358–360, September 11, 1903). This paper clearly showed what a pragmatist Trotter was ("Is one engineer out of five hundred interested in learning... [what a logarithm is?]"), and how broad-minded as an editor he must have been to accept Heaviside's mathematically dense papers.
31. On his 65th birthday, in February 1899, Preece was compelled by law to retire from the GPO. He remained influential for the next five years, however, as a paid consultant at 400 pounds a year.
32. First published in 1888, this London publication carried the subtitle "A Weekly Journal of Instruction, Literature, Interesting Facts, Amusing Anecdotes, and Jokes."
33. Bicycling was a *passion* for Heaviside in the 1890s and 1900s, as it was for literally millions on both sides of the Atlantic. An excellent paper that conveys a sense of the almost hysterical nature of the sport in those times is J. C. Whorton, "The hygiene of the wheel: An episode in Victorian sanitary science," *Bulletin of the History of Medicine*, vol. 52, pp. 61-88, 1978.
34. Heaviside's words have been taken from a preliminary draft he prepared, not from the actual letter he mailed. Many years later Raphael had a slightly different version of this episode, saying *he* had gotten the "The Lord will provide" on a postcard. But it was Carter who responded to Oliver, echoing the *carte blanche* condition,

and so I suspect Raphael's memory failed him here—see his "'The Electrician', 1897 to 1906," *The Electrician*, vol. 87, p. 618, November 11, 1921.

35. *The Electrician*, vol. 46, pp. 701–702, March 1, 1901.
36. EMT 3, p. 89. The death of FitzGerald had a practical side to it, too, for Oliver. As he rather coldly put it in a letter to his Boston correspondent (B. A. Behrend), dated April 5, 1918: "Unfortunately he died when I most needed him. He could and would have exerted himself to procure financial assistance of importance." For more on the scientific significance of FitzGerald's death, see B. J. Hunt, "The Maxwellians," Ph.D. dissertation, The Johns Hopkins University, Baltimore, MD, 1984, pp. 337–340.
37. *Journal of the Society of Arts*, vol. 50, p. 21, November 22, 1901. Preece was now the President of the Society!
38. The first commercially successful loaded cable was, in fact, a *continuously* loaded one, but it was not a Field's cable. The cable was designed by the Danish engineer C. E. Karup, and in 1902–1903 four such submarine cables, with soft iron wire spirals around the central copper conductors, were laid across the Baltic.
39. *The Electrical World*, vol. 32, p. 675, December 24, 1898.
40. J. Brittain, "The introduction of the loading coil: George A. Campbell and Michael I. Pupin," *Technology and Culture*, vol. 11, pp. 36–57, January 1970 (with additional information from the replies of various correspondents in the subsequent issue of October 1970, pp. 596–603). See also N. H. Wasserman, *From Invention to Innovation: Long-Distance Telephone Transmission at the Turn of the Century*, Baltimore, MD: The John Hopkins University Press, 1985.
41. Heaviside learned a few things from Stone, too. It was Stone, for example, who invented the "pre-emphasis" circuit to counter the frequency-dependent amplitude response of a telephone cable by inserting, between the transmitter mouthpiece and the input terminals to the cable, a special circuit with the *inverse* response. Stone called this a "counterdistortion" circuit.
42. A. M. McMahon, *The Making of a Profession: A Century of Electrical Engineering in America,* New York, NY: IEEE Press, 1984, p. 54. After leaving American Bell, Stone became caught up in the new wonder of "wireless," but he didn't forget his Heaviside. His paper at the 1904 St. Louis International Electrical Congress ("The theory of wireless telegraphy"), for example, used Oliver's operator methods from EMT 2, and referred to EP 2 for the physics of Maxwell's theory—see *The Electrician*, vol. 54, pp. 134–139, November 11, 1904.
43. M. I. Pupin, "Wave propagation over non-uniform cables and long distance air lines," *Transactions of the American Institute of Electrical Engineers,* vol. 17, pp. 445–507 (with discussion on pp. 508–512), May 19, 1900.
44. It was Campbell who derived the exact mathematics of where to insert coils of specified electrical parameters into any given circuit, something even Heaviside hadn't done (and which Pupin did only in a crude and approximate way at a later date). The only commercial coils *ever* built were all based on Campbell's work.
45. M. I. Pupin, *From Immigrant to Inventor*, New York, NY: Charles Scribner's, 1923, pp. 332–336. Pupin's self-serving statements were accepted at face value by historians, for decades, until Professor Brittain's paper (Note 40) set the record straight.
46. M. I. Pupin, "Propagation of long electrical waves," *Transactions of the American Institute of Electrical Engineers*, vol. 16, pp. 93–94, March 22, 1899.
47. From the Foreward to *The Collected Papers of George Ashley Campbell*, New York, NY: AT&T, 1937.
48. W. H. Preece, "On the Pupin mode of working trunk telephone lines," *The Electrical Engineer (London)*, vol. 40, pp. 237–238, August 16, 1907 and pp. 260–263, August 23, 1907, including discussion by Lodge, S. P. Thompson, and Preece.
49. Preece was, apparently, confused by the sign of *reactive* energy (positive for inductive and negative for capacitive) and didn't realize that this sign is related to the phase difference between current and voltage in the two kinds of reactances. When energy is going *into* the magnetic field of the cable's inductance it is coming *out* of the electric field of the cable's capacitance. The sign refers to the *direction* of energy flow, not to the energy itself. Preece's error is equivalent to asserting that if Bill gives Bob ten pounds (and if we call this a positive exchange), then Bob giving Bill ten pounds means there is such a thing as *negative* money!
50. We have here, I think, a psychological puzzle. For Preece to make such statements in 1907, either he must have been a very mean spirited, proud man who absolutely wouldn't admit he had been wrong all the many past years, or he had to be technically dense to an almost unbelievable degree. If the former is the case, it is astonishing that a proud man like Preece could continue to make a public fool of himself, even in the face of direct, open criticism from Thompson, and bold ridicule from Lodge.
51. *The Electrician*, vol. 40, p. 832, April 15, 1898. Thompson's defense was not that his critic was wrong in believing in tangential wireless rays, but rather that he had been misquoted, i.e., Thompson claimed he didn't believe in long-range wireless himself—see his letter in reply in *The Electrician*, vol. 40, p. 866, April 22, 1898.
52. *The Electrician*, vol. 43, pp. 35–36, May 5, 1899.
53. W. H. Preece, "Aetheric telegraphy," *Journal of the Society of Arts,* vol. 47, pp. 519–525 (including discussion), May 5, 1899. It was Carter, in fact, in the discussion, who pressed Preece on this question of waves bending around the Earth, and Carter's editorial note (Note 52) was his considered reply to Preece.

54. G. Marconi, "A note on the effect of daylight upon the propagation of electromagnetic impulses over long distances," *Proceedings of the Royal Society,* vol. 70, pp. 344–347, June 10, 1902.
55. Lord Rayleigh, "On the bending of waves round a spherical obstacle," *Proceedings of the Royal Society*, vol. 72, pp. 40–41, May 1, 1903.
56. W. H. Eccles, "Wireless communication and terrestrial magnetism," *Nature*, vol. 119, p. 157, January 29, 1927.
57. Sir William Henry Eccles (1875–1966), best known today to most electrical engineers (at least to "elderly" ones, like the author) as co-inventor of the Eccles–Jordan circuit, now usually called the "triggered bi-stable multivibrator" or "flip-flop," commonly used in counting circuitry.
58. There *is* a letter from the publisher, George Tucker, dated February 21, 1902, in which he said he was "much bewildered at the varying opinions I hear of the merits and demerits of Mr. Marconi's Wireless Telegraphy," and he then asked for Heaviside's opinion. Oliver responded, but evidently didn't say much (it was on a postcard), as a second Tucker letter of February 28 said he was "still in the dark." Tucker also worried about how Marconi's work was causing "investors in important British stocks [presumably the long-distance cable companies] to suffer a depreciation of their investments."
59. Reprinted in EMT 3, pp. 331–346. For this article Oliver received a payment of fifteen pounds and sixteen shillings and an invitation to a great dinner bash (for 500!) at the elegant Hotel Cecil in London, given in honor of all the contributors by *The Times*, publisher of the *Encyclopedia*. As might be expected, he took the money and ignored the dinner.
60. A. E. Kennelly, "On the elevation of the electrically-conducting strata of the Earth's atmosphere," *Electrical World and Engineer*, vol. 39, p. 473, March 15, 1902.
61. C. S. Gillmor, "Wilhelm Altar, Edward Appleton, and the magneto-ionic theory," *Proceedings of the American Philosophical Society*, vol. 126, pp. 395–440, 1982.
62. W. H. Eccles, "On the diurnal variations of the electric waves occurring in nature, and on the propagation of electric waves round the bend of the Earth," *Proceedings of the Royal Society*, vol. 87, pp. 79–99, August 1912. In this paper Eccles said Heaviside made his initial suggestion in 1900 (not 1902), but this must be an error.
63. W. H. Eccles, "Atmospheric refraction in wireless telegraphy," *The Electrician*, vol. 71, pp. 969–970, September 19, 1913.
64. *The Electrician*, vol. 75, p. 169, May 7, 1915.
65. J. A. Fleming, "On atmospheric refraction and its bearing on the transmission of electromagnetic waves round the Earth's surface," a paper read to the Physical Society of London and abstracted in *The Electrician*, vol. 74, pp. 152–154, November 6, 1914.
66. There was nothing *personal* in this wish to avoid the reflecting layer hypothesis, no desire to avoid it because of Heaviside's name. Such a layer, at the turn of the century (and indeed until *much* more advanced radio techniques were developed), was simply "unobservable" by direct means and thus demanded an act of faith, something all good scientists would generally like to avoid when wearing their scientist hats.
67. G. N. Watson, "The diffraction of electric waves by the Earth," *Proceedings of the Royal Society*, vol. 95, pp. 83–99, October 1918.
68. G. N. Watson, "The transmission of electric waves round the Earth," *Proceedings of the Royal Society*, vol. 95, pp. 546–563, July 1919.
69. J. Larmor, "Why wireless electric rays can bend round the Earth," *Philosophical Magazine*, vol. 48 (6th series), pp. 1025–1036, December 1924.
70. C. S. Gillmor, "The history of the term 'ionosphere'," *Nature*, vol. 262, pp. 347–348, July 29, 1976.
71. R. A. Fessenden, "The predetermination of the radiation resistance of antennae," *The Electrician*, vol. 61, pp. 650–651, August 7, 1908.
72. H. G. J. Aitken, *Syntony and Spark: The Origins of Radio,* Princeton, NJ: Princeton University Press, 1985. See, in particular, the entire chapter devoted to Marconi, pp. 179–297.
73. "Telegraphing without wires," *McClure's Magazine*, vol. 8, pp. 383–392, March 1897.
74. Aitken (Note 72), p. 211.
75. "Wireless telephony," *The Electrician*, vol. 45, p. 773, September 14, 1900.
76. *The Electrician*, vol. 59, p. 140, May 10, 1907. The Radio-Telegraph Committee was then considering the ratification of the International Convention on Wireless Communication at Sea, which broke the lengthy monopoly the Marconi Company had enjoyed on ship-to-shore and ship-to-ship communications. This quote shows that Preece, as late as 1907, still did not appreciate the distinction between his *induction*-field system, and Marconi's *radiation*-field system.
77. In 1905 Biggs reproduced the backs of menu cards from a hotel dinner party, signed by all those present (including Oliver's brother, Arthur) and given to the guest of honor as a memento. I am speculating here, but from the location of their signatures it would appear that Biggs and Preece sat next to one another, not likely if the two still retained strained feelings—see *The Electrical Engineer (London)*, vol. 36, p. 654, November 10, 1905.
78. "Oliver Heaviside: A personal sketch," *The Heaviside Centenary Volume*, London: IEE, 1950, p. 94.
79. NB 18:316.
80. A. Pais, "How Einstein got the Nobel prize," *American Scientist*, vol. 70, pp. 358–365, July-August 1982.

The prize in 1912 was worth about 140,000 Swedish kronor, equivalent to what, at that time, a senior professor in a good university might earn in *twenty years*.

81. Years later, after Heaviside's death, Homefield was made into a commercial enterprise, the Killester Hotel.
82. Searle knew what he was talking about. He suffered a very serious nervous breakdown himself in July 1910, which kept him absent from Cambridge until October 1911. The trauma of this episode is shown in the last line of the Preface to his book *Experimental Harmonic Motion*, written in April 1922: "Above all, I must give thanks to God for giving me the restoration of health that has enabled me to write this book." In a letter to Oliver he once expressed what he called "Searle's Law" of nervous breakdowns: "Time of Recovery = μ (Fatigue)^2." Heaviside thought Searle's problem was due to "a racketty life."
83. This is from the "Idylls of the King"—see, for example, *The Poems of Tennyson*, London: Longman, 1969, p. 1741. The words are those of Guinevere, who speaks them (line 655) after King Arthur learns of her affair with Lancelot. Literary peasant that I am, I confess I have not the slightest idea *why* Heaviside used it—perhaps a reader will enlighten me.
84. *The Electrician*, vol. 70, pp. 767–768, January 24, 1913.
85. *The Philosophical Magazine*, vol. 32 (6th series), pp. 600–602, December 1916.
86. Heaviside's comments are quoted from B. R. Gossick, "Where is Heaviside's manuscript for volume 4 of his *Electromagnetic Theory*?", *Annals of Science*, vol. 34, pp. 601–606, 1977.
87. "The romance of electricity—Death of a pioneer," *The Daily Telegraph,* pp. 11–12, November 7, 1913.
88. When the War broke out Burton resigned from his post at Oxford and went to work at the Royal Aircraft Establishment at Farnborough. He died there in an accident on February 3, 1917.
89. *Nature*, vol. 92, pp. 322–324, November 13, 1913.
90. *The Electrician* ran a black-bordered obituary of its "beloved chief" (who as of 1906 owned the entire operation), as well as a full-page photograph (vol. 76, p. 554, January 21, 1916). The writer(s) of the notice thought it important to mention Tucker's long-time support of Oliver's work, and related an amusing anecdote: "On one occasion he endeavored to persuade the author to send in the remainder of his manuscript by threatening to break up the type [this must be in reference to EMT 3], and a reply came back that 'Mr. Tucker would do nothing so foolish'."
91. R. Appleyard, *Pioneers of Electrical Communication*, London: Macmillan, 1930, pp. 240–241. The literary reference of Adullam is from the Old Testament, as in I Samuel 22:1, II Samuel 23:13, and I Chronicles 11:15. The biblical reference given in Appleyard's book is incorrect.
92. K. W. Wagner, "Oliver Heaviside," *Elektrische Nachrichten-Technik*, vol. 2, pp. 345–350, November 1925.
93. Harry Robert Kempe (1852–1935), assistant to Preece for many years, who himself became Principal Technical Officer and the Electrician to the GPO. I am only speculating now, but I suspect there was a link between Kempe and Oliver, through Arthur. In his paper read at the 1894 B.A. meeting at Oxford ("Signaling through space," *The Electrician*, vol. 33, pp. 460–463, August 17, 1894), Preece used a formula for the mutual inductance of the geometry of a square coil and a wire parallel to one side. This formula is credited to Kempe in an appendix of Preece's paper, but both its derivation and final form bear a striking resemblance to a page in one of Oliver's research notebooks (NB 17:72). The date on the page is May 1890, and Oliver noted he did it at Arthur's request. Once Arthur had the analysis, did Kempe see it?
94. The "friend" must have been Searle, who was at Farnborough from 1917 to 1919, as Burton was killed more than a year before Heaviside wrote (see Note 88).
95. Not to be outdone by the AIEE, the Secretary (A. N. Goldsmith) of the Institute of Radio Engineers wrote to the President of the IEE (Frank Gill) on April 17, 1923 to report that "The Board of Directors of the [IRE] had brought to their attention through a prominent French radio engineer [Bethenod] the unfortunate circumstances under which Dr. Oliver Heaviside is now living." The IRE Board was, in fact, prepared to award him money, but because of his well-known hostility and nasty responses to other such attempts, it was reluctant to do so without first consulting with the IEE. So, ironically, it was Oliver's own prickly personality that ended up, in this case, doing him in.
96. *Why* Heaviside thought vol. 4 "would do very well" is a mystery—the first three certainly hadn't. And he wanted an advance of *one thousand* pounds!
97. Searle (Note 78), p. 95.
98. E. J. Berg, "Oliver Heaviside: A sketch of his work and some reminiscences of his later years," *Journal of the Maryland Academy of Sciences*, vol. 1, pp. 105–114, 1930.
99. Searle (Note 78), p. 8.
100. Sir George Lee, *Oliver Heaviside and the Mathematical Theory of Electrical Communications*, London: Longman, Green, and Co., 1947, p. 17.
101. Ibid, pp. 29–30.
102. Berg, who in 1913 replaced Steinmetz as head of Union College's electrical engineering department, had been teaching Heaviside's methods since at least 1917. As he wrote to a former student in January 1918, "it is wonderfully beautiful and enables one in a very simple way to calculate transients in almost any network"—see the unpublished Ph.D. dissertation by R. R. Kline, "Charles P. Steinmetz and the development of electrical engineering science," University of Wisconsin-Madison, August 1983, pp. 394–395.
103. Heaviside was almost certainly referring to the article by J. R. Carson, "The guided and radiated energy in wire

transmission,'' *Journal of the AIEE,* vol. 43, pp. 908–913, 1924. There is some irony in this, of course, because after Oliver's death Carson became one of the great champions of Heaviside!

104. ''Bobby'' was Constable Henry Brock, who went well out of his way to help Heaviside in his everyday activities, including the fetching of food. His daughter would occasionally help straighten up Homefield. Oliver wrote many letters to the Brock family, but none of them seems to have survived. Brock died in 1947.

105. *The Electrician*, vol. 94, p. 186, February 13, 1925.

106. ''Around the world by telephone,'' *Bell System Technical Journal*, vol. 14, pp. 542–543, July 1935.

13
Epilogue

... the hand of death ... has struck down one whose genius, perspicacity and clear sightedness into fundamentals entitle him to be ranged with the greatest of philosophers, Archimedes, Newton, Kelvin and Faraday. It need hardly be said that we refer to Oliver Heaviside ... . Heaviside, to electrical engineers at any rate, was something more than a genius, ... he was a legend ... . Among much uncertainty nothing is surer than that Heaviside's name will live for evermore in the gallery of scientific heroes.

— Enthusiastic, but flawed, editorial prediction in *The Electrician* on Heaviside's passing

One picture of Oliver Heaviside in old age depicts him as having slipped decidedly below his intellectual prime, which is consistent with the record of his frequent suffering from emotional and mental strain. According to another picture, however, old age was a period in which Heaviside was exceptionally creative. Which picture is authentic?

— Professor B. R. Gossick, 1977

The Legend Grows

Two years after Heaviside's death, with a payment to his family of just 120 pounds, Oliver's personal library, research notebooks, and a great deal of correspondence came into the possession of the Institution of Electrical Engineers (IEE), London. Other books and letters were acquired by the Institution in 1936 from sources outside of the family. He was already something of a legend as an eccentric wizard of electricity and mathematics, and the writing of Rollo Appleyard (1867–1943), who had access to the IEE archives and knew Trotter of *The Electrician,* as well as the occasional story from Searle at Cambridge, contributed new anecdotes as the first years passed. These little tales were always about his personality and his battles, however, and the legend grew that he was a genius who had been cruelly unappreciated while he lived, but awareness of exactly *what* he had done slowly faded.

Heaviside, himself, through his 1922 letters to J. S. Highfield, President of the IEE, contributed to the legend.

The *technical relevance* of Heaviside's work also gradually faded as technology blossomed, with only his distortionless circuit theory continuing to this day to be of vital, continuing interest, and every now and then it is even associated with him. His vector analysis and vector formulation of Maxwell's theory, once considered odd and enormously difficult, have become so much a part of "basic knowledge" that they form a curious example of how *too much* success can lead to historical obscurity just as well as can failure!

His operational calculus, however, did live on for a bit, and for a while even became part of the education of the more theoretically trained electrical engineers and physicists. Even so, a great sense of mystery continued to surround its use. The situation was described by Norbert Wiener[1]:

> This Heaviside calculus had not as yet been given a thoroughly rigorous justification, but it had worked for Heaviside and those of his followers who had absorbed the spirit of his theory sufficiently to use it intelligently.

and[2]:

> When I returned [from France, in 1920] to MIT I found that the electrical engineers were beginning to count on me for resolving the very serious logical doubts which were attached to the new and powerful communication techniques of Oliver Heaviside.

One of the MIT electrical engineers Wiener worked with was Vannevar Bush, who wrote a book[3] for electrical engineers on Heaviside's operational methods. Other books on Heaviside for electrical engineers were written by John Carson[4] (AT&T), Louis Cohen[5] (George Washington University), and Ernst Berg[6] (Union College). All of these books were, literally overnight, made obsolete and of interest only to book collectors and historians with the 1937 publication of a book[7] by the German mathematician Gustav Doetsch. In his book Doetsch showed how, with the Laplace transform, the sometimes uncontrollable power of Heaviside's operators could be replaced with a systematic, routine method and, in fact, the Laplace transform is now just another tool for all modern electrical engineers. Doetsch's attitude toward Heaviside's mathematics was condescending, which no doubt added to the Heaviside legend. Doetsch wrote,[8]

> The formula relating to the partial fraction development is in symbolism known as "Heaviside's expansion theorem" and this is customarily surrounded by a mystical shimmer. Heaviside ... was self-taught and a passionate advocate of the experimental method in mathematics ... . Through him, the symbolic method of differential operators (which in truth had existed long before him) became popular in the technical literature. How Heaviside himself arrived at the "expansion theorem" is not quite clear, but probably through physical considerations. The formula itself, of course, had long been known in pure mathematics.

With the sneer of an immaculate purist who can't understand those who like to get their hands dirty, he also wrote[9] of the attempts to spread the experimental spirit of Heaviside's work:

> A vast amount of literature, throughout very incompetent mathematically, has attached itself to Heaviside's works ... .

And so, with the rise of Laplace techniques, the last direct link from Heaviside to the modern engineer was broken. This symbolic severing as the 1930s ended was made real when even his papers were removed from London. With the coming of the Second World War, the IEE, fearing destruction[10] of its priceless archives, including the Heaviside Collection, looked for a way to protect them. Fortunately, Mr. W. T. Eyton had room for it all in the cellars of his home near St. Asaph in North Wales. Unfortunately, however, his cellars were rather damp, and when the archives were returned to London in November 1945, after nearly five years of subterranean storage, not a little water and mildew damage had been suffered. This would prove to be of importance a few years later. After the war years, Heaviside's name appears to have surfaced, just once, on each side of the Atlantic. In America an essay[11] was devoted to his witty and unpretentious writing style, and to his personality. In England Sir George Lee wrote a much longer, more technical monograph[12] on his life and work. After that Heaviside's name faded below the noise level and he was at last really and truly forgotten.

Then came 1950, the centenary of his birth, and the IEE decided to host a celebration in his honor. A lucky decision it was as Searle, who was still at Cambridge, had only four years left to live and it gave him a formal occasion at which once more to reminisce about his long-dead friend. As he put it, he was the last living scientific man in England who had seen Oliver (and certainly the last on Earth to have met Oliver, Hertz, *and* Maxwell). But having even more impact was the participation of Mr. Henry J. Josephs, a mathematical physicist with the Post Office and the author of a book[13] using Heaviside's mathematics to solve electric circuit problems. Josephs had first looked at the Heaviside papers in 1928, but didn't really sit down and give them a prolonged and careful study until their return to London after the war. He presented his findings at the 1950 Centenary, and what he had to say caused a tremendous flurry of excitement.

Heaviside Profiled in *Time* Magazine!

The Centenary celebration resulted in several short essays[14] in the popular-science press in Heaviside's memory which confined themselves to straightforward historical accounts. Josephs' paper,[15] however, was a dramatic declaration that Heaviside had been a creative thinker practically up to the time of his death, and that scattered all through his unpublished papers are amazingly wonderful and new mathematical theorems and physical insights. He claimed that Heaviside, according to the partially rotted, difficult to read papers from the Welsh cellars, had produced a unified field theory combining gravity and electricity, and had also made bold new strides into the statistical quantum description of Nature! Josephs made no mention of Heaviside's own obvious distress at his declining ability to avoid "war-worm mistakes," nor of the deterioration of his mental state that is equally obvious from a study of his letters. These same letters give no indication he even *discussed* any of these revolutionary new ideas with his correspondents, something he had always done before. And also, we have Professor Berg's own statement, direct from Oliver, that "He never discussed his writings. He had forgotten all about them." and that it was not statistical quantum theory that interested him in his last years, but rather the obtaining of "Milkmaids Pure Cream with no Preservatives."

What Josephs claimed to have found was nothing less than (or at least parts of) the long-lost fourth volume of *Electromagnetic Theory*! It had been thought by some that Heaviside had actually completed this work, but that his house had been burgled soon after the announcement of his death on the radio. This story was mentioned by Sir George Lee in his address[16] during the Centenary, and repeated years later in a BBC radio talk in 1976, but Professor B. R. Gossick has reported[17] convincing evidence (or rather the lack of it) from a study of local newspaper files and police records that no such robbery took place. In any case, whatever the origin[18] of the papers cited by Josephs, it was his announcement of Heaviside's "unified field theory" that captured the imaginations of many. According to Josephs, Oliver

> ... imagined that all space was filled with energy-tubes, moving in straight lines according to Newton's first law, and in all directions with the speed of light. A single planet alone in space would be subject to a rain of these galactic energy-tubes from all directions at once and so would remain still. But two planets in space would screen each other from the galactic rays coming in particular directions. Consequently the galactic energy-density in the space between the two planets would be reduced and so they would be urged towards each other.

There is a lot more of this sort of thing in Josephs' Centenary paper, all of the same flavor. Is it really possible Heaviside believed this? The concept itself is the sort of thing a young person, bright enough to begin wondering about gravity, might think of, and while appealingly attractive in its simplicity, it is also quite easy to show it must be wrong. Richard Feynman, for example, briefly discussed[19] a version of this theory and very quickly showed the fatal flaw. It is hard to believe, if Heaviside was the elderly genius Josephs claimed, that Oliver would have missed the obvious trap.

But the suggestion that mildewed papers in the damp and gloomy earth of Wales had held a secret that had eluded the great Einstein for decades is wonderfully romantic, of course, and hard to resist. The popular press loved it. First to notice was the British monthly *Discovery*,[20] which quoted Josephs in an interview as saying,

> Einstein's analysis, like Heaviside's, clearly shows that the ideas of curved space are unnecessary for the development of a unified field theory that excludes atomic phenomena. Apparently this is not true if the unified field theory has to be extended to include atomic structure of mass characteristics. Einstein does not appear to have discussed the problem of complete unification; but Heaviside's unpublished notes show that he spent several years studying this problem and actually arrived at a solution. This solution may be regarded as Heaviside's greatest intellectual achievement.

This bizarre assertion, which ignores Heaviside's comment in 1922 to the President of the IEE that "practically nearly all my original work was done before 1887 and is contained in my *Electrical Papers*" is a grave disservice to both Einstein (who without doubt must have laughed if he heard about it) and Heaviside (who deserves credit for what he *did* do, not for the work of others). So outrageous was it all that *Time* quickly picked the story up and opened its piece[21] with the line "What pot of gold does science hope to discover at the end of the cosmic rainbow?" The magazine must have thought the gold more likely to be brass, as it concluded its comments with a parody of Josephs' own words:

> Scientists, whose pet theories radiate through the universe in all directions with the speed of light, are not likely to be unified in a hurry by Heaviside's [more correctly, I believe, Josephs'] unified field theory.

Josephs published one more paper[22] on what he continued to claim were sensational Heaviside discoveries, and then the matter dropped for a while. But not for long; in 1957 Heaviside again burst into view, this time as the direct result of a peculiar habit he practiced more than half a century before.

Formulas under the Floor

> On the 9th November 1957, Sir Edward Appleton received a letter which began "I have a sackful of Oliver Heaviside's papers in my garage ... ."

So began a new paper[23] by Josephs describing an astonishing discovery made by Mr. Harold Saunders, a chemistry teacher, while visiting a friend in Paignton. His friend, a Mr. Howard, happened to be the new manager of the Paignton branch of Barclays Bank, which in turn happened to own the house in which Oliver had lived during the years 1889–1897. Mr. Howard told Mr. Saunders that up in the attic, beneath the floorboards, was a large number of pieces of paper covered with calculations, as well as correspondence (including some postcards from Oliver Lodge). These papers, which filled three mail sacks, were removed from the attic and sent to the IEE in London. Mr. Saunders also reported a conversation with a workman who had originally found the papers and had told him that "he remembered jettisoning a heap of papers" before being asked to save the rest. How and why these papers came to be where they were found is a mystery, but one likely possibility is that Oliver stuffed them there himself, in an early attempt at modern attic insulation, to stop drafts. If so, we have silent testimony of just how much importance *Heaviside* attached to these papers!

The 1957 discovery re-ignited Josephs' interest in Heaviside to the point that he also published a new technical paper[24] in which he again made claim for a Heaviside unified field theory. Even later came another (and last) paper,[25] continuing with these claims, but Professor Gossick has provided, I think, convincing arguments[26] making such assertions highly doubtful.

Last Words

Certainly Heaviside had his thoughts about gravity, but from all that's available it appears he had no better luck with it than did any of the other pre-relativity physicists, including Maxwell. In July and August of 1893 Heaviside attempted to draw an analogy[27] between gravity and electromagnetism, with the key link being the localization of energy in a field. There *is* a crucial difference, however. If one brings two like charges together, energy is required to overcome the repulsion, and it is this energy that goes ''into the field'' (giving a positive field energy density in space). Two masses, on the other hand, *attract* each other and it takes energy to keep them apart, leading to the (strange) result of a *negative* field energy density. This result, implying the presence of *less* than no energy in space, so bothered Maxwell he gave gravity up as beyond 19th century physics.[28] Heaviside, too, reached the same conclusion: ''... it must be confessed that [negative energy density] is a very unintelligible and mysterious matter.''

Heaviside based his analogy on an ether (''It is as incredible now as it was in Newton's time that gravitative influence can be extended without a medium ...''), and reached the conclusion that gravity effects most likely propagate ''immensely fast,'' probably much faster than the speed of light. This is all anti-relativistic, but of course Heaviside can hardly be faulted on this as he was writing twelve years before Einstein's Special Theory was announced. This is all that Heaviside published on gravity, and it had no influence. Oliver Lodge liked it, but, as an old ''ether man'' himself, Heaviside was just preaching to the choir in his case.

Heaviside fought to the end for what he felt was his due in the loading coil dispute. If he had felt he had something new to say about gravity he would have published it, or at least have written to his many correspondents about it. He did neither.

In 1959 a curious book appeared, a *novel*[29] of all things, written by Norbert Wiener, the ''father of cybernetics.'' All the names were changed, but it is in fact a fictionalization of Heaviside's adult life as told through the events of the Campbell–Pupin loading coil episode (which is altered to be a ''stolen'' invention in feedback control). The book is perfectly awful *as a novel* (some of the dialogue is funny enough to be read aloud at a party), but most *technical* people would probably enjoy it because Wiener brings a lot of technical details into the story. Heaviside was tremendously admired by Wiener (Pupin was not), and the book was Wiener's attempt at justice for his late hero. The best one can say about it, perhaps, is that Wiener meant well. One passage,[30] however, does I think ring true:

> I have often reflected on the spirit in which [Heaviside] did his scientific work. Fundamentally, like all creative scientists, he did it because he couldn't help it. There were ideas pent up in him which demanded expression at any cost. He developed scientific ideas as naturally as a poet writes or a bird sings.

Heaviside should be remembered for his vectors, his field theory analyses, his brilliant discovery of the distortionless circuit, his pioneering applied mathematics, and for his wit and humor. The memory of what this enormously gifted man accomplished can only be blemished by the misguided appropriation of the work of others. Heaviside, a man of great personal

Torquay remembers a curious citizen.

integrity when it came to priority in technical matters, would have insisted that this is the only proper path to follow and anything else would simply be wrong. In our modern times, when the acquisition of fame, glory, and material wealth have become consuming goals for so many, such integrity may mark Oliver as a distinctly strange and naive person. But as Searle fondly remembered him during the Centenary,[31]

> Of course he was a first-rate oddity—he was Oliver.

After Heaviside's death, Oliver Lodge wrote,[32] "Heaviside lived an independent, self-contained life: and no doubt his insight into nature ... must have given him moments of sincere pleasure." And now, with O.H. gone to the Other Place with Preece, all the "pure" Cambridge mathematicians, Gibbs, Biggs, Kelvin, FitzGerald, Tait, Snell, Thompson, Lodge, Hughes, Perry, Poynting, Bromwich, Trotter, Searle, and all the rest of those he battled with and against, perhaps he has at last found what he never really did on this Earth. In the Other Place, where everyone is equal in every way.

Notes and References

1. N. Wiener, *I Am a Mathematician: The Later Life of a Prodigy,* Cambridge, MA: MIT Press, 1970, p. 78.
2. N. Wiener, *Ex-prodigy: My Childhood and Youth,* Cambridge, MA: MIT Press, 1972, p. 281.
3. V. Bush, *Operational Circuit Analysis,* New York, NY: Wiley, 1929.
4. J. Carson, *Electric Circuit Theory and the Operational Calculus,* New York, NY: McGraw-Hill, 1926.
5. L. Cohen, *Heaviside's Electrical Circuit Theory,* New York, NY: McGraw-Hill, 1928.
6. E. Berg, *Heaviside's Operational Calculus: As Applied to Engineering and Physics,* New York, NY: McGraw-Hill, 1929.

7. G. Doetsch, *Theorie und Anwendung der Laplace-Transformation,* New York, NY: Dover, 1943 (first published in Berlin by Springer-Verlag, 1937). In the same year (1937) an explicit application of the modern approach to electrical engineering problems appeared—see L. A. Pipes, "Laplacian transform circuit analysis," *Philosophical Magazine,* Seventh Series, vol. 24, pp. 502–511, September 1937. Pipes' paper builds on an earlier one which showed the general approach of using the Laplace transform to solve linear, constant-coefficient differential equations (with arbitrary initial conditions) excited by a step input (the function once famous as the "Heaviside step")—see B. van der Pol, "A simple proof and an extension of Heaviside's operational calculus for invariable systems," *Philosophical Magazine,* Seventh Series, vol. 7, pp. 1153–1162, June 1929.
8. Ibid, p. 337.
9. Ibid, p. 421.
10. The IEE building on Victoria Embankment at Savoy Place was, in fact, never hit by German bombs, although it wasn't for a lack of trying. Windows were more than once blown out by near misses.
11. C. M. Hebbert, "Oliver Heaviside—Humorist," *Journal of the Franklin Institute,* vol. 241, pp. 435–440, June 1946.
12. Sir George Lee, *Oliver Heaviside and the Mathematical Theory of Electrical Communications,* London: Longmans, Green and Co., 1947.
13. H. J. Josephs, *Heaviside's Electric Circuit Theory,* New York, NY: Methuen, 1950. This is a curious book, being a very nice treatment of Heaviside's *original* approach to operators. It is, however, a throwback to the spirit of the books by Bush, Carson, Cohen, and Berg in the 1920s. When first published in 1946, therefore, nine years after Doetsch's Laplace transform book, Josephs' book was already hopelessly out-of-date.
14. W. Jackson, "Life and work of Oliver Heaviside," *Nature,* vol. 165, pp. 991–993, June 24, 1950; P. E. Halstead, "Oliver Heaviside and his influence on modern radio research," *American Scientist,* vol. 38, pp. 610–611, October 1950; Sir Robert Watson-Watt, "Oliver Heaviside: 1850–1925," *The Scientific Monthly,* pp. 353–358, December 1950.
15. H. J. Josephs, "Some unpublished notes of Oliver Heaviside," in *The Heaviside Centenary Volume,* London: IEE, 1950, pp. 18–52.
16. Sir George Lee, "Oliver Heaviside—The man," in *The Heaviside Centenary Volume,* London: IEE, 1950, pp. 10–17.
17. B. R. Gossick, "Where is Heaviside's manuscript for volume 4 of his *Electromagnetic Theory*?", *Annals of Science,* vol. 34, pp. 601–606, 1977.
18. At the end of his essay Professor Gossick made it plain, I think, where *he* believes many of these papers came from and I agree with him.
19. R. Feynman, *The Character of Physical Law,* Cambridge, MA: MIT Press, 1965, pp. 37–39.
20. A. K. Astbury, "Heaviside's lost manuscript," *Discovery,* vol. 118, pp. 267–268, August 1950.
21. "Discovery in a cellar," *Time,* vol. 56, pp. 64, 66, August 14, 1950.
22. H. J. Josephs, "Unpublished work of Heaviside," *Electrical Review,* vol. 155, pp. 9–12, 1954. The story of the "burgled house" is repeated in this paper, which also puts forth claims that the 1886 debate with Hughes involved something Josephs calls "cross-effects and interference" (it was about the skin effect, as discussed in Chapter 8), and that Heaviside thought "sub-atomic particles are the result of the contraction and condensation of galactic energy tubes." No documentation has ever been presented to support these statements.
23. H. J. Josephs, "History under the floorboards," *Journal of the IEE,* vol. 5, pp. 26–30, January 1959.
24. H. J. Josephs, "The Heaviside papers found at Paignton in 1957," *Proceedings of the IEE,* vol. 106, part C, pp. 70–76, January 1959.
25. H. J. Josephs, "Postscript to the work of Heaviside," *Journal of the IEE,* vol. 9, pp. 511–512, September 1963.
26. B. R. Gossick, "Heaviside's 'Posthumous Papers'," *Proceedings of the IEE,* vol. 121, pp. 1444–1446, November 1974.
27. EMT 1, pp. 455–466. See also the editorial "Gravity and the ether," *The Electrician,* vol. 31, pp. 340–341, July 28, 1893.
28. P. C. Peters, "Where is the energy stored in a gravitational field?", *American Journal of Physics,* vol. 49, pp. 564–569, June 1981. One might think the same objection could be raised for the situation of bringing two *unlike* charges together, because they also attract. To get these two unlike charges, however, one has to separate them from two *other* charges (the universe is electrically neutral, while *all* mass is positive—or so we believe today) and when *all* the charge interactions are taken into account, a negative electromagnetic field energy density never occurs. Heaviside showed how all this puzzled him when he wrote in July 1893 (EMT 1, p. 455), "To form any notion at all of the flux of gravitational energy, we must first localize the energy. In this respect it resembles the legendary hare in the cookery book. Whether this notion will turn out to be a useful one is a matter for subsequent discovery. For this, also, there is a well-known gastronomical analogy."
29. N. Wiener, *The Tempter,* New York, NY: Random House, 1959.
30. Ibid, p. 115.
31. G. F. C. Searle, "Oliver Heaviside: A personal sketch," in *The Heaviside Centenary Volume,* London: IEE, 1950, pp. 93–96.
32. *The Electrician,* vol. 94, pp. 174–175, February 13, 1925.

Index

Credits

The photographs on the following pages have been reprinted courtesy of:

AIP Center for History of Physics (New York, NY): 52
AIP Niels Bohr Library (New York, NY): 190, 195
American Association for the Advancement of Science (Washington, DC): 23 (Copyright 1966)
Burndy Library (Norwalk, CT): 35 (top and bottom), 36, 40
Dover Publications (New York, NY): 7, 102, 192
Hodder & Stoughton Ltd. (London): 31 (from: *Kelvin the Man,* A. G. King, 1925)
IEEE Center for the History of Electrical Engineering (New York, NY): 19, 80 (original source: *Telegraphic Journal & Electrical Review,* vol. 8, p. 19, 1880), 111 (original source: Burndy Library), 247 (original source: AIP Niels Bohr Library, Zeleny Collection)
Institution of Electrical Engineers (London): frontispiece, xiv, 14, 16 (top and bottom), 18, 21, 44, 48, 53, 60, 104, 122, 144, 152, 156, 164, 188 (left and right), 246, 265, 268, 285, 291, 296, 308
The London Regional Transport (London): 61
The Mansell Collection (London): 5
The Royal Society (London): 123, 224, 228